Studies in Systems, Decision and Control

Volume 405

The series "Studies in Systems, Decision and Control" (SSDC) covers both new developments and advances, as well as the state of the art, in the various areas of broadly perceived systems, decision making and control–quickly, up to date and with a high quality. The intent is to cover the theory, applications, and perspectives on the state of the art and future developments relevant to systems, decision making, control, complex processes and related areas, as embedded in the fields of engineering, computer science, physics, economics, social and life sciences, as well as the paradigms and methodologies behind them. The series contains monographs, textbooks, lecture notes and edited volumes in systems, decision making and control spanning the areas of Cyber-Physical Systems, Autonomous Systems, Sensor Networks, Control Systems, Energy Systems, Automotive Systems, Biological Systems, Vehicular Networking and Connected Vehicles, Aerospace Systems, Automation, Manufacturing, Smart Grids, Nonlinear Systems, Power Systems, Robotics, Social Systems, Economic Systems and other. Of particular value to both the contributors and the readership are the short publication timeframe and the world-wide distribution and exposure which enable both a wide and rapid dissemination of research output.

Indexed by SCOPUS, DBLP, WTI Frankfurt eG, zbMATH, SCImago.

All books published in the series are submitted for consideration in Web of Science.

More information about this series at https://link.springer.com/bookseries/13304

Kofi Kissi Dompere

The Theory of Problem-Solution Dualities and Polarities

Information-Decision-Choice Foundations of the Unity of Knowing and the Unity of Science

Kofi Kissi Dompere
Department of Economics
Howard University
Washington D.C., WA, USA

ISSN 2198-4182 ISSN 2198-4190 (electronic)
Studies in Systems, Decision and Control
ISBN 978-3-030-90281-0 ISBN 978-3-030-90279-7 (eBook)
https://doi.org/10.1007/978-3-030-90279-7

This Springer imprint is published by the registered company Springer Nature Switzerland AG
The registered company address is: Gewerbestrasse 11, 6330 Cham, Switzerland

Preface

This monograph is epistemically special in its orientation as it is concerned with the relational structures of information, knowledge, decision-choice processes of problems and solutions in theory and practices relative to investment and capital accumulation processes by the way of the logic of the economic theory of production specified in the spaces of input-output and production-consumption dualities. It is about explanatory and prescriptive conditions of diversity and unity principles of knowing and information-knowledge systems under the variety input-output processes of horizontal and vertical supply-demand chain dynamics. The monograph is a continuation in the sequence of epistemic works on the theories of *info-statics, info-dynamics, entropy* and their relational connectivity to information, language, knowing, knowledge, cognitive practices relative to variety *identification problem-solution dualities* and *variety transformation problem-solution dualities* in all areas of knowing. It is about economic-theoretic approach in understanding the diversity and the unity of knowing and science through decision-choice actions over the space of problem-solution dualities and polarities in relation to elements in the space of incentive-disincentive dualities embedded in the space of real cost-benefit dualities. In this approach, all human endeavors are viewed as variety transformations where all neuro-decision-choice actions are viewed in terms of neuro-decision-choice balances of opposites in socioeconomic variety transformations such that the opposites are mutually constraining with a change in relation, property or both.

The problem-solution dualities are argued to connect all areas of knowing including science and non-science, social science and non-social science into unity with diversities under neuro-decision-choice actions, where individual and collective creativities play important determining roles to set variety solutions against variety problem over the space of problem-solution dualities. The concept of diversity is defined and explicated to connect with the tactics and strategies of neuro-decision-choice actions over the space of problem-solution dualities. The concepts of problems and solutions are defined and explicated in the space of relativity rather than in the space of absoluteness. All neuro-decision-choice activities are viewed in terms of the logic of the economic theory of production over the space of real cost-benefit dualities with identities of real costs and real benefits broadly defined in terms of characteristic-signal dispositions. In this epistemic process, the variety identities may be distorted through the decoding of variety signal dispositions to create false interpretations of the characteristic dispositions that may deviate from the true variety characteristic dispositions leading to misidentifications of varieties.

There are two important structures to pay attention to in the processes of knowing. One is a movement from the ontological space to the epistemic in identifying ontological varieties. The other is intra-epistemological movements, where propaganda, misinformation and disinformation may occur in the space of source-destination dualities with an intent to influence the inputs and the directions of neuro-decision-choice for a source advantage. In terms of intellectual production, the variety misidentification may put us on a possible wrong path of knowing and information-knowledge production to create increasing real costs over the real benefit of knowing in the space of mistake-correction dualities. In terms of general transformation dynamics, the misinterpretations may lead to a path of neuro-decision-choice actions to create increasing costs over real benefit in the space of success-failure dualities viewed in terms of elements in the space of progress-retrogress dualities within the space of individual-collective dualities in the same society or within the space of domestic-international dualities.

All analytical and non-analytical concepts are specified and defined in the space of relativity to provide an infinite framework and space of variety comparisons, differences, commonness, similarities and ranking under a preference ordering in establishing sequences of visions, goals, objectives for selections and missions under the socioeconomic principles of non-satiation. All definitions follow the conceptual framework of the theory of definitions in establishing the relevant characteristic-signal dispositions required for linguistic meanings and identities within the same language. The concepts of diversity, the unity of knowing and the unity of science are structured around the nature of variety problem-solution dualities over the space of static-dynamic dualities. It is argued that the diversity and unity of knowing and science are explainable through neuro-decision-choice actions over the infinite space of problem-solution dualities to provide an expanding information-knowledge system.

The interactions between the system of neuro-decision-choice actions and the space of variety problem-solution dualities define the meaning of life and are explainable through the logic of economic theory of production, where diversities are created within the general unity of production to meet the diverse cross-sectional preferences in time and changing time series of preferences of intrageneration and intergeneration dynamics. The way these interactions take place over the space of the problem-solution dualities defines the essence of life in diversity and unity of knowing for continual socio-natural interactions. Similarly, the manner in which these interactions take place specifies the individual and social creativities relative to individual and social visions within the space of imagination-reality dualities that helps to define the space of possibility-impossibility dualities linked to the space of possible world-impossible world dualities which may help the understanding of cognitive illusions, innovations, anticipations, curiosity and other behavioral characteristics over the space of quality-quantity dualities.

The monograph is used to explore the explanatory and prescriptive pathways that may contribute to the interdisciplinary education in teaching and learning, and foster innovatively collaborative research in knowing and information-knowledge production, especially, for the understanding of the give-and-take sharing modes of categories of knowing through neuro-decision-choice actions over the space of problem-solution dualities. It is, then, argued that any problem-solution process is a variety destruction-replacement process, where every variety problem is simultaneously an offspring of a variety solution and a parent to a variety problem in a never-ending process to an unknown destination that drives the formation of visions with goals and objectives as sequential missions and actions. From the logic of economic theory of production, the theory of intellectual production is divided into two of the theory of intellect investment as flows and the theory of intellectual information-knowledge accumulation as stocks. The theory of intellectual investment is the theory of knowing, and the theory of intellectual information-knowledge accumulation is the theory of intellectual capital in dynamic system. The theories of intellectual investment and information-knowledge accumulation constitute the general theory of intellectual stock-flow disequilibrium dynamics over the space of diversity-unity dualities to establish the conditions of the space of fundamental-applied dualities and the general input-output dualities relative to the system of consumption-production processes of cognitive existence.

The theory of knowing is explained and presented as an investment in intellectual flows from problem solving, and the theory of information-knowledge system is explained and presented as intellectual capital accumulation of success-failure outcomes of decision-choice actions over the space of the problem-solution dualities and polarities. The conceptual system is developed with the fuzzy paradigm of thought over the space of relativity that is characterized by the principle of opposites composed of duality and polarity, where every variety is characterized by a series of dualistic-polar conditions, where knowledge resides in ignorance and ignorance resides in knowledge in the space of variety ignorance-knowledge dualities. The theory of knowing and the theory of knowledge, just like the theories of basic economic investment and capital accumulation, are simultaneously developed as the theory of production, where behind every variety is a characteristic-signal disposition to reveal its identity. The intellectual production is a variety problem solving which is about variety destruction-replacement process leading to neuro-decision-choice outcomes in the space of ignorance-knowledge dualities contained in the space of imagination-reality dualities where cognitive illusions, innovations, anticipations, curiosity and other behavioral characteristics instead of belonging to the space of imagination may creep into the space of realities over the space of quality-quantity dualities.

The neuro-decision-choice actions over the space of problem-solution dualities and the space of variety destruction-replacement processes are guided by the assessments of real cost-benefit conditions and rationality of cognitive agents with information-knowledge of quality control as provided by relationally entropic tests within the interactive spaces of certainty-uncertainty dualities, doubt-surety dualities and decidability-undecidability dualities as may be abstracted from the space of possibility-probability dualities connected to the space of necessity-freedom dualities, where possibility houses necessity and initial conditions of categorial conversion, and probability houses the freedom and sufficient conditions from the social philosophical consciencism to realize variety imaginations in the space of reality. The information-knowledge development is set to distinguish the initial spaces of ontological space in which nature operates and the epistemological space in which cognitive agents operate to know and take advantage of ontological varieties.

The Monograph

Given the interactive conditions among the space of decision-choice activities, the action space and the space of problem-solution dualities as explaining diversity and unity of categorial knowing and science, the monograph is set to provide answers to several questions. These questions come to us as a set of primary questions and a set of derived questions. They center around the role of how neuro-decision-choice actions over the space of problem-solution dualities bring about variety knowing and how the variety knowing processes generate information-knowledge accumulation irrespective of the category of knowing. The process of knowing and information-knowledge accumulation proceed from the space of acquaintance to the space of descriptions in the epistemological space as a derivative from the ontological space as sub-spaces of the ontological-epistemological dualities. The outcomes from the space of acquaintance are the results of observations and encounters, while the outcomes from the space of descriptions are the results of logical operations with a paradigm of thought. The outcomes of the processes of knowing and information-knowledge accumulation are the results of neuro-decision-choice actions, where information-knowledge processes are also input-output processes to generate an organic self-correction system based on subjectivity of individual and collective creativities irrespective of the areas of knowing.

The space of information-knowledge input-output processes is continually expanding under socio-natural decision-choice activities creating increasing possibilities with necessity and anticipations, and expanding the space of probability with freedom and expectations making easier for the discovery of other varieties in the space of actual-potential dualities through the neuro-decision-choice activities over the space of imagination-reality dualities. In other worlds, the knowing processes and information-knowledge processes are not only interdependent but the results of neuro-decision-choice actions irrespective of the category of knowing in give-and-take sharing modes. Every area of knowing is a category of problem-solution dualities under decision-choice actions passing through the spaces of uncertainty-certainty dualities, doubt-surety dualities and decidability-undesirability dualities, where the outcomes are nothing more than experiences, and information-knowledge accumulation is either justified or unjustified experiences. The input-output processes are composed of intellectual investment processes through knowing and the intellectual capital stock processes through information-knowledge accumulation under the general entropy for quality control.

The neuro-decision-choice processes are established by variety preference ranking based on variety real cost-benefit dispositions, and variety selections based on the real cost-benefit rationality in relation to a goal-objective element or a set of goal-objective elements in visionary or non-visionary conditions in the space of imagination-reality dualities. The definition of real cost as a real opportunity cost is to link the decision-choice actions of economic production and neuro-decision-choice processes over the space of production-consumption dualities and the principle of resource scarcity to the conditions of physical processes of organic matter-energy equilibrium in accordance with the first law of thermodynamics, where neither matter nor energy can be created or destroyed but can have inter-categorial and intra-categorial transforms with the same matter such that socioeconomic variety production and socio-natural variety transformations are in continual disequilibrium dynamics which is different from the conditions of market equilibrium analysis.

The real opportunity cost relates to the first and fifth laws of info-dynamics, where the previous variety cost information is always indestructible to expand the space of the potential varieties while a new variety information-knowledge item is created to expand the space of reality as a sub-space of the space of actual varieties, and hence the amount of information-knowledge accumulation is continually in disequilibrium dynamics to satisfy the second and fourth laws of info-dynamics [4]. In its simplest form, the first law of info-dynamics states that neither information nor knowledge as established by characteristic-signal dispositions can be destroyed even though the variety can be destroyed by transformation in support of the principle of continual creation within the matter-energy equilibrium. The amount of information-knowledge accumulations in the observable and non-observable universe is continually expanding through variety destructions, transformations, dividedness and differentiations within the space of matter-energy stock-flow equilibrium.

Information is continually being created through categorial conversions from one variety to another and from one form to multiple forms by variety differentiations, or from one process to another or multiple processes, or from one state to another or multiple processes with continual changes, moves, controls and accumulation relative to relation and properties. The epistemic reflection projection conditions that the information-knowledge accumulation is always created from some varieties induced by categorial conversions based on socio-natural technologies to create appropriate transformational processes to bring about new information and possible knowing while retaining the old information in the space of ignorance-knowledge dualities [1–4]. Comparative analytics of info-statics and thermodynamic are undertaken to show the information property of matter-energy configuration.

The comparative analytics presents the differences and similarities between info-dynamics and thermodynamics:

(1a) Commonness: Information is a property of mater-energy configuration without which matter is indistinguishable from energy and energy is indistinguishable from energy; (1b) information is non-existence without matter-energy configuration;

(1c) the awareness of matter-energy configuration and the identities of matter and energy are made possible by information as seen in terms of variety characteristics. (2a) Differences: Matter-energy configuration is always in equilibrium state in the sense that matter-energy is in flow equilibrium as well as in stock equilibrium.

(2b) Information-knowledge configuration is always in disequilibrium state in the sense that the ontological information-knowledge configuration is continually in flow and stock disequilibrium and this ontological information-knowledge disequilibrium is transformed to the epistemological information-knowledge process from the potential space to the space of the actuals through the neuro-decision-choice activities over the space of imagination-reality dualities transforming variety imaginations to variety realities under fuzzy-stochastic conditionality satisfying the defined cognitive transversality conditions from the journey from ignorance space to the space of knowledge.

In a simple reflection, we may say that thermodynamics is about energy transformations within matter-energy equilibrium configuration, while variety energy transformation destroys one variety energy and replaces it with another variety energy by categorial conversion or transformation, where the previous variety energy enters into the potential space with the possibility of retransformation to the space of the actuals, where it is not available. The info-dynamics are about variety transformations of characteristic dispositions of states, processes, relations and properties within the matter-energy equilibrium, while the information about the transformed variety remains in the space of actual and the information about the new varieties becomes additions to the existing information-knowledge stocks. In this sense, the transformed varieties no longer belong to the space of the actuals, their characteristic dispositions, however, are available while the characteristic dispositions of the new variety energies become part of the space of the actuals and addition to the information-knowledge accumulation. In other words, the variety energy or the variety matter disappears by transformational process in the space of actuals and then enters the space of potential through cognitive transforming activities in the space of imagination-reality dualities.

The conditions of characteristic dispositions of all transformed varieties remain in the space of the actuals as part of telescopic past and present, while their potential for reentry into the space of the actual remains in the telescopic future. The thermostatics is about the studies of heat energy equilibrium within the thermodynamics, while the info-statics is about the studies of variety identification within the info-dynamics of the information-knowledge stock-flow disequilibrium dynamics. It must be understood in passing that the formations of new galaxies in the observable and non-observable universe are ongoing and support the theory of info-dynamics while their observations through the space of acquaintance-description dualities support the theory of info-statics. The discoveries of new galaxies in the existence and in the making cannot be used to justify the idea of expanding universe under matter-energy stock-flow equilibrium.

The quality control of information-knowledge items obtained from neuro-decision-choice actions over the space of problem-solution dualities is dealt with under entropic conditions over the space of ignorance-knowledge dualities with acceptance conditions defined over the space of variety doubt-surety dualities through the conditions from the space of certainty-uncertainty dualities under the journey from possibility to probability as we traverse through the space of ignorance-knowledge dualities.

The theory of problem-solution duality and polarity is used to deal with (1) the understanding of problem-solution dynamics as the unifying force of knowing in all areas of human endeavor, (2) the understanding of infinite problem-solution process, (3) the understanding of preferences as shaping knowing and information-knowledge development, (4) the understanding of cost-benefit conditions as defining the incentive space for problem-solving dynamics, neuro-decision-choice actions and transformations relative to change in variety relation, change in variety properties and simultaneity of both, (5) the understanding of relations among the neuro-decision-choice actions, vision, goals, objectives, knowing, knowledge and ignorance, (6) the understanding of an epistemic relation between the spaces of actual-potential dualities and the imagination-reality dualities and (7) the multiplicity of relation among a number of sub-spaces of dualities and polarities contained in the space of imagination-reality dualities.

The monograph is, thus, devoted to the development of a general framework to deal with categorial diversities, interdependency of knowing, information-knowledge systems through the dynamics of the space of problem-solution dualities to understand the human decision-choice behavior over and the space of ignorance-knowledge dualities. The neuro-decision-choice activities are linked to the meaning and essence of human life, creation of new institutions due to the rise of new problems and transformation of social organizations through dualistic-polar conflicts. In this respect, the relational condition among possibility, anticipations, possible world, impossible world, necessity and imagination are examined as we move from the potential space to the space of actuals. Similarly, the relational conditions among probability, expectations, freedom, risk and failure are examined as we finally enter the space of reality as a sub-space of the space of actuals, where the epistemic journey through the space of imagination-reality dualities that contains hope, creativity, beliefs and other qualitative characteristic that are driven by the principle of non-satiation generating sub-spaces of peace-war dualities, kindness-wickedness dualities, good-evil dualities and many of such human dualistic-polar behaviors to abstract advantage from conditions of the space of real cost-benefit dualities.

The monograph is an attempt to find answers to the following broad general questions in the space of ignorance-knowledge dualities in relation to epistemic conditions over the space of doubt-surety dualities under cognitive limitations, limitationality and limitativeness as seen from the space of certainty-uncertainty dualities as cognitive agents travel from the imagination space as a sub-space of potential space to the possibility space, from the possibility space to probability space and from the probability space with cognitive capacity limitations to the space of the realities as a sub-space of the space of the actuals and then to the potential space in complex processes of decision-choice actions to create variety transformation processes over the epistemological space within the space of ontological-epistemological dualities.

The ontological transformation processes are substantially different from the epistemological transformation processes. The natural processes are such that variety transformations are directly defined in the space of actual-potential dualities. The epistemological transformations pass through the sub-space of imagination-reality dualities which contains three important sub-spaces of dualities to effect the epistemological transformation within the space of actual-potential dualities.

The spaces are (a) the space of imagination-possibility dualities, (b) the space of possibility-probability dualities and (c) the space of probability-reality dualities, where the space of realities is a sub-space of the space of the actuals and the space of imaginations is a sub-space of the space of potentials. All these sub-spaces are contained in the space of imagination-reality dualities as a sub-space of the space of actual-potential dualities which is improper sub-space of ontological-epistemological dualities. This rectangular process in knowing and information-knowledge system is explainable by cognitive capacity limitations that generate the space of necessity-freedom dualities over the space of cost-benefit dualities which is supported by an imperfect information structure. The cognitive neuro-decision-choice processes are such that cognitive agents cannot epistemologically travel directly on the pathways from the space of potential to the space of the actuals in terms of operating directly within the space of actual-potential dualities. That direct pathways from the potential space to the space of actuals are defined by variety neuro-decision-choice processes that are supported by perfect information structures which are also perfect knowledge structures which require no quality control process because the outcomes are the identities. These questions are provided for the examinations in the process of understanding of the diversity and unity in knowing, information-knowledge processes and neuro-decision-choice processes over the variety spaces of varieties and the problem-solution dualities as:

(1) In what space can one find the activities of knowing and decision-choice actions?
(2) Similarly, in what space can one find the activities of information-knowledge production and decision-choice actions?
(3) What are the similarities and differences between knowing and knowledge, and between the theory of knowing and the theory of knowledge?
(4) Who are the subjects of the activities of knowing and information-knowledge production?
(5) Who and what are the objects under search, research, knowing and information-knowledge production and what roles do problems and solutions play in the diversity and unity of knowing and information-knowledge production?
(6) In what space can one find problems and solutions and what are the relationships between problems, solutions and states within the space of static-dynamic dualities?
(7) What are the relational structures among neuro-decision-choice actions, information-knowledge states, problem-solution elements and socio-natural varieties?

(8) How are these questions related to diversity and unity of knowing, diversity and unity of science diversity and unity of non-science and the fundamental-applied dualities in general, knowing, science and knowledge system as viewed in the input-output space with input-output supply chain?

These are some fundamental questions in cognitive search for answers to problems of meaning, variety existence and variety changes as they relate to environmental sustainability, human survivability, human capacity to produce and reproduce forms of new ideas in relation to dualistic-polar conditions of nature and society, construction and destruction, war and peace, conflict and resolution, tension and change, stagnation and progress, stability and chaos, energy and power, ignorance and knowledge and many others with dualistic-polar relations. The monograph attempts to deal with these questions where neuro-decision-choice actions are the center of these input-output processes over the space of problem-solution dualities.

The Organization of the Monograph and the Chapter Summaries

The monograph is organized with a preamble, prologue, preface, acknowledgment and seven chapters which are concluded with an epilogue. Each chapter deals with a specific set of analytical difficulties within the development of the theory of the problem-solution duality and polarity in support of our general understanding of the principles of diversity and unity of knowing, science and general cognition.

There is a *preamble to the monograph.* The preamble is composed of series of quotations from important works of different authors in philosophy, science, mathematics, economics, decision-choice theories and other relevant areas to the information-knowledge questions about the general science and the art of knowing under cognitive capacity limitations that generate questions over the spaces of certainty-uncertainty dualities and projected to the space of doubt-surety dualities. The preamble is used to initialize the idea that all information-knowledge systems are generated by knowing actions irrespective of the area of human endeavor through neuro-information-decision-choice processes over the space of problem-solution dualities within the space of ontological-epistemological dualities. Again, the objective of the preamble is to introduce the reader to the conflict zones of knowing and to point out the epistemic zones of irreducible disagreements in knowing as intellectual investment flows and information-knowledge system as intellectual capital accumulation enterprise in the general space of production-consumption dualities. Here, the conflict zone of knowing will reveal the conditions of paradoxes, ill-posed problems and many other inconsistencies of information-knowledge accumulation.

The *prologue* is tightly structured to follow the preamble. The prologue introduces a set of essential epistemic propositions in the theoretical system on problem-solution duality and polarities and its relationships with investments and capital accumulation in intellectual and variety spaces. Based on these essential propositions, the concepts of methodology, techniques and methods for discussing the propositions are defined, explicated and analyzed to provide conditions of their differences and commonness in theoretical and applied analytics. The techniques and methods are toolbox collections in any methodology of interest for knowing, thinking and reasoning over the space of the problem-solution dualities. The prologue is also used to introduce the concepts of exact and inexact information structures and how they relate to the space of absoluteness and the space of relativity to affect the methodology of thinking and reasoning over the space of decidability-undecidability dualities. The prologue points to the path of the theoretical journey and the parameters of containments and deviations regarding the applicable areas over the space of question-answer dualities and polarities to establish diversity and unity principles in knowing, science and general information-knowledge accumulation.

The *preface*: The prologue is followed by a preface which introduces the motivation and conditions for writing the monograph, what it is about and how the conceptual framework of the monograph is connected to other previous monographs by the same author on information-knowledge processes and general conditions of categories of knowing. The general conceptual frame makes explicit the structure of the theory of problem-solution duality and polarity and how the structure constitutes the complex foundation of knowing and information-knowledge accumulation in all areas of cognitive activities over the space of fundamental-applied dualities in knowing. The analytical connectedness of these monographs is found in the consistent use of the principle of opposites with relational continuum and unity, inexact information with vagueness and incompleteness, and fuzzy paradigm of thought with its mathematics, logic and corresponding toolbox of techniques and methods of analysis, where knowing is seen as an intellectual investment flow in a discovery of knowledge and information-knowledge accumulation is seen as an intellectual capital stocks of discoveries, providing cognitive agents with intellectual factories of production and consumption in the general input-output space and neuro-decision-choice actions over the space of problem-solution dualities. Together, these monographs present the meaning and essence of my works, research agenda and possible contribution to the organization of knowing, research and teaching, where the system of information-knowledge stock-flow disequilibrium dynamics is seen as part of the system of economic productions and related to continual neuro-decision-choice dynamics over the space of problem-solution dualities that resides in temporary equilibrium states and permanent disequilibrium dynamics. In a sense, the interrelated works provide foundations for explaining the continual process of transformation of varieties in terms of parent-offspring replacement processes at the levels of both the fundamental and the applied knowledge on the dynamics of emergence of the new and the disappearance of the old, where the end is in the beginning and the beginning is in the end in a continual past-present-future telescopic connectivity as well as taking a critical look at the logic of economic theory of production in the understanding of intellectual stock-flow disequilibrium dynamics.

Chapter 1 sets the stage in understanding the similarities and differences between the theory of knowing and the theory of knowledge from an approach of neuro-decision-choice activities over the spaces of problem-solution dualities and the input-output dualities. The differences and similarities of the theory of knowing and the theory of knowledge are presented by the way of definitions and supported by explicative analytics. The theory of knowing is developed as the unity of the theory of info-statics directed to the neuro-decision-choice activities over the space of *identification problem-solution dualities* and the theory of info-dynamics directed to the neuro-decision-choice activities over the space of *transformation problem-solution dualities* both in relation to *what there was* (past), *what there is* (the present) and *what would be* (the future). The chapter sets the epistemic stage to state a set of questions and discuss the conditions for their answers within the space of variety problem-solution dualities in terms of production and quality control, where production relates to constructionism and quality control relates to reductionism in the space of constructionism-reductionism dualities such that most of the existing theories of knowledge are simply about the discussions of quality control methods and hence belong to the reductionism dynamics, where the concept of justified belief in defining knowledge is not analytically useful in the understanding of knowing, variety production and quality control processes. The chapter is, then, used to introduce several analytical concepts and their analytical representations to prepare the discussion on the pathway connectivity in the chapter that follows.

Chapter 2 presents a conceptual pathway analysis of the theory of knowing and the theory of knowledge as neuro-decision-choice activities in the space of ontological-epistemological duality, where knowing and information-knowledge production proceed from the space of imaginations as a sub-space of the potentials to the space of possibility to the space of probability and to the space realities as a sub-space of the space of actuals in a continual search for information and knowledge through the neuro-decision-choice activities in support of general neuro-decision-choice actions over problem-solution dualities relative to the spaces of variety identifications and transformations. The problem-solution dualities establish the principle of diversity and the unity of knowing, the principle of diversity and the unity of information-knowledge accumulation as well as the principle of diversity and the unity of science. The concepts of ontology and epistemology, and the corresponding spaces are defined and explicated to be useful to economic-theoretic approach to the understanding of knowing as production of intellectual investment flows and information-knowledge accumulations as intellectual capital stocks. The methods of the conceptual pathway analytics are introduced and discussed to provide an understanding of the unbreakable rectangular connectivity of the epistemic processes of information-knowledge stock-flow disequilibrium dynamics.

Chapter 3 advances the concept of epistemic space. It is then defined and explicated for the understanding of fuzzy spaces through the principle of opposites in relational continuum and unity in the development of the theory of knowing as the theory of intellectual investment flows and the theory of information-knowledge accumulation as the theory of intellectual capital stocks. In this framework, the theory of intellectual investment-capital process is seen as the theory of production of intellectual stock-flow disequilibrium dynamics, where the intellectual information-capital stocks provide continual capital-service input flows into the general neuro-decision-choice space relative to the space of fundamental-applied dualities, and where the intellectual investment flows provide a continual updating of intellectual stocks in the space of quality-quantity dualities. The relevant spaces and sub-spaces are introduced leading to the definition and explication of the epistemic space and the development of the theory of epistemic space for the understanding of both the theories of knowing and knowledge as seen in terms of neuro-decision-choice activities over the spaces of problem-solution dualities, input-output dualities and production-consumption dualities under the principle of opposites with relational continuum and unity.

Chapter 4 presents the theory of problem-solution dualities as the first part of the general theory of opposites, where the second part is the theory of problem-solution dualities which will be taken up in the next chapter. It discusses the conditions of the nature of problems relative to solutions in dualistic-polar structures with relational continua, unity and give-and-take-sharing modes in the space of fundamental-applied dualities to provide conditional connectivity to the development of the theory of polarities. The concepts of ontological and epistemological spaces are introduced, defined, explicated and connected to the space of problem-solution dualities. The connectivity of these concepts to the roles they play in the knowing as intellectual investment flows and information-knowledge system as intellectual capital accumulation stocks is then explained. The space of incentive-disincentive dualities governing the neuro-decision-choice processes is introduced and linked to the space of cost-benefit dualities in order to establish the concepts of primary-derived spaces and parent-offspring processes relative to the elements in the problem-solution dualities as transformational elements where problems are transformed into solutions to give rise to new problems over the multiplicity of cost-benefit or destruction-construction enveloping that establishes the space of problem-solution genetic paths of socio-natural variety dynamics. The space of problem-solution dualities is established over the space of imagination-reality dualities which is a sub-space of actual-potential dualities that constitutes the space of ontological-epistemological dualities.

Chapter 5 is used to present the conditions of epistemological space for the understanding of neuro-decision-choice activities in knowing as intellectual investment flows in understanding the nature of cognitive actions over the space of problem-solution dualities and how such actions relate to information-knowledge accumulations as intellectual capital stocks, the service flows of which become inputs into neuro-decision-choice activities. It presents a theoretical understanding of the space of problem-solution dualities as embedded in the space of imagination-reality dualities which is a sub-space of the space of actual-potential dualities. The concepts and phenomena of problem, solution and the space of problem-solution dualities are defined through the conditions of cost-benefit dualities over the space of relativity. The concepts of criterion, criteria space and neuro-decision-choice space are verbally and algebraically introduced and defined with explanations to their relationships to knowing and information-knowledge processes. The connectivity of these concepts to the space of problem-solution dualities is specified and explained. The important and prominent role played by conditions of cost-benefit duality are made explicit and linked to the criteria space. The problem-solution process is argued to link the knowledge by acquaintance to knowledge by description through a general paradigm of thought irrespective of whether an area is defined as science or non-science, where the classifications of areas of knowing and problem-solution dualities are viewed in terms of fuzzy sets with fuzzy-set closures and fuzzy-stochastic conditionality. The chapter also presents analytical connectivity within the space of ontological-epistemological dualities and how the ontological space is the primary identity existing with information-knowledge equality and the epistemological space is a derived identity existing with information-knowledge inequality.

Chapter 6 presents discussions on the meanings of the concepts and phenomena of category, categorial conversion and their uses in the understanding of knowing, knowledge and the principles of diversity and unity that they may entail as an entry point to the understanding of the interactive behaviors of problem-solution structures as parent-offspring dynamics. The concepts are introduced to deal with transformations and changes within the space of quantitative-qualitative dualities as they relate to info-statics and info-dynamics of varieties in all areas of knowing as intellectual investment flows and information-knowledge accumulation as intellectual capital stocks. Alongside, the theory of categorial conversion is a visitation to the theory of philosophical consciencism and the discussions on its relation to neuro-decision-choice processes, education, preference formation, varieties of human progress, and increasing individual and social abilities in the space of variety creation-destruction dualities as seen in terms of individual and collective curiosities over the space of imagination-reality dualities and continual variety transformations. The theories of categorial conversion and philosophical consciencism are mutually interdependent, inter-supportive and inseparable, just like the theories of info-statics and info-dynamics are mutually interdependent, inter-supportive and analytically inseparable. Similarly, the theories of categorial conversion and info-statics are mutually creating while the theories of philosophical consciencism and info-dynamics are also mutually interdependent, creating and inseparable. The frameworks of both the theories of categorial conversion and info-statics are shown to be linked to the frameworks of possibility and necessity, while the frameworks of the theories of philosophical consciencism and info-dynamics are shown to be linked to the frameworks of probability and freedom over the spaces of neuro-decision-choice activities and the action space as they relate to conditions over the spaces of identification-transformation and imagination-reality dualities within which the concepts and the phenomena of phantom, ill-posed, unresolved, unintended and other such problems are defined and discussed in relation to paradigms of thought and cognitive capacity limitations in knowing. The knowing processes and information-knowledge process are shown to be cognitive journeys between variety knowing and variety knowledge in the space of ignorance-knowledge dualities contained in the space of imagination-reality dualities as a sub-space of actual-potential dualities.

Chapter 7 contains discussions on the concepts and phenomena of problem-solution polarities and the role they play in the understanding of knowing as intellectual investment flows and information-knowledge accumulation as intellectual capital stocks in order to establish the principles of diversity, relational continuum and unity that they entail. The relational structure between duality and polarity is made explicit in an epistemic organicity, where the diversities are in a multiplicity of give-and-take sharing mode. The poles of the polarities are established by the residing duality such that the negative (positive) duality defines the negative (positive) pole while the dominance of the negative or the positive pole reveals the identity of the polarity in inter-categorial transformations. The categorial conversion establishes the external conditions for dualistic-polar change, while the philosophical consciencism establishes the internal conditions for dualistic-polar change over the space of variety quantitative-qualitative dualities relative to the space of statics-dynamics dualities with extensive discussions on the variety dynamics in knowing as production of intellectual investment flows and information-knowledge accumulation as production of intellectual capital stocks.

The *epilogue* concludes the monograph with views and reflections for further research and development in the understanding of the interactions of the spaces of neuro-decision-choice actions, problem-solution dualities, knowing and information-knowledge stock-flow disequilibrium dynamics relative to matter-energy stock-flow equilibrium dynamics under the principles of diversity, unity and opposites with relational continua. The directions of the role that institutions and organizations play in knowing and the knowledge processes are discussed in relation to information-knowledge processes as a system of input-output processes under a system of neuro-decision-choice actions with *cognitive capacity limitations* composed of *information-knowledge capacity limitations* and *paradigmatic capacity limitations*, where goals, objectives and visions are defined and expressed in the space of imagination-reality dualities as well as the manner in which the space of imagination-reality dualities is contained in the space of actual-potential dualities for the understanding of the general human behavior and cognitive actions over the space of production-consumption dualities. The epilogue is concluded with discussions on the directions for the examination of imagination, reality, belief, hope, belief, curiosity and the relevant methodology in knowing and information-knowledge stock-flow dynamics through the space of problem-solution dualities, where categorial inter-relationalities for the development future research universities are discussed.

The direction of the theory of problem-solution dualities and polarities points to an epistemic idea that in the neuro-decision-choice actions over the space of information-knowledge accumulation, knowing always comes against difficult conditions set by new problem-solution dualities to challenge our methods and techniques of construction-reduction dualities within an accepted methodology, where adjustments are made to abstract new variety solutions to new variety problems under dualistic-polar conditions. Our research, learning and teaching in the knowing process through neuro-decision-choice actions over the space of problem-solution dualities must reflect how new experiential experiences from the spaces of success-failure and doubt-surety dualities overcome the limits of epistemic familiarity, give rise to the development of new techniques and methods and develop new thinking and reasoning toward new means of obtaining knowledge by continuously creating consistent tools to enhance the individual and collective cognitive abilities for dealing with new and anticipated problem-solution dualities. Our global research, teaching and learning must also instruct us of the disequilibrium dynamics of our knowledge system where old problem-solution dualities will give way and give rise to new problem-solution duality in the parent-offspring process such that *learning by doing* must be seen as an important part in knowing irrespective of the area of human endeavor. In this respect, the gains in the family of fundamental information-knowledge systems are to widen the space of successes of the family of applied information-knowledge systems in terms of learning by doing.

In this respect, the epilogue is devoted to the initial discussions on the debate on the concept of unified sciences and the cognitive telescopic future which are done through the theory of knowing and how the theory of knowing must be seen in terms of neuro-decision-choice actions over the spaces of problem-solution and production-consumption dualities. The concepts of unified sciences, knowing and knowledge are used to establish the intellectual and research framework that extends to the concept of the theory of unified applied sciences at all levels of neuro-decision-choice actions, engineering sciences broadly defined to include all forms of human creation over the production-consumptions space by shaping the culture of intellectual change and transformation over the general space of problem-solution dualities, where all neuro-decision-choice actions draw their guidance from the philosophical consciencism at the generation time point. The concepts of unified sciences and knowing are further extended to the general unity of knowledge over the epistemological space, where the fundamental information-knowledge system and the applied information-knowledge system involving non-science, sciences, engineering, non-engineering, explanatory sciences, prescriptive sciences and the whole information-knowledge system find meaning and cognitions in human existence with neuro-decision-choice system under the dynamics of individual, collective and social preferences. The proposition pointed out in the epilogue is that the unity of knowledge and unified sciences must find expressions in information as the third demission of universal existence.

The direction of the theory of problem-solution dualities and polarities points to an epistemic idea that in the neuro-decision-choice actions over the space of information-knowledge accumulation, knowing always comes against difficult conditions set by new problem-solution dualities to challenge our methods and techniques of construction-reduction dualities within an accepted methodology, where adjustments are made to abstract new variety solutions to new variety problems under dualistic-polar conditions. Our research, learning and teaching in the knowing process through neuro-decision-choice actions over the space of problem-solution dualities must reflect how new experiential experiences from the spaces of success-failure and doubt-surety dualities overcome the limits of epistemic familiarity, give rise to the development of new techniques and methods and develop new thinking and reasoning toward new means of obtaining knowledge by continuously creating consistent tools to enhance the individual and collective cognitive abilities for dealing with new and anticipated problem-solution dualities. Our global research, teaching and learning must also instruct us of the disequilibrium dynamics of our knowledge system where old problem-solution dualities will give way and give rise to new problem-solution duality in the parent-offspring process such that learning by doing must be seen as an important part in knowing irrespective of the area of human endeavor.

In this respect, the gains in the family of fundamental information-knowledge systems are to widen the space of successes of the family of applied information-knowledge systems in terms of learning by doing. The epistemic limits behave differently under perfect information and the classical paradigm and under inexact information and fuzzy paradigm. In the theory of knowing as a theory of intellectual investment flows and the theory of information-knowledge accumulation as a theory of intellectual capital stocks, we must always distinguish between potentialization and imagination, between actualization and realization and how they relate to neuro-decision space of actions and then to the space of incentive-disincentive dualities as contained in the space of cost-benefit dualities. The variety potentializing is achieved through an imaginational destructive process of the variety actual, while the variety actualization is achieved through a creative realization process of variety reality in the space of intellectual stock-flow dynamics as it relates to the dynamics of the spaces of production-consumption dualities, input-output dualities, construction-destruction dualities and supply-demand dualities.

Washington D.C., USA Kofi Kissi Dompere

Contents

Preamble

Noise, sound, song, speech: These also reveal a cause-and-effect world to which we are so accustomed that the phenomenon no longer affects us. Yet, this phenomenon is formidable once we train our attention on it. Symbolic or hieroglyphic writing that becomes ideogrammatically and finally alphabetical is similarly veritable magic, since it allows mam to think and live a world unto himself, and to transmit to a reader his thoughts, sensations, experiences or will by means of simple signs without any other direct personal communication, without even the help of the spoken world.

In writing as in speech, we see an evolution or, more correctly, a "transformation of means" corresponding to the growing complexity of the causes and aims of the transmission.

This transformation of means does not at all signify an evolution toward perfection, for an evolution can equally go toward an imperfection as it signals only a transformation in conformity with the time and ambient conditions.

Had the means of transmission of consciousness remained pure, the spoken word would be a noise or a modulated and rhythmic sound without complication of grammar and syntax, expressing everything by rhythm only, by the intensity and the sequence of sounds and their variations: The writing would be made of images representing objects and conventional ideogram for abstract [1].

Elementary explications concerning numbers lead one to infer a cosmic character for number and a special mentality for its comprehension. How can contemplation of Number shed light on cosmic laws? This can be done only by considering the necessities immanent to numbers. There is no doubt about it. That is why number belongs to the "Eternal World" of Plato's Timaeus. Number is that non-created and immutable world; thus, the science of number is the science of unity: It shows us intelligibly that all proceeds from unity and returns to it through diversity. This diversity is precisely our world created in the image of the Eternal World's example [1].

The development of natural science and philosophy has shown that it is impossible to explain the whole wealth of the relationship between theory and experience and answer many important questions posed by the progress of science (particularly the question of the character of scientific revolutions) if one confines oneself to formal-logical analysis in considering these problems. A far more comprehensive approach embracing all the essential elements in man's search for knowledge is required [2].

Physics is regarded as the leader of modern natural science. It owes its dominant role to the fact that it is concerned with the primary fundamental elements and properties of matter, and also that it has achieved the highest degree of organization of knowledge and thus reflects most strikingly the basic features of modern science. Consequently, the style of thought characteristic of physics exerts a very considerable influence on the style of scientific thought in general. The question of the status of physics, its ideas and methods interest scientists in many fields. Further progress in chemistry and the whole complex of the biological sciences would be unthinkable in isolation from the development of physic. So, we are faced with the complex task of investigating the interrelation and interdependence of the various levels of cognition, the form and structure in which knowledge is expressed, the name of scientific theory, its logical structure and the sources of its development [2].

This duality manifests itself in all areas and on all levels of being as a major organizing force. In order to understand God, ourselves, the world and life, we must be able to identify, understand and live in harmony with the dualizing shaping forces of life [3].

The Tree of Life shows that the creative process of the world is based on a plan in which all the things in the world are modifications of one and the same material substance and being. Although they are different in their needs, mode of existence and appearance, they are all part of "One Whole." The equilibrium between the oneness at the top and difference on the bottom must be maintained [3].

It is important to acknowledge the dynamic nature of the disciplinary character of scholarship. What we regard as entrenched disciplines today have changed considerably in the past and continue to do so. New ideas and concepts continue to explode forth at ever-increasing pace. We have ceased to accept that there is any coherent or unique form of wisdom that serves as the basis for new knowledge. We have simply seen too many instances in which a new concept has blown apart our traditional views of the field. Just as a century ago, Einstein's theory of relativity and the introduction of quantum mechanics totally revolutionized the way that we thought of the physical world; today, speculation about dark matter and quantum entanglement suggest that yet another revolution may be under way. The molecular foundations of life have done the same to the biomedical sciences [4].

There is a definite hierarchy of academic prestige—or perhaps better stated, an intellectual pecking order—within the university. In a sense, the more abstract and detached a discipline is from "the real world," the higher its prestige. In this ranking, perhaps mathematics or philosophy would be at the pinnacle, with the natural sciences and humanities next, followed by the social sciences and the arts. The professional schools fall much lower down the hierarchy, with law, medicine and engineering followed by the health professions, social works and education at the bottom. Clearly, within this culture of academic snobbery, the distinction of basic ("curiosity-driven" or Baconian) versus applied ("mission-oriented" or Newtonian) research becomes significant, perhaps tracing back to the Humboldtian ideal of pure Wissenschaft.

In reality, however, the progress of basic knowledge from the library or the laboratory to societal application is far from linear, and the distinction between basic and applied research is largely in the eye of the beholder (Sonnerr & Holton, 2002). Furthermore, there is yet another mode of research that represents a conscious combination of basic and applied research so-called Jeffersonian science. Such research aims at providing the fundamental knowledge essential to address a key social priority [4].

Scientific progress has been two-dimensional. First, the range of questions and problems to which science has been applied has been continuously extended. Second, science has continuously increased the efficiency with which inquiry can be conducted. The products of scientific inquiry then are (1) a body of information and knowledge which enables us better to control the environment in which we live, and (2) a body of procedures which enables us better to add to this body of information and knowledge.

Science both informs and instructs. The body of information generated by science and the knowledge of how to use it are two products of science. As already indicated, we will not be concerned here with the body of information and knowledge which it has generated, that is, not with the specific theories, laws and facts that have been developed in the various physical, life and behavioral sciences. Instead, we will be concerned with the procedures by which science generates this body of knowledge and the process of inquiry.

The procedures which characterize science are generally referred to as "tools, techniques and methods." The common inclination to use these three terms interchangeably conceals some distinctions which are important to understand in discussing scientific procedures [5].

The passer of human development is therefore seen to be an extension, enlargement and acceleration of the pattern of biological development, operating through mutation and selection. Selection is ecological interaction constantly creating new niches and destroying old ones; mutation takes the form of invention, discovery, expansion of sphere and human non-genetic structures. Niches open up, and sometimes are filled, sometimes not, depending on the capacity of the system for mutation; each successful mutation opens some niches and closes others. The pattern jogs along in an immensely complex interaction of things, organization and people, with biological, meteorological, environments, structures and population [6].

In some sense, every empirical researcher is reporting the results of an experiment. Every researcher who behaves as if an exogeneous variable varies independently of an error term effectively views their data as coming from an experiment. In some cases, this belief is a matter of a priori judgment, in some cases it is based on auxiliary evidence and inference, and in some cases, it is built into the design of the data collection process. But the distinction is not always as bright and clear. Testing that assumption is a recurring difficulty for applied econometricians, and the search always continues for variables that might better qualify as truly exogeneous to the process under study. Similarly, the growing popularity of explicit experimental methods arises in large part from the potential for constructing the proper counterfactual [7].

Some Notes on the Preamble in Relation to Dualistic-Polar Conditions in the Knowing and the Information-Knowledge Process

The quotations given in the preamble deserve some reflections by the author and those who may have the opportunity to read this monograph and the related monographs such as the theories of info-statics, info-dynamics, fuzzy rationality, entropy and related monographs by the same author. *Science like any category of knowledge informs* to establish necessity over possibilities and imaginations from potentials and *instructs* to establish freedom over probabilities and realities from actuals. It is analytically useful to reflect on the idea that all information-knowledge systems are generated by knowing through neuro-decision-choice actions irrespective of the area of neuro-information-decision-choice processes over problem-solution duality within the space of ontological-epistemological duality. It must be clear that the information system generated by knowing through the acceptable paradigm of thought composed of methodological toolbox and analytical toolbox, for realizing the knowing actions, and how to use the acquired knowledge systems are themselves the results of knowing through neuro-decision-choice actions over the space of problem-solution dualities, where the space of problem-solution dualities is always mapped onto the space of question-answer dualities. From the reflections on these quotations and the author theories of info-statics, info-dynamics, general entropy, knowledge square and related monographs, the concern, here, is on the understanding of interactive processes among neuro-decision-choice actions, problem-solution dualities and question-answer dualities for which the knowing is involved in the creations of stock-flow disequilibrium dynamics of information-knowledge system in the interactive unity and diversity, where the understanding of social outcomes in forward and backward motions is dependent on inseparable interactions between neuro-decision-choice actions and the elements of problem-solution dualities.

It is useful to keep in mind that the spaces of problem-solution dualities and question-answer dualities are continually expanding involving information-knowledge systems to understand the environment of existence. The expansion of the space of problem-solution dualities generates new techniques in both methodological toolbox and analytical toolbox for the expanding area of knowing. It also generates the need for the development of techniques and methods for mimicking nature in creative processes where the neuro-decision-choice processes over the spaces of the problem-solution dualities and information-knowledge systems continually destroy old forms and create new forms through information-knowledge discoveries and method-technique innovations. In these processes, there are reductions of individual and collective cognitive capacity limitations by expanding the area of collective intelligence through effective social organizations of neuro-decision-choice actions over the individual and collective spaces of problem-solution dualities, where the knowing processes inform and instruct actions over the space of the problem-solution dualities to provide evolving fundamental and applied information-knowledge systems needed to generate input-output processes over the space of interactions between neuro-decision-choice actions, problem-solution dualities and individual and collective preference orderings.

The understanding of the behaviors of the elements in the space of interactions among the spaces of neuro-decision-choice actions, the problem-solution dualities, the question-answer dualities and the information-knowledge systems requires the acknowledgement of the dynamic nature of categories of areas of knowing as captured by the dynamic notions of epistemic spaces in relation to the epistemic field with relational and information fields which presents the character of categories of problem-solution dualities and the scholarships that may be required of them. The introduction of the concept of dynamic field of cognition will help to explain the expanding areas of categories of problem-solution dualities and the corresponding areas of knowing, where there are continual destruction-replacement process of problem-solution dualities and question-answer dualities, such that the variety destruction-replacement process constitutes the universal epistemic basis for the understanding of the stock-flow dynamics of intra-categorial and inter-categorial information-knowledge accumulation in a continual disequilibrium processes.

The intra-categorial and inter-categorial structures are based on the principle of opposites with relational continuum and unity such that the progress of complexity of sciences, Industrial Revolution, agricultural revolution, medical revolution and others would be unconceivable in separable dualistic-polar epistemic conditions under the principle of excluded middle, where diversities exist in disconnected modes and varieties reside in the space of absoluteness. The ideas in this monograph are centered around relational continua of diversities and unity of categories in knowing and cognition within the information-knowledge production through neuro-decision-choice actions over the space of problem-solution dualities, where every variety resides in dualistic-polar conditions under the universal principle of opposites with relational continuum and unity in the space of relativity. Our world of neuro-decision-choice actions over the space of problem-solution dualities, composed of fundamental and applied problem-solution dualities to generate fundamental and applied information-knowledge systems, operates in the space of relativity with error correction processes toward perfections which reside in the space of absoluteness as the collections of ultimate states of universal affairs, where the principle of opposites in dynamic existence gives way to the principle of uniqueness with no change or possibility of change under static conditions over the space of quality-quantity dualities, where quality does not reside in quantity but quality is quantity and quantity is quality in the state where static is dynamic and dynamic is static.

In the epistemic system of knowing and information-knowledge development over the static-dynamic space and variety problem-solution processes, it is useful to see that duality and polarity manifest themselves in all areas of variety and on all levels of existence as powerful organizing factors in understanding of the operative forces of the duals in duality and the poles in polarity for the development of information-knowledge system under the stock-flow disequilibrium dynamics in order to also understand the socio-natural conditions and the corresponding environment, its contents and the variety creative processes, the universe and conditions of existence through knowing and information-knowledge production that provide the input-output structures for the management and controls of social setups relative to the established natural environment, dynamics which can be affected by the variety behaviors as well as affect the variety behaviors and survivals. It is also useful to understand that forces are generated from tension produced by dualistic-polar conflicts, where general dynamics are understood through our search for dualistic-polar structures which shape the forces of variety transformations in the space of quality-quantity dualities. Additionally, it must be understood in all zones of thought that conditions of variety transformations may not point to a journey toward a zone of perfection as desired; they may also point to the opposites, toward a zone of imperfection as undesired under a preference scheme under the principle of opposites. It is through this logic of the principle of opposites and input-output dynamics that one can understand the results of ontological dynamics such as climatological and geomorphological changes and problems of variety extensions.

Here, knowing is seen as the science of unity in intellectual investment where diversity in knowing begins from the organic structure of unity and returns to it through neuro-decision-choice actions over the space of categorial problem-solution diversities in the information-knowledge accumulation under the stock-flow disequilibrium dynamics of intellectual capital stock. The theory of knowledge should not be the identification problem of knowledge, but it should include the journey over the space of ignorance-knowledge dualities and the test of epistemic transversality conditions of variety ignorance-knowledge dualities and acceptances where the epistemic transversality conditions show when and where we exit from the space of a dual and enter into the space of another dual of the same duality. It may also be pointed out that the concepts of relativity, the space of relativity, absoluteness and the space of absoluteness are very important in the development of the theory of knowing, the theory of information-knowledge system, paradigms of thought, language development, definition and explication of terms and dictionary of words and terms. It must be understood that our neuro-decision-choice systems find meaning in the space of relativity that allows variety comparison and order to be established, and not in the space absoluteness that deprives variety comparison. The quotations in this preamble will provide us with foundations of necessity and freedom to examine the variety knowing and information-knowledge accumulation processes from the economic theory of production with input-output processes in relation to the space of intellectual investment-capital dualities.

References

[1] Schwaller de Lubicz, R.A.: The Egyptian Miracle: An Introduction to the Wisdom of the Temple, Rochester (Vermont) Inner Traditions International (1985)
[2] Fedoseyer, P.N. et al.: Philosophy in USSR: Problems of Dialectical Materialism, Progress Pub. Moscow (1977)
[3] Amen, Ra Un Nefer: Metu Neter Vol.1, Khamit Corp., Bronx, NY (1990)
[4] Weber, Luc E., James J. Duderswtadt (Eds.): Reinventing the Research University, Economica, London (2004)

[5] Ackoff, R.L.: Scientific Methods: Optimizing Applied Research Decisions, John Wiley, New York (1962)
[6] Boulding, Kenneth E.: A New Theory of Societal Evolution, Sage Pub. Beverly Hills Ca, (1978)
[7] Harrison, Glenn W., John A. List: "Field Experiments," J. Econ. Lit., Vol. 42, Dec. (2004)

Prologue: Epistemics of Past-Present Research Directions on Information-Knowledge Systems

I. Introductory Reflections

The prologue introduces and discusses the analytical and epistemic points of entry of the current monograph in relation to my previous works and their connections to the existing works on information-knowledge systems and our understanding of the problem-solution process of general and specific cognitive actions under neuro-decision-choice actions. The current monograph is thus special in its orientation toward the understanding of the information-knowledge process in the evolution of socio-natural existence of matter-energy conditions and the diversity and unity that they connote in socio-natural variety identifications and transformations as seen in the space of problem-solution dualities and translated into information-knowledge systems. The unity of knowing and the diversity of information-knowledge structures are seen in terms of decision-choice activities over the space of problem-solution dualities and polarities, where diversity relates to categories of problem-solution dualities and unity relates to inter-categorial dependency in give-and-take-sharing processes in the processes of intellectual investments and information-knowledge accumulations as disequilibrium stock-flow dynamics.

The general principle in this epistemic work is that all specific areas of knowing, whether science or non-science, exist as diversity in decision-choice strategy over the space of problem-solution dualities and polarities which are united in the results of information-knowledge accumulation for the unity of understanding of universal matter-energy processes and for their social utilizations toward individual social goals, objectives and visions over the space of individual-collective dualities. The monograph is devoted to present conditions of the diversity and unity principles of knowing and science through a complex system of arguments projects the idea that all decision-choice activities are about variety identifications and transformations, where the energies are generated by creative conflicts, contradictions and tensions in destruction-replacement processes under the cost-benefit rationality as seen in the space of ignorance-knowledge dualities over the space of ontological-epistemological dualities that contains the space of actual-potential dualities with a sub-space of imagination-reality dualities. This monograph is the fourth of my tetralogy on information-knowledge phenomena and their understanding of socio-natural decision-choice actions to bring about knowing and information-knowledge production of variety identifications and transformations where a change in conditions creates a vacuum with replacement of new conditions. At social level, a change of conditions means a creation of conditions of variety solution to a variety of a problem or problems to establish new conditions for new social environment. At the level of nature, a change of conditions means a destruction of conditions of existing variety environment which is incompatible with existing elements in order to establish new conditions of variety natural compatibility of the element-environment relationality.

The sequence of the monographs is—(1) The Theory of Info-Statics: Conceptual Foundations of Information and Knowledge [1], (2) The Theory of Info-Dynamics: Foundations of Information-Knowledge Dynamics [2] and (3) A General Theory of Entropy: Fuzzy Rational Foundations of Information-Knowledge Certainty [3]. The basic framework of these theories on information-knowledge phenomena was established in another monograph entitled: The Theory of the Knowledge Square: The Relational Foundations of the Knowledge Production Systems [4] and supported by further developments in previously interdependent monographs [5–7] with theoretic development of cost-benefit analysis [8, 9], where the central link is neuro-decision-choice actions as guided by subjective preferences, where the real cost-benefit conditions are within the *asantrofi-anoma principle* which is explained in relation to a universal condition that every variety is presented a cost-benefit duality where the benefit dual cannot be selected without the cost dual providing the foundations for the development of the general theory of cost-benefit analysis toward information-decision-choice interactive processes.

These interdependent theories deal with some fundamental information-knowledge *problem-solution dualities* in intra-variety and inter-variety systems on one hand, and intra-categorial and inter-categorial systems on the other hand as they relate to information-decision-choice interactive processes in the matter-energy and the space-time dynamics over the space of quality-quantity dualities for socio-natural productions which relate to the maintenance and transformations of varieties, given the structure of variety identifications. Information-knowledge systems are seen in terms of the space of general social production-consumption dualities through social decision-choice activities over the space of variety problem-solution dualities, where there is no absolute problem or solution. Every variety existence is seen in terms of problem-solution dominance in dualistic-polar conditions over the space of relativity specified in the space of negative-positive dualities and translated into variety real cost-benefit dualities under given individual and collective preferences defined in intra-generational and inter-generational time structures. The set of essential epistemic propositions in the theoretical system on problem-solution duality and polarities is given below.

The essential propositions:

1. Every variety exists under a complex dualistic-polar structure of negative-positive, problem-solution, cost-benefit and static-dynamic states, where the identity of the variety is revealed by a relativity of dualistic-polar dominance of its characteristic dispositions, where the epistemic structure and analytics are over the space of relativity and not over the space of absoluteness.
2. Every neuro-decision-choice activity is a relational action to set a variety solution against a variety problem or to maintain the existing solution against a potential variety in the space of variety problem-solution dualities by changing the problem-solution relation structure.
3. Every information-knowledge item is an input-output variety of decision-choice activity over the spaces of problem-solution dualities and polarities relative to an element in the goal-objective space, where any variety is claimed to be a problem or solution in terms of its relative characteristic disposition as seen in the space of relativity. As an output, it is a construction from destruction; as an input, it is a destruction from construction under the space of destruction-construction dualities.
4. The meaning of life finds expressions in neuro-decision-choice activities over the space of problem-solution dualities, while the essence of life finds expression in the way that neuro-decision-choice activities are exercised over the space of problem-solution dualities.
5. The neuro-decision-choice actions and the problem-solution processes are telescopically infinite backward as well as telescopically infinite forward relative to the present state under the *principle of social non-satiation* toward perfection from the space of relativity toward the space of absoluteness, where the dynamic of variety problem-solution is infinite such that the variety solution always generates variety problems which in turn generate possibility of new variety solutions to equip the information-knowledge system with the properties of self-exiting and self-correction.
6. Collectively and individually, our neuro-decision-choice actions are guided by the time trinity (the *sankofa-anoma rationality* of past-present-future structure). Here, the flying bird has its head turned backward, knowing the information of the current position, retrieving information from the past regarding success-failure duality in order to determine its future direction and action.

II. The Methodological Toolbox

The methodological tools needed to establish the principles of unity and diversity in knowing and science must be elements that see information-knowledge generation as neuro-decision-choice actions under opposing forces with conflicts, contradiction, tension, dominance, friction and many other events in relational continua and unity, and in infinitely continual processes over the space of static-dynamic conditions that reside in internal and external interdependent modes with relational continuity and unity relative to the dynamic space of variety problem-solution dualities.

The relevant methodological tools in this theoretical construct over the space of constructionism-reductionism dualities are: (1) the principle of opposites, (2) the space of problem-solution dualities and polarities with relational continua and unity and (3) the fuzzy paradigm of thought which is composed of dualistic-polar conditions in relational continua, unity and opposites with give-and-take-sharing interdependencies in the space of relativity. These methodological tools allow one to deal with both the fuzzy and classical information structures as well as internalize the neuro-decision-choice agent with subjective judgments as part of the process of knowing through neuro-decision-choice actions over the space of true-false or correct-wrong dualities. The notion of subjective judgment cannot be discounted away from neuro-decision-choice activities over the space of problem-solution dualities irrespective of whether the space is defined over the category of science or non-science. The methodological tools, as indicated here, allow one to identify variety opposites in relative and not in absolute terms that may be represented in the static states by conditions of fuzziness and in dynamic states by conditions of fuzzy processes where every variety exists in dualistic-polar state in time and over time.

Given the fuzzy paradigm of thought with its logic and mathematics as constituting the essential methodological frame, and the principle of opposites composed of duality and polarity with relational continuum and unity, one must then relate them to the fundamental issues of problem-solution duality and polarity by establishing the idea that every variety identity is defined by its relative characteristic disposition in dualistic-polar conditions, where the duals and the poles are always in a simultaneous existence. These fundamental issues of problem-solution duality and polarity must also be related to the classical paradigm of thought with its logic and mathematics, where the set of classical analytical processes with excluded middles over the space of absoluteness with perfect information structure is contained in the set of fuzzy analytical process over the space of relativity with imperfect information.

In this methodological process, one speaks of dualistic-polar conditions of matter-energy elements as varieties. The direction of research into the understanding of knowing and the development of the information-knowledge system based on neuro-decision-choice activities over the space of problem-solution dualities is such that our main concern is shifted away from variety specificities of knowing to categorial generalities of knowing in terms of unity and diversity as well as shifted from information-knowledge specificity to information-knowledge generality over the space of diversity-unity duality. Here, we are not concerned with a particular set of techniques and methods which is relevant to setting a variety solution against a variety problem in categories of problem-solution dualities, since each category may require a different toolset of techniques and methods to deal with a specific set of dualistic-polar conditions that relates variety solutions to variety problems under the principle of opposites in the space of relativity.

In all attempts with regard to knowing as intellectual investment and to information-knowledge system as intellectual capital accumulation, it is epistemically important to distinguish between *methodology* and *methodological tools* on one hand, and *analytics* and *analytical tools* on the other hand in pursuing the understanding of the process of knowing and information-knowledge generation through neuro-decision-choice actions over the spaces of problem-solution dualities and input-output dualities. In the cognitive model of reality represented here, a distinction is made between the concept of a *methodological toolbox* and that of an *analytical toolbox*; furthermore, the space of reality is seen as sub-space of the space of actuals. A question, therefore, arises as to what the differences and similarities are. The importance and the essence of the differences find expressions in conditions of the theory of definition and explications to reveal their characteristic dispositions [4, 10, 11]. Let us always keep in mind that definitions are the means to establish the characteristic dispositions of varieties through a language in the source-destination communications, and explications are refinements from the family of ordinary languages (FOL) to the family of abstract languages (FAL) to restrict the domain of linguistic understanding and applications within the communication process in relation to messages from the variety signal dispositions where any language is a system of codes in sound, visual and other forms for the elements in the space of source-destination dualities. The distinction between the space of actual and the space of reality is useful to understand anomalies in knowing such as imagination, possible world and impossible world paradoxes, phantom problems, unsolvable problem and other such anomalies in the space decision-choice actions. Similarly, the distinction between the space of imagination-reality dualities and the space actual-potential dualities is useful in understanding the natural mathematics, social transformations, variety productions and their relationships to the theory of knowing as a process and the theory of the information-knowledge system as the output of the knowing process as well as an input into the knowing process.

Definition IIA: Methodology and Methodological Toolbox

The *methodological toolbox* encompasses the general cognitive framework of information structure, constructionism-reductionism duality, the unacceptability-acceptability dualities connected to the spaces of variety true-false, problem-solution and question-answer dualities, and laws of thought in guiding *thinking* to create epistemic algorithms to guide reasoning in knowing as well as the development of information-knowledge systems that contain accepted and unaccepted information-knowledge items to reconcile individual and collective conflicts over the spaces of true-false dualities, problem-solution dualities and question-answer dualities. The collection of the elements of the methodological toolbox is the paradigm of thought that sets rules and controls to guide individual and collective knowing as well as the path of the social information-knowledge system. It is the methodology of knowing, information-knowledge accumulation and strategic neuro-decision-choice actions over the space of problem-solution dualities.

Definition IIB: Analytics and Analytical Toolbox

The *analytical toolbox* is composed of methods and techniques for the development of thinking as an algorithm over the space of constructionism-reductionism duality where constructionism is a forward creation and reductionism is a backward creation within a paradigm of thought, to deal with experiential information (knowledge by acquaintance) as the primary category of knowledge, and knowledge by description as the derived category of knowledge with a paradigm of thought. The analytical toolbox is always defined within a paradigm of thought as the methodology. It is a *meta-paradigm* as a set of techniques and methods to accomplish the decision-choice activities within the space of constructionism-reductionism dualities with a defined nature of information under an organic paradigm. As a meta-paradigm under an organic paradigm, its development and acceptability are always under the rules and behaviors as established within a paradigmatic framework for knowing and information-knowledge development, to create the conditions for over the spaces of computability-uncomputability dualities, predictability-unpredictability dualities, prescriptibility-unprescriptibility dualities over the space of decidability-undecidability and doubt-surety dualities.

A note about these definitions will be helpful for the enhancement of the epistemic understandings of the similarities and differences between the *methodological toolbox* and the *analytical toolbox* in relation to methodology and methods. Methodology is the primary framework for thinking, while the set of methods and techniques is a derived framework of instruments within the primary framework of thinking that provides tools for executing an acceptable thinking process to guide reasoning to conclusions over the space of the problem-solution dualities which relates to the spaces of question-answer dualities and true-false dualities. The derived framework of instruments includes experiential information and language representations, multitudes of algorithms, assumptions and rules of application for reasoning in the family of categories of problem-solution dualities. The language representations include the family of ordinary languages (FOL) and the family of abstract languages (FAL) that represent real and nominal variety characteristic dispositions as abstracted from the ontological variety signal dispositions over the space of variety quality-quantity dualities for source-destination communications.

II A: There are Two Organic Paradigms of Thought of Interest

IIA.1) The classical paradigm of thought with (a) exact information structure composed of linguistic clarity and substantial quantitative characteristics in the space of quality-quantity duality, (b) the principle of opposites, where the variety opposites exist in a way such that the dualistic characteristic dispositions exist with relationally dualistic-polar separations and excluded middle, where the varieties cannot exist with the simultaneity of their opposites in the same space and at the same time, (c) problems and solutions, and questions and answers existing in a space of absoluteness to create relational non-interdependencies and (d) a paradigm of thought with an excluded middle, non-acceptance of contradiction which is composed of conditions of relational separation, disunity and opposites with no-give-and-take-sharing mode and relational interdependencies in the space of subjective-objective dualities over the space of absoluteness, where every variety exists in absolute characteristic disposition for identity. The summary of the classical laws of thought will be stated and discussed in relation to knowing and information-knowledge accumulation as neuro-decision-choice activities over the space of economic production-consumption dualities and the space of input-output dualities.

IIA.2) The fuzzy paradigm of thought with (a) inexact information composed of vagueness and linguistic unclarity with substantial qualitative characteristics in the space of quality-quantity dualities, (b) the principle of opposites where varieties exist in relational continuum and unity and varieties exist and are defined by the dualistic-polar characteristic dispositions, (c) problems and solutions and questions and answers exist in the space of relativity to create the space of problem-solution and question-answer dualities and polarities, where a problem (solution) is defined by the relative dominance of the problem (solution) dual and (d) a paradigm of thought which is composed of dualistic-polar conditions in relational continua, unity and opposites with give-and-take-sharing interdependencies in the space of subjective-objective dualities over the space of relativity, where all varieties exist in relative characteristic dispositions for identity. The summary of the fuzzy laws of thought will be stated and discussed in relation to knowing and information-knowledge accumulation as activities of economic production, where neuro-decision-choice activities are journeys between dualistic-polar structures.

As explained, the set of conditions of the classical paradigm is contained in the set of conditions of the fuzzy paradigm, and hence the set of elements of the classical methodological toolbox is contained in the set of elements of the fuzzy methodological toolbox. The thinking algorithms of the classical paradigm externalize the subjectivity and rational value judgments of the users from the knowledge by acquaintance to knowledge by description, where the possibility space that houses the variety necessity is given, and the probability space that houses freedom is constructed. The thinking algorithms of the fuzzy paradigm internalize the subjectivity and rational value judgments of the users from the knowledge by acquaintance to knowledge by description, where the possibility space that houses necessity and initial conditions is constructed to give rise to the probability space to generate the conditions for the exercise of freedom. The thinking framework defining algorithms for reasoning over both the spaces of absoluteness and relativity must consider the absolute and relative categories as projected by the conditions of $[0, 1]$, (0.1) and $\{0,1\}$ where $((0.1) \subset [0, 1])$, $(\{0,1\} \subset [0, 1])$ and $(\{0,1\} \not\subset (0, 1))$ as well as $[0, \infty]$, $(0.\infty)$ and $\{0,\infty\}$, where $((0.\infty) \subset [0, \infty])$, $(\{0,\infty\} \subset [0, \infty])$ and $(\{0,\infty\} \not\subset (0, \infty))$ in the spaces of relativity and absoluteness. These mathematical representations of information relate to conditions of nothingness-somethingness dualities where nothingness resides in somethingness and somethingness also resides in nothingness, and to the case where we concern ourselves with the excluded middle of $((0, 1)$ and $(0, \infty))$. The information-knowledge connectivity relative to continual transformation of matter-energy equilibrium to information-knowledge disequilibrium and their relational continuum and unity are discussed under paradigms of thought in [2, 5, 12, 13].

III: The Principle of Opposites, Neuro-Decision-Choice-Action and Knowing

The principle of opposites is an essential indestructible framework to understand destruction-substitution processes in self-exciting and self-correction systems such as information-knowledge systems and self-organizing systems such as a society and related ones. There are two ways of conceptualizing the principle of opposites. One way is where the opposites are cast in relational continua and unity with give-and-take interdependent sharing modes that maintain their interdependencies for their mutual existence and change, where the opposites simultaneously exist in one variety to present a duality or a polarity or both. The other way is where the opposites are cast in relational separations with excluded middles without give-and-take interdependent sharing modes that maintain their mutual separation and non-interdependencies for their independent existence, where the change of an opposite is independent of the other opposite.

The principle of opposites reveals itself as dualities and polarities to generate the existence of dualistic-polar conditions under relational continua and unity or generate existence of dualistic-polar conditions under relational separation and excluded middle. Under the principle of opposites, we have a system of dualities and polarities, the mutual interactions of which induce universally internal variety transformations in terms of problem-solution dualities and polarities in all areas of knowing to generate continual information-knowledge accumulation and in areas of decision-choice actions to generate continual variety transformations in the negative-positive progressive states as well as affect the cost-benefit relational conditions. The combined conditions of negative and positive duals of a variety duality allow the identification of variety identity. In the knowing process, this is the *identification problem-solution duality* in all areas of knowing and it is used as a condition for the principles of unity and diversity in knowing. The combined conditions of negative and positive duals of variety duality and negative and positive poles of variety polarity allow the identification of variety transformations. Within the knowing process of identification of the processes of internal changes of variety dispositions and the development of information-knowledge accumulation is the *transformation problem-solution duality and polarity* in all areas of knowing.

The transformation variety problem-solution dualities and polarities are used as the conditions in establishing the principles of unity and diversity in all areas of knowing and information-knowledge accumulation. The principle of opposites also relates to paradigms of thought with neuro-decision-choice actions defined over the space of true-false dualities with dualistic-polar relational continuum and unity, where one speaks of degrees of truth and falsity in the space of true-false dualities within the space of relativity or dualistic-polar separation, where one speaks of absolute truth and absolute false in the space of true-false dualities within the space of absoluteness. The principle of opposites is connected to variety information structure, paradigm of thoughts and neuro-decision-choice actions. All neuro-decision-choice actions are defined in the space of statics or in the space of dynamics. The space of statics holds the conditions of variety identification-non-identification dualities, while the space of dynamics holds the conditions of variety identification-non-identification dualities as well as a variety transformation-non-transformation dualities or both that relate to either changes in relations or changes in properties or changes in both over the space of quality-quantity dualities. In this respect, the whole of human actions through is related to identification problem-solution dualities and transformation problem where variety transformations are induced by real cost-benefit rationality under the psychosocial principle of non-satiation.

IV. The Problem-Solution Dualities and Polarities

Given the principle of opposites as composed of a sub-principle of *duality* and a sub-principle of *polarity* in an integrated mode, we now turn our attention to the meanings of the problem-solution duality and the problem-solution polarity as they provide conditions for the diversity and unity of knowing and by logical extension, the unity of fundamental science and applied science. When one selects dualistic-polar relational separation with the excluded middle under the principle of opposites, some analytical difficulties tend to arise, making it more difficult to use the principle of opposites as an epistemic tool to unite knowing in all areas of knowledge. At the conditions of dualistic-polar relational separation and the principle of the excluded middle, some problems arise as to the relationship between duality and polarity and between the duals of the dualities and the poles of the polarities. At the levels of problem-solution spaces, the principle of the excluded middle places every problem and every solution in absolute states, where variety problems exist independently of variety solutions in the space of decision-choice activities over the action space where every variety is described by the classical exact information-knowledge system. These seem not to be found in the experiential information for destruction-construction processes.

At the level of dualistic-polar relational continua and unity, the duals of any duality are connected in a relational give-and-take process. Similarly, the poles of any polarity are also connected in a relational give-and-take process for their mutual existence and destruction. The dualities and the corresponding polarities exist in mutual give-and-take interdependency where every pole of a polarity has a residing duality which determines the identity of the pole as well as the identity of the polarity by the conditions of dualistic-polar dominance. Translated at the levels of problem-solution spaces, the principle of relation continua places every problem and every solution in relative states where there is variety problem-solution duality in the space of neuro-decision-choice activities within the action space where every variety is described by a fuzzy information-knowledge structure. Any variety in the space of acquaintance is determined to be a problem or a solution by its relative dominance in dualistic-polar conditions. This notion seems to be consistent with the experiential information as an input into destruction-construction processes in all areas of knowing through decision-choice actions. Furthermore, the space of the problem-solution dualities is an image reflection of the space of question-answer dualities. The fuzzy information-knowledge system is central in the development of the monograph with the proposition that the classical information-knowledge system is contained in the fuzzy information-knowledge system in the sense of set inclusion or space inclusion. Similarly, all activities over the epistemological space take place within the space of relativity-absoluteness duality where there is a continual journey from the relativity to absoluteness, providing cognitive systems with self-exiting and self-correction toward elements in the space of goals, objectives and visions under the principles of knowing and information-knowledge production.

V. Paradigms of Thought in Knowing and Decision-Choice Actions

We have discussed some aspects of the concepts of methodology, methodological toolbox, analytics and analytical toolbox as they relate to the concepts of organic paradigm and micro-paradigms in knowing and information-knowledge development. The processes of knowing and information-knowledge development not only are economic production but encompass the totality of economic activities of human existence, where the knowing process is an intellectual investment and a flow, the information-knowledge system is a collection of intellectual capital accumulations and a stocks, and the thinking-reasoning processes are organic technologies with micro-technologies generated by neuro-decision-choice actions under cognitive capacity limitations of many forms. The knowing process and the development of the information-knowledge system about universal awareness proceed from acquaintance to acceptance based on neuro-decision-choice actions, where acquaintance relates to knowledge by acquaintance or experiential information or epistemological information, and acceptance relates to knowledge by description or knowledge by laws of thought.

The development of knowledge by acquaintance or experiential information structure, and knowledge by description or knowledge by laws of thought are defined in a space of acceptance-rejection dualities with intense social conflicts in the space of true-false dualities under organizational capacity limitations. The resolution of the elements in the conflict zone of agreement requires a socially agreed-upon framework of information-knowledge development and conditions of acceptance in relation to neuro-decision-choice activities of stock-flow conditions over the space of social existence. One of the concepts that generates linguistic confusion in the space of general neuro-decision-choice activities operating in the action space of social existence is the *paradigm of thought*. The paradigm of thought is an organic framework for thinking in the development of algorithms for general reasoning which relates to all decision-choice activities in an individual and collective action space governed by a non-conflict zone of individual preferences and a conflict zone of collective preferences over the space of problem-solution dualities and polarities. It is useful in this prologue to examine some aspects of the nature of the paradigm of thought and its effects on knowing, the information-knowledge system, the diversity and unity of knowing, science and neuro-decision-choice activities of human actions.

The knowing activities, information-knowledge development and decision-choice activities are defined over the problem-solution dualities and cover all areas of human existence and actions, guided by a paradigm of thought. In the path of research that is being followed, a paradigm of thought is a generally agreed-upon model of thinking for the development of variety information-knowledge from acquaintance to acceptance in order to minimize the conflict zone of individual judgments and collective acceptance in the space of decision-choice actions. For the purpose of epistemic clarity, paradigms are grouped as *external* and *internal paradigms*. The external paradigms relate to the general framework of thought in processing information-knowledge systems from acquaintance and transform them into knowledge by description or knowledge by laws of thought through reasoning. The external paradigm is the organic paradigm that relates to the general framework of decision-choice actions over the space of problem-solution dualities. It is the primary category of the processes of knowing and information-knowledge production.

The internal paradigms relate to frameworks of methods and techniques of thinking to create algorithms within the organic paradigm for reasoning to follow the paradigmatically acceptable reasoning in processing information-knowledge systems from acquaintance, and transform them into knowledge by description or knowledge by laws of thought through reasoning over the space of decidability-undecidability dualities. The internal paradigm is the micro-paradigm that relates to the specific framework of decision-choice actions over the space of problem-solution dualities. It is the derived category of the processes of knowing and information-knowledge production from the primary paradigm. The set of concepts of organic, external and primary category of paradigms relative to the set of concepts of micro-internal and derived category of paradigms is important if we are to understand the differences and appropriate roles of the methodological toolbox and analytical toolbox in the knowing process, information-knowledge accumulation and decision-choice actions over the space of problem-solution duality without epistemic confusion in the decision-choice actions over the space of problem-solution dualities. The sequence of knowing involving decision-choice actions is such that there is experiential information, followed by the organic paradigm that gives rise to the derived paradigm within the construction-reduction duality. Hence, there is the *primary category of paradigms* and *derived category of paradigms*, where the primary category of paradigms gives rise to the primary category of paradoxes in methodology and methodological toolboxes, while the derived category of paradigms gives rise to the derived category of paradoxes in methods, techniques and analytical toolboxes.

There are two types of external paradigms constituting the organic paradigms of thought or primary category of paradigms, and corresponding to them are two internal paradigms of thought consisting of derived paradigms of thought or micro-paradigms of thought. There is the fuzzy *paradigm of thought* which relates to the fuzzy experiential information-knowledge system in the space of relativity, and there is the *classical paradigm of thought* which relates to the classical experiential information-knowledge system in the space of absoluteness. The movement from either the classical paradigm of thought to the fuzzy paradigm of thought or from the fuzzy paradigm of thought to classical paradigm of thought will be called an *inter-paradigm shift* or simply paradigm shifting with a change in the logic of thinking and reasoning as well as a change of the conditions of verification and acceptance of elements over the spaces of true-false dualities and decidability-undecidability dualities relative to decision-choice actions over the space of problem-solution dualities.

Within the classical and the fuzzy paradigms, there are methods and techniques for dealing with intra-dualistic-polar relationships of categories of problem-solution dualities, inter-dualistic-polar relationships among categories of problem-solution dualities and all forms of decision-choice activities with truth table conditions for acceptance over the space of decidability-undecidability dualities. The truth table, corresponding to the classical laws of thought for the logic of thinking, reasoning and decision-choice actions, is called *classical truth table*, where contradictions are not accepted to maintain the relational separation and excluded middle over the spaces of subjectivity-objectivity dualities and absoluteness. The truth table, corresponding to fuzzy laws of thought for logic of thinking, reasoning and decision-choice action, is called the *fuzzy truth table*, where contradictions are accepted to effect relational continua and unity in the existentially nominal and real varieties over the spaces of subjectivity-objectivity dualities, quality-quantity dualities and relativity. The fuzzy paradigm in the space of relativity provides us with movements between duals of duality, poles of polarity and opposites under the principle of continuum and unity. The classical paradigm in the space of absoluteness, on the other hand, has no logical mechanism to connect the duals of the duality, poles of polarity or the opposites under the principle of excluded middle and separation.

A change in the formulation of variety problem-solution duality, techniques and methods of dealing with relationships among problem-solution dualities within categories may be viewed as an *intra-paradigm shift* or shifting of a micro-paradigm within an organic paradigm. Thus, Kuhn's concept of a paradigm is the classical one, and the concept of paradigm shifting is an intra-classical-paradigmatic in nature which relates to problem-solution formulation, techniques and methods of decision-choice activities within the classical paradigm over the space of problem-solution dualities with dualistic-polar separation under the principle of the excluded middle in the space of absoluteness [14–17].

The knowing process of a number of approaches on information-knowledge systems and decision-choice actions such as Zadeh and Brouwer is a fuzzy paradigm, and the shifting of the rules of thought is inter-paradigmatic in nature shifting from the classical paradigm to the fuzzy paradigm of thought, where the truth table and reasoning are governed by conditions of *relativity* but not *absoluteness* and every conclusion has a qualification of fuzzy conditionality. The fuzzy paradigm of thought is developed from the space of relativity such that truth is truth with fuzzy conditionality, while the classical paradigm of thought is developed from the space of absoluteness of nominal and real varieties and categorial varieties such that truth is truth without a conditionality. The micro-paradigms or the derived paradigms from within the organic paradigm contain not only the laws of thought of the organic paradigm but also the possible abstract languages such as mathematics. These concepts will become clearer as the development of the monograph proceeds. We must hold on at this point that we have (1) fuzzy information with vagueness, inexactness and incompleteness and (2) non-fuzzy information representing the classical information with exactness, clarity and incompleteness.

Currently, the dominant paradigm of thought in all neuro-decision-choice actions, development and application of fundamental and applied decisions is the classical paradigm of thought with its logic and mathematics, where all information-knowledge systems in all areas of learning, teaching and research are assumed to be of a classical nature by simply discounting the conditions of vagueness and inexactness with approximation whether such approximation is appropriate or not, especially in the processing of the information-knowledge system and decision-choice actions over the space of problem-solution dualities. In fact, the development and the use of the theory of approximation within the classical paradigm need re-examination in relation to imperfections and cognitive capacity limitations. Similarly, the gains of the fuzzy paradigm of thought with its logic, mathematics and the corresponding truth table must be examined in relation to the general information-knowledge systems and neuro-decision-choice actions in the space of problem-solution dualities in all areas of knowing. These are the elements of the past-present path of research on information-knowledge systems, paradigms of thought and decision-choice actions over the space of problem-solution dualities to generate the disequilibrium dynamics of the stock-flow processes of knowing.

Under the principle of opposites, we regard the fuzzy paradigm of thought in the spaces of truth-false dualities, problem-solution dualities, decision-choice actions, decidability-undecidability dualities, static-dynamic dualities and variety identification-transformation dualities, as constituting a comprehensive epistemic system of thinking and reasoning with micro-paradigms in dealing with elements in the space of problem-solution dualities in the process of knowing and information-knowledge accumulation in all areas of knowing. No area of knowing is exceptional to these conditions just like no area of knowing is exceptional to the organic and derived paradigms of thought or exceptional to experiential information as initializing the path of knowing.

VI. Information Structures and Paradigms of Thought

There are two types of information structures. There is the *inexact information structure* where variety characteristic dispositions are defined over the *space of relativity* where subjectivity and contradictions are allowed, and there is the *exact information structure* where variety characteristic dispositions are defined over the *space of absoluteness* with no allowance for subjectivity and contradictions. These types of information structure are important in establishing a distinction and similarity between the fuzzy and classical paradigms of thought as they form the input for paradigmatic processing to derive the knowledge by description or the knowledge by laws of thought which is the output in the paradigmatic input-output processes. The other distinguishing factor is the nature of the behavior of these paradigms over the spaces of quality-quantity and subjectivity-objectivity dualities where decision-choice conflicts of preferences arise within the space of individual-collective dualities. The type of information to be processed provides the framework of the development of a primary paradigm as the processing module within which micro-paradigms are developed as a toolbox for implementing the processes within the space of ignorance-knowledge dualities with an input-output quality control. Alternative to the classical paradigm, the fuzzy paradigm relates to both exact and inexact information structures and operates over the space of relativity, while the classical paradigm relates only to the exact information structure and operates over the space of absoluteness.

VII. Decision-Choice Activities, Certainty-Uncertainty Duality and Problem-Solution Duality and Polarity

Decision-choice activities are the innate property of cognitive agents, where every socio-natural variety of knowing resides in the space of ignorance-knowledge dualities that is superimposed on the space of problem-solution dualities, where every decision-choice activity is a function of an element within the action space in relation to the spaces of identification problem-solution dualities and transformation problem-solution dualities in which ignorance is a problem and knowledge is a solution in relative proportions, in the sense that, for every solution disposition, there is a corresponding problem disposition defined in a dualistic-polar structure in relative proportions in continuum and unity under the principle of opposites, and where every neuro-decision-choice action is a change in relation or a change in property as understood under the system of qualitative and quantitative motions.

Every decision-choice activity is an action on an element in the space of static-dynamic dualities for either *variety identification*, *variety transformation* or both to set a variety solution against a variety problem and a variety problem against a variety solution, and to set surety against doubt in an infinite sequence of occurrences over the spaces of decidability-undecidability, certainty-uncertainty and risk-benefit dualities. Every variety identification is the result of decision-choice action guided by some comparative analysis of characteristic-signal dispositions under a criterion of dualistic-polar dominance. Every variety transformation is the result of decision-choice action guided by some subjective criteria of preference. Every decision-choice action is also a solution to a variety problem where hope operates in the space of risk-benefit dualities and every solution is a parent to a new problem as well as being an offspring to a predecessor problem in a dualistic-polar process of varieties over an input-output space, and where every decision-choice action passes through the space of decidability-undecidability dualities for a comparative analysis and cognitive action in disequilibrium dynamics. The inputs of decision-choice activities are information-knowledge elements defined over the space of quantity-quality dualities from acquaintance, where the outputs are information-knowledge elements which also act as inputs into decision-choice activities over all categorial endeavors in the general space of production-consumption dualities.

In thinking about theories of knowing and knowledge in all areas of varieties and categorial varieties of science and non-science, three interdependent spaces of the concepts of problem and solution must be considered. They are the *space of problem-solution dualities*, the *problem space* which is a collection of varieties where the problem duals dominate the solution duals, and the *solution space* which is a collection of varieties, where the solution duals dominate the problem duals as seen from some elements in the general goal-objective space as a sub-space of the potential space. The definitions and explications of the concepts of these spaces find expressions in the general relative space and not in the general absolute space of nominal and real varieties. The information-knowledge system, composed of knowledge by acquaintance and knowledge by description, is imperfect due to cognitive capacity limitations at the levels of acquaintance and paradigm of thought for thinking and reasoning in the space of quality-quantity dualities of all areas of knowing and information-knowledge production. The information-knowledge imperfections composed of qualitative and qualitative imperfections create the space of certainty-uncertainty dualities that affects the enveloping path of success-failure outcomes of decision-choice activities as a uniting factor of all areas of knowing and decision-choice actions that must traverse over the spaces of decidability-undesirability and risk-benefit dualities.

In this way, it seems clear that the only factors that establish the principles of unity and diversity in knowing and science are information, decision-choice activities, the spaces of variety problem-solution dualities and the primary paradigm of thought. The variety problem-solution dualities reveal themselves as variety *identification problem-solution dualities* and *variety transformation problem-solution dualities*. Every variety problem-solution duality in any area of knowing is either an identification, transformational problem or both. These conditions provide a cognitive framework for the development of the current monograph on unity of knowing and science from the point of view of conditions of the decision-choice actions on the space of problem-solution dualities as it relates to spaces of input-output and production-consumption dualities. Within each primary paradigm, there are categorial micro-paradigms under continual development as seen in relation to the corresponding category of problem-solution dualities, where the categorial conditionality of acceptance may also vary under internal and external disequilibrium dynamics in knowing and information-knowledge production. Within each primary paradigm, the changes in the derived paradigm are about techniques and methods where the primary paradigm sets boundaries. Within all paradigms, the acts of thinking and reasoning are mediations in dualistic-polar structure under asantrofi-anoma conditions, cost-benefit rationality and the sociopsychological principle of non-satiation.

VIII: The Asantrofi-Anoma Problem and Cost-Benefit Analysis Over the Space of Problem-Solution Dualities

The analytical concept of existence introduced the universe as a collection of past-present-future of matter-energy varieties with negative-positive dualistic-polar conditions defined over the space of variety identity and transformations for decision-choice activities. Each variety is an *asantrofi-anoma problem* over the decision-choice space in the sense that every variety is simultaneously composed of cost disposition (cost dual) and benefit disposition (benefit dual) such that every variety exists as a cost-benefit duality where the asantrofi-anoma problem is simply that there is a general real cost-benefit relationality where a selection of variety for its real benefit is also a selection of the variety real cost disposition. This gives rise to the asantrofi-anoma problem and rationality where a rational variety decision-choice action among a set of varieties requires an examination of intra-variety and inter-variety real cost-benefit conditions of elements in the set before a selection. The conceptual framework gives justification to the real cost-benefit analysis in rational decision-choice actions over the space of variety problem-solution dualities where the set of real cost-benefit relativity measures are generated for relational ranking of the varieties for decision-choice action, and where each variety constitutes a real opportunity cost relative to new variety benefit.

The asantrofi-anoma problem, real cost-benefit analysis and corresponding rationalities are foundational tools for all decision theories over the space of variety problem-solution dualities irrespective of the paradigm of thought used. Every variety is defined by negative and positive duals which are reflected by negative and positive characteristic dispositions and made available at the space of acquaintance that present the variety information specified as variety characteristic-signal dispositions where the characteristic dispositions are the contents and the signal dispositions are the messages. The positive dual and the negative dual reside in the space of cost-benefit dualities relative to some elements in the goal-objective space where negative or positive duals may be assessed as either a real cost or a real benefit by the decision-choice agent. In this respect, the asantrofi-anoma problem, principle and rationality are first defined over the space of negative-positive dualities where the negative dual and the positive dual are individually defined over the space of the cost-benefit duality decision-choice ranking under a preference scheme.

All static and dynamic decision-choice actions reflect either a benefit maximization constrained by a cost disposition or a cost minimization constrained by a benefit disposition. The nature of the real cost and real benefit dispositions will depend on the elements in the goal-objective space and the categorial transformation where generally by the *asantrofi-anoma principle*, one cannot choose the benefit disposition (dual) and leave the cost disposition (dual) of any variety defined in the quantity-quality space. The nature of the asantrofi-anoma rationality is to choose the variety with the highest real benefit dominance over the variety real costs to provide a solution to the asantrofi-anoma problem. The meanings of real cost and real benefit dispositions must be seen from the economic theory of the opportunity cost of varieties, where varieties reside in the space of actual-potential dualities. The concept of the space of actual-potential dualities containing the spaces of problem-solution dualities and decision-choice actions on varieties, where many decision-choice processes may be complex in the space of variety destruction-replacement processes. In this sense, real cost-benefit analysis is a general theory of decision-choice actions over the problem-solution space that unites all areas of knowing whether they are category of science or non-science, or category of social science or non-social science. The social and individual utilities of all areas of knowing are found in the results of decision-choice actions over the space of problem-solution dualities.

The existence of a variety problem simply implies that the variety cost disposition dominates the variety benefit disposition in relative terms. The existence of variety solution simply means that the potential variety benefit disposition dominates the potential variety cost disposition and hence the variety solution dominates the variety problem to allow variety destruction-replacement action. In the destruction of a variety problem, not only one does get rid of unwanted real cost disposition, but one forgoes the intrinsic real benefit disposition. In designing the variety solution, one brings in real benefit disposition as well as the intrinsic real cost disposition of the replacement. The dualistic-polar cost-benefit variety relation is a powerful way to understand the concepts of unintended consequence, continual transformation and insatiability problems in continual socio-natural transformation. It is also an epistemic powerful way to conceptualize the idea of beginning-end duality, nothingness-somethingness duality and finite-infinite duality in the understanding of the activities of the universe.

Let us keep in mind that the problems, solutions and decision-choice processes find expressions in the space of cost-benefit dualities, where variety problems and solutions are ranked with the cost-benefit conditions in the space of relativity. In other words, the whole of the working mechanism of neuro-decision-choice actions is to strike balances between system of variety costs and the system of variety benefits under the principle of opposites, where every variety is simultaneously cost and benefit as well as input and output. It is here that critical development of epistemics of cost-benefit analysis is required for the understanding of the conditions of the telescopic past in relation to the conditions of the telescopic present and to the conditions of telescopic future under the Sankofa-anoma information-time connectivity.

IX: The Principle of Opposites and the Concept of Relativity

The principle of opposites projects the notion that at the static state, every variety resides in opposites called duality with a combination of negative and positive characteristic dispositions for its identity, where the variety is called negative (positive) duality if the negative (positive) dual relatively dominates the positive (negative) dual, where the duals are defined by characteristic dispositions. The negative characteristic disposition is called the negative dual, and the positive characteristic disposition is called the positive dual. Each dual resides in cost-benefit duality where the dual is said to be cost or benefit depending on the relative dominance and relation to an element in the space of goals-objectives and vision. Similarly, every variety besides residing as duality simultaneously resides as polarity at dynamic states with negative and positive poles, where a pole is identified as negative (positive) if the residing duality is negative (positive) and the polarity is said to be negative (positive) polarity if the negative pole relatively dominates the positive pole. In this respect, the identity of any variety, either real or nominal, over the dualistic-polar spaces is claimed by the relativity of the duals and poles but not by the absoluteness of the duals. In the source-destination communication process, we have the concepts of *real relative identity* and *nominal relative identity* of varieties. The space of real relative identities corresponds to the space of real definitions in linguistic constructs. Similarly, the space of nominal relative identities corresponds to the space of nominal definitions in linguistic constructs.

The epistemic structure of the principle of opposites is very useful to the understanding of dualistic-polar negation and the development of neuro-decision-choice strategies to create favorable conditions in the space of success-failure dualities within the knowing and information-knowledge stock-flow disequilibrium dynamics, where there are more successes than failure where the social culture of philosophical consciencism is to set hope against despair, good against evil, peace against war, benefit against cost and positive against negative by manipulating the relevant characteristic dispositions of duals for specific directions of variety categorial conversions under information-knowledge-decision interactive processes. This will also allow us to create a category of problem-solution dualities and polarities useful in the space of thinking-reasoning dualities. The understanding of the principle of opposites will help in the development of transitional processes within the space of imagination-reality dualities as a sub-space of actual-potential dualities, and also in terms of the relational meanings among imagination, fantasy, dreams, goals, objective and visions over the space of subjective preferences. The cost-benefit duality and the space of cost-benefit dualities provide us with general and specific understandings of elements in the space of constraint-unconstraint dualities on knowing and information-knowledge production over the space of neuro-decision-choice actions in relational conflicts between the cost duals and benefit duals within the space of static-dynamic dualities.

Proposition I: Internal Characteristic Disposition

In the socio-natural existence, varieties exist as internal relativity of internal characteristic dispositions for identification decision-choice actions in determining variety identities in property and relation, and as external relativity of external characteristic dispositions for transformation decision-choice actions in determining changes in variety disposition in property and relation.

Proposition II: Identification-Transformation Condition

From the principle of opposites and the relativity of dualistic-polar conditions, every variety problem-solution duality is either an identification problem-solution duality or a transformation problem-solution duality or both. The identification problem-solution duality is a property-identification or a relation-identification or both at the static state, and the transformation problem-solution duality is a change in property or change in relation or both at the dynamic state. Every problem or solution is determined by the internal and external relativity conditions in the space of relativity.

X: Possibility, Imagination, Reality, Possible Worlds, Creativity and Anomalies in Knowing and the Knowledge Production System

In conceptualizing knowing as an intellectual investment process and information-capital production as an intellectual capital accumulation process, we also introduced into the epistemic structure by-products as outcomes in the space of the input-output dualities in the supply-demand chain dynamics as one journeys from the potential space through the space of imagination to the space of actuals through the space of realities over the epistemological space. The epistemological space represents all neuro-decision-choice actions over the space of static-dynamic dualities. It is within this space that we have sub-spaces of acquaintance, descriptions, cognitive journeys, true-false conflicting claims, epistemic complexities and epistemic tensions that provide a system of continual feedback modes over the space of intellectual stock-flow dynamics.

In the epistemological space, we have the space variety imagination-reality dualities which gives to the space of variety possibility-impossibility dualities, possible world-impossible world dualities, clarity-confusion dualities, representation-misrepresentation dualities, disinformation-misinformation dualities and the rise of errors, mistakes and many conflicting understandings and beliefs over the space of individual-community dualities that gives rise to tensions to generate elements in the energy-power dualities under the cognitive self-exiting and self-correction system for a continual knowing and information-knowledge production within the socio-natural production in knowing with information-knowledge disequilibrium dynamics and matter-energy equilibrium under primary epistemic complexities. Further complications arise as cognitive and epistemic complexities in the information-knowledge stock-flow disequilibrium dynamics. These complications are housed in the space of imagination-reality dualities that generates the space of possibility-impossibility dualities, some elements of which may reveal themselves as paradoxes, phantom problems, ill-posed problems and many other problems in the information-knowledge stocks. It is here that we see a relational connectivity of hope, creativity aspiration, audacity, motivation curiosity and the rise of the space of possible world-impossible world dualities in the space of cognitive activities and human progress over the space of negative-positive dualities under the principle of non-satiation in cognitive existence. The monograph will be devoted to discussions of these concepts in relation to intellectual stock-flow dynamics over the space of equilibrium-disequilibrium dualities and set directions for future research.

References

[1] Dompere, Kofi K.: The Theory of Info-Statics: Conceptual Foundations of Information and Knowledge, Springer (Series: Studies in Systems, Decision and Control, Vol. 112), New York (2017)

[2] Dompere, Kofi K.: The Theory of Info-Dynamics: Rational Foundations of Information-Knowledge Dynamics, Springer (Series: Studies in Systems, Decision and Control, Vol. 114), New York (2017)

[3] Dompere, Kofi K.: A General Theory of Entropy: Rational Foundations of Information-Knowledge Dynamics, Springer (Series: Studies in Systems, Decision and Control, Vol. 114), New York (2019)

[4] Dompere, Kofi K.: The Theory of the Knowledge Square: The Fuzzy Rational Foundations of Knowledge-Production Systems, Springer, New York (2013)

[5] Dompere, K. K.: Fuzzy Rationality: Methodological Critique and Unity of Classical, Bounded and Other Rationalities, (Studies in Fuzziness and Soft Computing, vol. 235), Springer, New York (2009)

[6] Dompere, Kofi K.: Epistemic Foundations of Fuzziness, (Studies in Fuzziness and Soft Computing, vol. 236), Springer, New York (2009)

[7] Dompere, Kofi K.: Fuzziness and Approximate Reasoning: Epistemics on Uncertainty, Expectation and Risk in Rational Behavior, (Studies in Fuzziness and Soft Computing, vol. 237), Springer, New York (2009)

[8] Dompere, Kofi K.: Cost-Benefit Analysis and the Theory of Fuzzy Decisions: Identification and Measurement Theory (Series: Studies in Fuzziness and Soft Computing, Vol. 158), Springer, Berlin, Heidelberg (2004)
[9] Dompere, Kofi K.: Cost-Benefit Analysis and the Theory of Fuzzy Decisions: Fuzzy Value Theory (Series: Studies in Fuzziness and Soft Computing, Vol. 160), Springer, Berling, Heidelberg (2004)
[10] Robinson, R.: Definition, clarendon Press, Oxford (1950)
[11] Gorsky, D.R.: Definition, Progress Publishers, Moscow (1974)
[12] Dompere, Kofi K.: The Theory of Categorial Conversion: Rational Foundations of Nkrumaism in socio-natural Systemicity and Complexity, Adonis-Abbey Pubs. London (2016–2017)
[13] Dompere, Kofi K.: Polyrhythmicity: Foundations of African Philosophy, Adonis and Abbey Pub, London (2006)
[14] Kuhn, T.: "The Function of Dogma in Scientific Research," in Brody, Baruch A. (ed.) *Reading in the Philosophy of Science*, Englewood Cliffs, NJ., Prentice-Hall, pp. 356–374 (1970)
[15] Kuhn, T.: The Essential Tension: Selected Studies in Scientific Tradition and Change, University of Chicago Press, Chicago (1979)
[16] Lakatos, I.: The Methodology of Scientific Research Programmes, Vol 1, Cambridge University Press, New York (1978)
[17] Lakatos, I., Musgrave A. (eds.), Criticism and the Growth of Knowledge, Cambridge University Press, New York (1979). Holland, 1979, pp. 153–164

Acknowledgements

The theories of knowing and information-knowledge intellectual stock-flow processes, viewed in terms of the theory of neuro-decision-choice rationality in the space of input-output dualities, irrespective of how rationality is conceived and interpreted, affect all areas of human thought and the laws of thought that may guide neuro-decision-choice behaviors toward fundamental and applied systems in managing all social setups in time and over time under the Sankofa telescopic past-present-future traditions, where all variety transformation decision-choice actions are governed by the *asantrofi-anoma* conditions of internal and external cost-benefit variety relations with relational continua on the bases of knowing and information-knowledge conditions that demand the constructs of the theory of knowing and the theory of information-knowledge system through the theory of problem-solution processes in a relational unity of *denkyem-mereku-funtum-mereku* tradition of dualistic-polar conflicts with relational interconnectedness and resolutions over the spaces of variety production-consumption and input-output dualities under no-free lunch conditions. This approach has benefited from economic theory of production-consumption dualities and cost-benefit analysis and the adinkra system of symbolic thoughts involving *adinkramatics*, *anansimatics*, *adinkralogy*, *anansiology* and conditions of *paremiology*.

There are many cognitive pathways to the construct of the theory of input-output processes as production-consumption processes. One pathway may be conceived from the viewpoint of theory of production-consumption processes with quality control through *epistemic transversality* conditions by entropic analytics. Any of these paths to the theory of knowing and information-knowledge production, therefore, is about the discovery of problem-solution processes in human actions, and the explanation as to how these production-consumption processes manifest themselves in information-knowledge structures and as inputs into neuro-decision-choice processes that allow the universal object set to be reflected in human mind relative to social and natural processes. The greatest danger to the discovery of the problem-solution process over the space of fundamental-applied dualities and the understanding of applications that may be required of it in the use of social transformation processes is ideological, religious and political credulity without the conditions of the constraints of space of scientific truth-falsity duality. In this respect, we offer great thanks to the gains in philosophy and psychology.

This ideological and scientific credulity finds expression in the classical paradigm with the principle of exactness and absolute truth that cement the foundations of current scientific knowledge over the space of absoluteness. The danger may be diminished by developing cognitive habits of different forms of the mendacity in order to develop a more robust foundation of fuzzy paradigm composed of its logic and mathematics with the principle of opposites composed of duality and polarity over the space of relativity. In this respect, in the development of the fuzzy paradigm, the enhancement of its logic and the expansion of its mathematical domain, we must be thankful for all researchers and scholars who have freed themselves from the ideological grip of the classical paradigm with Aristotelian principle of excluded middle of the duals of duality in order to work on the frontier of fuzzy phenomena in all areas of the knowledge enterprise. This monograph has benefited from their contributions in the global search over the spaces of question-answer dualities and problem-solution dualities through some aspects of mathematical representation and reasoning.

Special thanks go to friends and foes whose positive and negative encouragements, respectively, have made this work challenging and enjoyable to the finish. My thanks also go to all my graduate students in my courses in economic theory in open and closed systems, mathematical economics and cost-benefit analysis for allowing me to introduce them to new techniques and logic of thinking in problem-solution conflicts in an attempt to free them from the rigid logical structure of the classical system with the principle of excluded middle with non-acceptance of contradictions. I express great appreciation to Ms. Jasmin Blackman for providing proofreading advice that has allowed this work to be clearer and easier to read. This monograph has also benefited from the works of all the authors in the references for cognitive conflicts and resolutions. I also express thanks to all those who will be interested to read this monograph and offer some new directions or continue the path of thinking and reasoning.

A Proposal

Statement of Aims and Rationale behind the Monograph

(a) The *aim* of this monograph is to provide a self-contained analytical framework in terms of philosophy and mathematics for the understanding of the conditions that provide a categorial diversity and unity of knowing and science where the conditions are abstracted from the continual interaction between space of neuro-decision-choice actions and the space of problem-solution dualities. The general explanatory-prescriptive process relating to decision-choice behavior over the problem-solution dualities under dualistic-polar conditions is the *theory of problem-solution dualities and polarities*. It presents a case where information-knowledge systems are derived from neuro-decision-choice actions over the general space of problem-solution dualities. The aim is to show that human history, human progress, culture and civilization are generated by decision-choice outcomes over the space of success-failure dualities and to connect the analysis to diversity and unity of knowing and science.

(b) The *objective* of the monograph is to develop a logically consistent unified explanatory and prescriptive theory to explain the general categorial diversity and unity of knowing and science not as induced by methods and techniques but as induced by neuro-decision-choice input-output process through the behavioral conditions of the space of problem-solution dualities in the space of static-dynamic dualities. Techniques and methods are derivatives from the nature of variety problem-solution dualities; however, one may categorize the variety problem-solution dualities in terms of the common techniques and methods that may be useful to individuals and the collective in the problem-solution situation but not in classificatory terms of diversity and unity of knowing and science. The epistemic process is to show the role that individual, collective and social preferences play on the space of problem-solution dualities through the space of variety cost-benefit dualities as facilitating continual productions of variety problem-solution dualities and seen in the transformations of variety characteristics over the space of acquaintance.

The other objective is to show that the principle of opposites, consisting of duality and polarity with relational continuum and unity, acts as powerful analytical tools in understanding the diversity and unity of knowing and science under the conditions of parent-offspring process, where the basic premise is that all socio-natural varieties exist in dualistic-polar states and transformations do not destroy the basic proposition of variety dualistic-polar universal existence. It is also to show that the fuzzy paradigm deals with all information structures within the space of exact-inexact dualities as well as within the space of absolute-relative dualities to make explicit the philosophical and analytical debate for information-knowledge representation and analyses among different mathematical and logical schools of thought with excluded or non-excluded middle.

The principal objective and supporting objectives are enhanced by placing the diversity and unity of knowing and science through the variety problem-solution dualities in an epistemic field with relational and information fields. The theory of problem-solution dualities and polarities is, thus, an attempt to develop a linking framework in support the theories of info-statics, info-dynamics and general entropy through the interactions of the space of neuro-decision-choice actions and the space of problem-solution dualities to deal with our understanding of conditions of diversity over the space of unity-disunity dualities. The objectives are also amplified by the use of dualistic-polar games as methods and techniques in the understanding of diversity and unity of knowing and science as one examines the epistemic progress through fuzzy-stochastic dynamic games between the duals of the variety problem-solution duality and between poles of variety problem-solution polarities in the conflict space of variety actual-potential existence and to provide an explanation for infinite problem-solution process that complements the postulate of non-satiation.

(c) The *rationale* for writing the book is derived from multiple epistemic sources of concern.

(1) There is a need for justifications of multidisciplinary programs in research, teaching and learning in our education systems toward restructuring the whole educational foundations for redefining research universities, teaching and learning universities and colleges in terms of problem-solution dualities.

(2) There is also a need to re-examine the research directions of the meaning and development of information-knowledge systems and place them in the spaces of production-consumption dualities and input-output dualities relative to interactive structures of the space of neuro-decision-choice systems and the dynamic conditions of input-output space, and to emphasize the nature of information-knowledge systems as continual outcomes of the historical process. This will require an analytical distinction and separation between the phenomenon of information transmission and the

phenomenon of the meaning and production of information-knowledge systems in relation to the space of source-destination dualities.

(3) There is a continual need to clarify the methodological platforms of the classical paradigm, the neoclassical paradigm and the fuzzy paradigm for their utilities and limitations in all analytics, decision-choice processes over the space of problem-solution dualities and the results that they produce.

(4) The need to provide an epistemic structure for the concepts of problem, solution and their interactions in affecting the direction of history. These needs lead to the basic rational which is to bring into focus the idea that all decision-choice actions over the space of space of problem-solution dualities are about information-knowledge production and processing as an input-output process in diversity and unity within the space of human socio-natural activities of decision-choice actions and source-destination transformations.

(5) The need to show the economic production morphology of knowing and information-knowledge accumulation, where knowing is an intellectual investment as a flow and information-knowledge accumulation, is an intellectual capital accumulation as stock in the space of neuro-decision-choice systems irrespective of the area of human action. The use of the intellectual investment as a flow and intellectual capital accumulation as stocks provides a way to allow the establishment of stock-flow disequilibrium dynamics of all information-knowledge systems with self-excitements, self-corrections and self-improvements and to link knowing and information-knowledge systems to the space of varieties which allow us to link knowing to learning with supervision and learning by doing as well as provide an explanatory and prescriptive links to knowing and knowledge accumulation.

(d) The *monograph*: Given the interactive conditions among the space of decision-choice activities, the action space and the space of problem-solution dualities as explaining diversity and unity of categorial knowing and science, the monograph is set to provide answers to several questions through the development of the *theory of epistemic field*. These questions come to us as primary and derived, and are center around the role of how neuro-decision-choice activities over the space of problem-solution dualities bring about knowing and how the knowing processes generate information-knowledge accumulation in disequilibrium dynamics irrespective of the category of knowing in terms of science and non-science, social science and non-social science. The stock-flow disequilibrium conditions are defined and explicated.

The process of knowing and knowledge accumulation process proceed from the space of acquaintances to the space of descriptions. The outcomes from the space of acquaintances are the results of diverse observations and diverse encounters, while the outcomes from the space of descriptions are the results of logical operations from a paradigm of thought. The outcomes of the processes of knowing and information-knowledge accumulation in both space of acquaintances and space of descriptions are the results of neuro-decision-choice actions, where information-knowledge processes are also input-output processes to generate organic self-correction system based on subjectivity of individuals and collectives over the spaces of certainty-uncertainty dualities, doubt-surety dualities and decidability-undecidability dualities within the space of diversity and unity irrespective of the areas of knowing or category of problem-solution dualities.

The space of information-knowledge input-output processes is continually expanding under socio-natural decision-choice actions over the space of problem-solution dualities creating increasing possibilities with variety necessities and widening the space of probabilities with variety freedoms, making easier for the discovery of other varieties in the space of actual-potential dualities as well as exercising preferences in the decision-choice space. In other worlds, the knowing processes and information-knowledge processes are not only interdependent but the results of neuro-decision-choice actions irrespective of the category of knowing in terms of mutually creating potential ideas and mutually destructing existing ideas. Every area of knowing in the space of actual-potential dualities is a category of problem-solution dualities connected to the space of actual-potential dualities under neuro-decision-choice actions passing through the spaces of uncertainty-certainty dualities, doubt-surety dualities and decidability-undesirability dualities, where outcomes are nothing more than experiences, and information-knowledge accumulation is a stock of either justified or unjustified experiences or both. The epistemic input-output processes are composed of intellectual investment processes through knowing and capital stock processes through intellectual-discovery accumulation of all categorial elements in the epistemic field. This epistemic structure is different from our traditional factory system of input-output processes of our basic economic production where investment-capital processes require other inputs in addition to intellectual information-knowledge inputs to bring about basic economic outputs.

The decision-choice processes are established by variety preference ranking based on variety cost-benefit dispositions, and variety selections made, based on the real cost-benefit rationality in relation to a goal-objective element or a set of goal-objective elements in visionary or non-visionary conditions. The definition of real cost as a real opportunity cost is to link the decision-choice actions of economic production and neuro-decision-choice processes over the space of production-consumption dualities to the conditions of physical processes of organic matter-energy equilibrium in accordance with the first law of thermodynamics, where neither matter nor energy can be created or destroyed but can have inter-categorial transforms with the same matter into various forms of variety matter. In other words, the real opportunity cost is related to the first law of thermodynamics. In the space of information-knowledge systems, the real opportunity cost relates to the first and fifth laws of info-dynamics, where the previous variety cost information is indestructible while a new variety information-knowledge item is created, and hence the amount of information-knowledge accumulation is continually in disequilibrium dynamics to satisfy the second and fourth laws of info-dynamics. In its simplest form, the first law of info-dynamics states that neither information nor knowledge as established by characteristic-signal dispositions can be destroyed even though the variety can be destroyed by transformation. The amount of information-knowledge accumulations in the universe is constantly expanding through variety destructions and differentiations. Information is continually being created through categorial conversions from one variety to another, or from one process to another, or from one state to another with continual changes, moves, controls and accumulations in terms of variety relations and places. The conditions are such that the information-knowledge accumulation is created from some variety induced by categorial conversion based on technology to create an appropriate transformational process to bring about a new variety with information and possible knowing while retaining the old information for knowing of the past variety.

There is comparative analytics which presents the differences and similarities between info-dynamics and thermodynamics:

(1a) **Commonness**: Information is a property of mater-energy configuration without which matter is indistinguishable from energy and energy is indistinguishable from matter.

(1b) Information is non-existence without matter-energy configuration.

(1c) The awareness of matter-energy configuration in different forms and their identities in matter-energy are made possible by information as seen in terms of variety characteristic dispositions and are known to cognitive agents through signal dispositions.

(2a) **Differences**: Matter-energy configuration is always in equilibrium state in the sense that matter-energy configuration is always in a flow equilibrium as well as in stock equilibrium where neither matter or energy can be created.

(2b) Information-knowledge configuration is always in disequilibrium state in the sense that the ontological information-knowledge configuration is continually in a flow-stock disequilibrium dynamics and this ontological information-knowledge disequilibrium dynamics is transformed to the epistemological information-knowledge process from the potential space to the space of the actual.

(3) The commonness and similarities of info-dynamics and thermodynamics should allow us to develop a universal theory of diversity and unity of neuro-decision-choice actions in information and matter-energy dynamics.

The theory of problem-solution dualities provides explanatory and prescriptive conditions for the understanding of the role of problem-solution dualities in our civilization and cultural dynamics and social change and transformation as seen from information-knowledge processes of cognitive actions.

(4) The conditions of the characteristic-signal dispositions are contained in the structures of info-dynamics and info-statics.

(e) The *abstract* provides a general summary of the contents of the book a thematic structure, where the topical developments of the argument are given in the *table of contents.*

The General Reflections About the Monograph

This monograph is epistemically special in its orientation as it is concerned with the relational structures of information, knowledge, decision-choice processes of problems and solutions in theory and practice. It is a continuation of the sequence of epistemic works on the theories of *info-statics, info-dynamics,entropy* and their relational connectivity to information, language, knowing, knowledge, cognitive practices relative to variety *identification problem-solution dualities* and *variety transformation problem-solution dualities* and human actions. It is about an economic approach in the understanding the diversity and unity of knowing and science through decision-choice actions over the space of problem-solution dualities and polarities. The problem-solution dualities are argued to connect all areas of knowing including science and non-science, social science and non-social science into unity with diversities under neuro-decision-choice actions. The concept of diversity is explicated to connect to the tactics and strategies of decision-choice actions over the space of problem-solution dualities. The concepts of problems and solution are defined in the space of relativity rather than in the space of absoluteness. The efforts, here, are not about information technology and its usage which are second-level processes in the sequential processes of thought. It is about the foundations of information technology where computes are viewed as mechanical support of the mind in storage, processing, person-machine, machine-person and machine-machine in the space of source-destination dualities in the space of relativity with distribution of degrees of acceptance in the spaces of true-false dualities, doubt-surety dualities and decidability-undecidability dualities.

All analytical and non-analytical concepts are specified and defined in the space of relativity to provide an infinite framework and space of comparisons, differences, commonness and ranking under individual, collective, community and social preference orderings for establishing sequences of visions, goals, objectives for selections under the socioeconomic principles of non-satiation. The concepts of diversity and unity of knowing and unity of science are structured around the nature of variety problem-solution dualities over the space of static-dynamic dualities. It is argued that the diversity and unity of knowing and science are explainable through neuro-decision-choice actions over the infinite space of problem-solution dualities. The interactions between the system of neuro-decision-choice actions and the space of variety problem-solution dualities define the meaning of life, and the way these interactions take place define the essence of life in diversity and unity of knowing.

The monograph is used to explore the pathways that may contribute to interdisciplinary education and foster innovatively collaborative research in knowing and information-knowledge production especially the understanding of the give-and-take sharing modes of categories of knowing through neuro-decision-choice actions over the space of problem-solution dualities. Unity in knowing and science is established by information as an input into the neuro-decision-choice modules for actions over the space of decidability-undesirability dualities, while diversity in knowing and science is established by information as an input in establishing differential categories of problem-solution dualities where the categorial variety is relevant for the construction of the needed methods and techniques for categorial actions. It is then argued that the problem-solution process is a variety replacement process where every variety problem is simultaneously an offspring of a variety solution and a parent to a variety problem in a never-ending process to an unknown destination where any problem gives rise to a solution which then gives rise to a problem and where every variety problem is also a variety solution relative to an element in the space of the goals and objectives.

The theory of knowing is presented as an investment in intellectual flow from problem solving, and the theory of information-knowledge system is presented as intellectual capital accumulations of success-failure outcomes of decision-choice actions over the space of the problem-solution dualities and polarities. The conceptual system is developed under the fuzzy paradigm of thought over the space of relativity that is characterized by the principle of opposites composed of duality and polarity where every variety is characterized by series of dualistic-polar conditions. The theory of knowing and the theory of knowledge are simultaneously developed as the theory of production of human actions, where behind every variety information is a characteristic-signal disposition. The production is a system of variety problem-solving actions which is about a system of variety destruction-replacement processes leading to outcomes in the spaces of ignorance-knowledge dualities, success-failure dualities and true-false dualities in terms of input-output processes with entropic conditionality. The decision-choice actions over the space of problem-solution dualities and the variety destruction-replacement processes are guided by preference ordering under real cost-benefit conditions and rationality of cognitive agents with information-knowledge of quality control as provided by a system of entropic relational tests within the interactive spaces of certainty-uncertainty dualities, doubt-surety dualities and decidability-undecidability dualities.

The theory of problem-solution dualities and polarities is an attempt to demonstrate the existence of diversity and unity of knowing and science by developing a theory of an epistemic field with relational and informational fields to allow a consistent development of the importance of interdisciplinary approach over the space of the problem-solution dualities and a unified knowledge system, where every decision-choice action over a variety problem-solution knowing is an addition to the information-knowledge accumulation over the epistemological space.

The important co-lateral contribution of the monograph is its methodological approach to the concepts of knowing and science in the monograph and the judicious use of the principle of opposites, *dualistic-polar* game theory, preference ordering, cost-benefit analytics and fuzzy paradigm of thought that together allow a linkage of the primary categories to an evolving matrix of infinite set of derived categories of problem-solution dualities in a chain of parent-offspring creative-destructive process in the space of problem-solution dualities to explain the never-ending problem-solution process on the path to perfection in the sense of understanding continual expansion of the space of problem-solution dualities and information-knowledge stock-flow disequilibrium dynamics. The space of the problem-solution dualities is also in disequilibrium dynamics in dualistic-polar stock-flow processes which has no end where an end will mean the end and essence of life as we know it.

The Synopses

A. For the general trust of the book, see the attached abstract.
B. For the main topics, see the table of contents.

Organization of the Monograph

The **monograph** is organized with a preamble, a prologue, a preface, an acknowledgment, seven chapters and epilogue.

The **preamble** is composed of series of quotations from important works of different authors in philosophy, science, mathematics, economics, decision-choice theories, information theory and other relevant areas to the information-knowledge questions about the general science and art of knowing under cognitive capacity limitations that generate questions over the spaces of certainty-uncertainty dualities and projected to the spaces of doubt-surety dualities and decidability-undecidability dualities. The preamble is used to introduce the disagreements and agreements on the idea that all information-knowledge systems are generated by knowing actions irrespective of the area of human endeavor through neuro-information-decision-choice processes over the space of problem-solution dualities within the ontological-epistemological duality relative to the space of actual-potential dualities and how one moves from the space of acquaintance to the space of description.

The **prologue** introduces a set of essential epistemic propositions in the theoretical system on problem-solution duality and polarities and how decision-choice actions are practiced. Based on these essential propositions, the methodology, techniques and methods for discussing the propositions are defined and analyzed to provide conditions of their differences and commonness. The techniques and methods are a toolbox collection in any methodology of interest.

The preface introduces the motivation for writing the monograph, what it is about and how the conceptual framework of the monograph is connected to other previous monographs by the same author on information, knowledge, data, evidence and decision-choice actions. The general conceptual frame makes explicit the structure of the theory of problem-solution duality and polarity and how it forms the foundations of knowing and information-knowledge accumulation as well as provides us with foundational links to information technology and relevant computer connections. The analytical connectedness of these monographs is found in the consistent use of the principle of opposites, fuzzy paradigm of thought with its mathematics, logic and corresponding toolbox of techniques and methods of analyses, where knowing is seen as an intellectual investment flow in the discovery of knowledge, and information-knowledge is seen as an intellectual capital accumulation of discoveries providing us with a system of intellectual factories of production and decision-choice actions over the space of problem-solution dualities. Together, these monographs present the meaning and essence of my works and research agenda, where information-knowledge stock-flow disequilibrium dynamics is seen as an economic production and related to continual neuro-decision-choice dynamics over the space of problem-solution dualities and where stock-flow equilibrium and disequilibrium dynamics are related to matter-energy states and information-knowledge states in the universal existence, where information stock-flow disequilibrium takes within the matter-energy stock-flow equilibrium.

Chapter 1 is devoted to introducing the reader to the theory of knowing and the theory of knowledge. The concepts of knowing and knowledge are defined in order to establish the differences, commonness, similarities and their connecting cords as a production chain, where the process of knowing is connected to the production of intellectual investment flows from the space of acquaintance to produce the output of experiential information of knowledge by acquaintance which becomes an input into a processing modulus to produce an acceptable variety output as information-knowledge for updating intellectual capital stocks. Comparative analytics are performed on the theory of knowing and the theory of knowledge to establish their relative positions in the economic-theoretic chain of production-consumption process.

Chapter 2 presents conceptual pathway analysis of the theory of knowing and the theory of knowledge as neuro-decision-choice activities in the space of ontology-epistemology duality, where the knowing process and information-knowledge production proceed from the space of imagination as a sub-space of the potential to the space of possibility to the space of probability and to the space of reality as a sub-space of the space of actuals in a continual search for information and knowledge through the neuro-decision-choice activities in support of the general neuro-decision-choice actions over problem-solution dualities relative to the space variety identifications and variety transformations. The problem-solution dualities establish the principle of diversity and the unity of knowing, the principle of diversity and the unity of information-knowledge accumulation as well as the principle of diversity and the unity of science. The concepts of ontology and epistemology and the corresponding spaces are defined and explicated to be useful to economic-theoretic approach for the understanding of the knowing process as the production of intellectual investment flows and the production of the information-knowledge system as the accumulation of intellectual capital stocks.

The chapter also introduces the concepts of primary structures and derived structures linking them to the general parent-offspring destruction-replacement process in transformations. The epistemic differences and similarities of the concepts of primary and derived are discussed in relation to diversity and unity of knowing and science. The forces of transformations are discussed in relation to diversity and unity of knowing and science where the variety characteristics are represented by the family of ordinary languages (FOL) and the family of abstract languages (FAL) in coding for storage, remembrance, sharing and analysis in relation to *the categorial diversity and unity principles of science* through knowing under dualistic conflicts to establish the framework transformation through polar conflict games of varieties. The emphasis is on the role of duality in the journey between ignorance and knowledge.

In Chapter 3, the concept of epistemic space is advanced, defined and explicated for the understanding of fuzzy spaces in the development of the theory of knowing as the theory of intellectual investment flows and the theory of information-knowledge system as the theory of intellectual capital stocks. In this framework, the theory of intellectual investment-capital process is seen as the theory of intellectual stock-flow dynamics, where the intellectual information-capital stocks provide capital-service input flows into the general neuro-decision-choice space and the intellectual investment flows provide continual updating of intellectual stocks in the space of quality-quantity dualities. The relevant spaces and sub-spaces are introduced leading to the definition and explication of the epistemic space and the development of theory of epistemic space for the understanding of both the theories of knowing and knowledge seen in terms of neuro-decision-choice activities over the space of problem-solution dualities.

Chapter 4 is used to present the theory of problem-solution dualities as the first part of the general theory of opposites where the second part is the theory of problem-solution dualities which will be taken up in the next chapter. It discusses the conditions of the nature of problems relative to solutions in dualistic-polar structures with relational continua, unity and give-and-take-sharing modes in the space of fundamental-applied dualities to provide conditional connectivity to the development of the theory of polarities. The concepts of ontological and epistemological spaces are introduced, defined, explicated and connected to the space of problem-solution dualities. And the connectivity of these concepts to the roles they play in the knowing process as intellectual investment flows and the production of information-knowledge system as intellectual capital accumulation stocks is then explained. The space of incentive-disincentive dualities is introduced and linked to the space of cost-benefit dualities in order to establish the concepts of primary-derived spaces and parent-offspring processes relative to the elements in the space of problem-solution dualities as transformational elements where problems are transformed into solutions to give rise to new problems over the multiplicity of cost-benefit or destruction-construction enveloping that establishes the space of problem-solution genetic paths of socio-natural variety dynamics. The space of problem-solution dualities is established over the space of imagination-reality dualities which is shown as a sub-space of actual-potential dualities that constitutes the space of ontological-epistemological dualities.

The essential elements in all categories of unity are grouped into *primary categories* that find meaning in the space of problem-solution dualities and *derived categories* that find meaning in the logical processes for epistemic maneuvering over the space of problem-solution dualities under the control and management of decision-choice actions to effect variety problem-solution replacement processes, the results of which go to update the universal information-knowledge accumulation *disequilibrium dynamics*. The instruments of thinking and reasoning and their roles in the cognitive journey are introduced, defined and explicated to set a stage for the development of the theory of epistemic fields in relation to epistemic spaces.

Chapter 5 is used to present the structural conditions of the epistemological space for the understanding of neuro-decision-choice activities in knowing as intellectual investment flow, cognitive actions over the space of problem-solution dualities and how such actions relate to information-knowledge accumulation as intellectual capital stocks, the service flows of which become inputs into neuro-decision-choice actions. It presents a theoretical understanding of how the space of problem-solution dualities is embedded in the space of imagination-reality dualities which is a subspace of the space of actual-potential dualities. The concepts and phenomena of problems, solutions and the space of problem-solution dualities are defined through the conditions of cost-benefit dualities over the space of relativity. The concepts of criterion, criteria space and neuro-decision-choice space are verbally and algebraically introduced and defined, and their relationships to knowing, information-knowledge processes and their connectivity to the space of problem-solution dualities are specified and explained. The important and prominent role played by conditions of cost-benefit duality is made explicit and linked to the criteria space. The problem-solution process is argued to link the knowledge by acquaintance to knowledge by description through a general paradigm of thought irrespective whether an area is defined as science or non-science, where the classifications of areas of knowing and problem-solution dualities are seen as fuzzy-set closure with fuzzy conditionality. The chapter also presents the nature of the analytical connectivity within the ontological-epistemological duality and how the ontological space constitutes the primary identity of all variety existence in the past-present-future telescopic structure functioning with information-knowledge equality constraints, while the epistemological space constitutes the derived identity of variety existence, in the a derived past-present-future telescopic structure functioning with information-knowledge inequality constraints.

Chapter 6 is devoted to the discussions on the meanings of the concepts and phenomena of category and categorial conversion and their uses in the understanding of knowing, knowledge and the principles of diversity and unity that they may entail. The concepts are introduced to deal with transformations and changes within the space of quantitative-qualitative dualities as they relate to info-statics and info-dynamics of varieties in all areas of knowing as intellectual investment flows and information-knowledge accumulation as intellectual capital stocks. Alongside the theory of categorial conversion is an introduction of the theory of philosophical consciencism and the discussions on its relations to neuro-decision-choice processes, education, preference formations, varieties of human progress, shifting of generational preferences and increasing individual and social abilities in the space of variety creation-destruction dualities as seen in terms of individual and collective curiosities over the space of imagination-reality dualities and continual variety transformations. The theories of categorial conversion and philosophical consciencism are mutually interdependent, inter-supportive and inseparable, just like the theories of info-statics and info-dynamics are mutually interdependent, inter-supportive and analytically inseparable. Similarly, the theories of categorial conversion and info-statics are mutually creating, while the theories of philosophical consciencism and info-dynamics are also mutually interdependent, creating and inseparable. The epistemic frameworks of both the theories of categorial conversion and info-statics are shown to be linked to the epistemic frameworks of *possibility* and *necessity*, while the epistemic frameworks of the theories of philosophical consciencism and info-dynamics are shown to be linked to the frameworks of *probability* and *freedom* over the spaces of neuro-decision-choice activities and the action space as they relate to the conditions established over the spaces of identification-transformation and imagination-reality dualities within which the concepts and the phenomena of phantom, ill-posed, unresolved, unintended and other problems are defined and discussed in relation to paradigms of thought and cognitive capacity limitations in knowing. The knowing and information-knowledge processes are shown to be cognitive journeys between variety ignorance to variety knowledge in the space of ignorance-knowledge dualities as a sub-space of the space of actual-potential dualities, where the journeys are constrained by information-knowledge inequality conditions with uncertainty and risks.

Chapter 7 contains discussions on the concepts and phenomena of problem-solution polarities and the role they play in the understanding of knowing as intellectual investment flows and information-knowledge accumulation as intellectual capital stocks in order to establish the principles of diversity, relational continuum and unity that they entail. The relational structure between duality and polarity is made explicit in an epistemic organicity, where the diversities are in a multiplicity of give-and-take sharing mode. The poles of the polarities are established by the residing dualities such that the negative (positive) duality defines the negative (positive) pole while the dominance of the negative or the positive pole reveals the identity of the polarity in variety transformations. The categorial conversion establishes the external conditions for dualistic-polar change, while the philosophical consciencism establishes the internal conditions for change over the space of variety quantitative-qualitative dualities relative to the space of statics-dynamics dualities with extensive discussions on the variety dynamics in knowing and information-knowing development.

The **Epilogue** concludes the monograph with views and reflections for further research and development in the understanding of the interactions of the spaces of neuro-decision-choice actions, problem-solution dualities, knowing and information-knowledge stock-flow disequilibrium dynamics relative to matter-energy stock-flow equilibrium dynamics under the principles of diversity, unity and opposites with relational continua. The directions of the role that institutions and organizations play in knowing and the knowledge processes are discussed in relation to information-knowledge processes as a systems of input-output processes and construction-destruction processes under a system of neuro-decision-choice actions with *cognitive capacity limitations* composed of *information-knowledge capacity limitations* and *paradigmatic capacity limitations*, where goals, objectives and visions are defined and expressed in the space of imagination-reality dualities as well as the manner in which the space of imagination-reality dualities is contained in the space of actual-potential dualities for the understanding of the general human behavior and cognitive actions over the space of production-consumption dualities, where preferences are governed by the principle of non-satiation over the space of varieties relative to the space of imagination-reality dualities. The epilogue is concluded with discussions on the directions for the examination of imagination, reality, belief, hope, goals, objectives, visions, curiosity and the relevant methodology in knowing and information-knowledge stock-flow dynamics through the space of problem-solution dualities under the principle of philosophical consciencism with the discussions on categorial inter-relationalities for the development of future research universities.

Professor Emeritus Kofi Kissi Dompere
Department of Economics, Howard University, Washington, D.C., 20059, USA
Home: 3822 18 ST. NE., Washington, D.C., 20018, USA
E-mails: kdompere@howard.edu with CC: domperekofi494@gmail.com

About the Author

Professor Emeritus Kofi Kissi Dompere is Professor Emeritus of economics at Howard University. He holds a doctoral degree in economics, master's degree in applied mathematics, master's degree in business administration in finance and bachelor's degree in economics and mathematics, all degrees from Temple University in Philadelphia. He has authored several scientific and scholarly works in economics, philosophy, decision theory and fuzzy mathematics. Some of the articles may be found at the following site https://howard.academia.edu/KOFIKISSIDOMPERE.

He is the winner of 2006 Cheikh Anta Diop Award for excellence in scholarship. He has a number of published monographs including: (1) A General Theory of Entropy: Fuzzy Rational Foundations of Information-Knowledge Certainty, www.springer.com (2019), (2) The Theory of Info-Dynamics: Rational Foundations of Information-Knowledge Dynamics, www.springer.com (2018), (3) The Theory of Info-Statics: Conceptual Foundations of Information and Knowledge, www.springer.com (2018), (4) The Theory of Categorial Conversion, www.adonis-abbey.com (2017), (5) The Theory of Philosophical Consciencism, www.adonis-abbey.com (2017), (6) Fuzziness, Democracy, Control and Collective Decision-Choice Systems: A Theory on Political Economy of Rent-Seeking and Profit-Harvesting, www.springer.com (2014), (7) Social Goal-Objective Formation, Democracy and National Interest: A Theory of Political Economy Under Fuzzy Rationality, www.springer.com (2014), (8) The Theory of the Knowledge Square: The Fuzzy Rational Foundations of the Knowledge Production System, www.springer.com (2013), (9) Fuzziness and Foundations of Exact and Inexact Sciences, www.springer.com (2013), (10) Fuzzy Rationality: A Critique and Methodological Unity of Classical, Bounded and Other Rationalities, www.springer.com, (11) Epistemic Foundations of Fuzziness: Unified Theories on Decision-Choice Processes, www.springer.com, (12) Fuzziness and Approximate Reasoning: Epistemics on Uncertainty, Expectation and Risk in Rational Behavior, www.springer.com, (13) African Union: Pan-African Analytical Foundations, www.adonis-abbey.com, (14) Polyrhythmicity: Foundations of African Philosophy, www.adonis-abbey.com (2006), (15) Cost-Benefit Analysis and the Theory of Fuzzy Decisions: Identification and Measurement Theory (2004) (www.springer.com), (16) Cost-Benefit Analysis and the Theory of Fuzzy Decisions: Fuzzy Value Theory (2004) (www.springer.com), (17) The Theory of Aggregate Investment in Closed Economic Systems (www.greenwood.com), (18) The Theory of Aggregate Investment and Output Dynamics in Open Economic Systems (www.greenwood.com), (19) Epistemics of Development Economics: Toward a Methodological Critique and Unity (Co-author with Dr. Ejaz) (www.greenwood.com, 1995). His teaching areas include economic theory (micro and macro), international economics (trade theory, monetary theory and theory of commercial policy), cost-benefit and project analysis, mathematical economics and econometrics at all levels. His current research focus is on the theories of knowing and information-knowledge development through the understanding of the phenomena of problem-solution and question-answer dualities as they relate to the theories of economic production-consumption, input-output processes under the conditions of cost-benefit dynamics, fuzzy paradigm of thought and the principle of opposites, mathematics and epistemics on fuzzy phenomena and their applications to neuro-decision-choice actions, mathematics of politico-legal economy and data analytics. He is a resource person for UNITAR (United Nations); economic consultant in the areas of cost-benefit analysis, national income accounts, development and macroeconomics; resource person, World Bank; and referee for Studies on African Economic Policy and Trade Negotiations. He served as Senior Technical Research Consultant for Cost-Benefit Analyst—Zero-Based Regulatory Review Project of Federal Highway Administration (USA) Basic Technologies International Annandale, VA. He also served as Consultant to consolidation of East and Southern African Regional Groupings into a common market organized by International Development Research Center—Canada. He served as Technical Consultant, Government of Botswana, USAID Washington, DC National Accounts, Macroeconomic Policy and Industrial Data Systems Design; Economic Consultant, World Bank Social Dimensions of Structural Adjustment—Ghana; and Technical Consultant and African Delegate, 24th Session, United Nations Statistical Commission, New York.

He is the founder of The Research Institute of Information Decision and Economic Sciences (RIIDES) in which he serves as Distinguished Research Professor. His partial research impact on the scientific and scholarly community may be assessed on http://worldcat.org/identities/lccn-n93122486/, and the partial list of monographs may be see on https://www.abebooks.com/book-search/author/kofi-kissi-dompere/, https://dl.acm.org/profile/81100240813, https://www.semanticscholar.org/author/K.-K.-Dompere/2608523.

He is an active participant on fundamental problems in physics FQXi Community. Under RIIDES, he is working on the frontiers of information-knowledge and decision-choice systems with monographs of (1) A General Theory of Entropy: Fuzzy Rational Foundations of Information-Knowledge Certainty and (2) Unity of Knowing and Unity of Science: Information-Decision-Choice Foundations in the Fuzzy-Stochastic Space.

He is the producer and host of a radio program "African Rhythms and Extensions" on WPFW 89.3 FM in Washington D.C., USA, www.wpfwfm.org.

E-mails: kdompere@howard.edu with CC: domperekofi494@gmail.com

Chapter 1
The Theory of Knowing and the Theory of Knowledge: Comparative Analytics

Abstract This chapter is devoted to introducing the reader to the theory of knowing and the theory of knowledge. The concepts of knowing and knowledge are defined in order to establish the differences, commonness, similarities and their connecting cords as a production chain, where the process of knowing is connected to the production of intellectual investment flows from the space of acquaintance to produce the output of experiential information of knowledge by acquaintance which becomes an input into a processing modulus to produce an acceptable variety output as information-knowledge for updating intellectual capital stocks. Comparative analytics are performed on the theory of knowing and the theory of knowledge to establish their relative positions in the economic-theoretic chain of production-consumption process.

1.1 Introduction to Knowing and Knowledge

In my monographs dealing with some important epistemic questions on information-decision-choice interactive processes of human endeavors, the idea of the *theory of knowing* was advanced and related to the variety identification problem, the variety transformation problem and the information-knowledge-certainty problem in the space of methodological constructionism-reductionism dualities as an analytical toolbox for an *epistemic search, discovery and acceptance of knowledge* [501, 670, 671, 832]. The discovery and acceptance of knowledge include decidability, computability, and predictability associated with knowledge by acquaintance and knowledge by description with a complex built in automation of the vehicle of knowing with self-exciting, mistake-properties and self-correction properties for providing a varying dynamic instruments of epistemic journey over an infinite space over the space of mistake-correction dualities, where the mistake-correction processes are under the structure of the ruling philosophical consciencism. In the development of the relational structure of the theory of knowing and the theory of information-knowledge structure, the concepts of information, data, knowledge, fact,

K. K. Dompere, *The Theory of Problem-Solution Dualities and Polarities*,
Studies in Systems, Decision and Control 405,
https://doi.org/10.1007/978-3-030-90279-7_1

evidence, and evidential things were verbally, analytically and algebraically introduced, defined, explicated and structured in a general interrelated framework for scientific, philosophical, technological and general understanding as neuro-decision-choice input-output processes in all areas of human endeavors and actions, in the sense that information is an input as well as an output, knowledge is also an input as well as output in the complete production-consumption processes where the production-consumption facilities are the general and specific neuro-decision-choice actions in socio-natural existence and the concept of progress is defined over the space of cost-benefit dualities.

These interrelated and interdependent concepts are set up as sequential derivatives from information which is considered as the primary categorial input obtained from the acquaintance space in active and inactive encounters in the search for socio-natural understanding and knowledge as the outcomes in the general input-output process with a system of *epistemic search engines*. The system of epistemic search engines is the system of collective and individual neuro-decision-choice processes in competition and cooperation that constitute the foundations of cognitive intellectual investment-capital enterprise for developing the information-knowledge system of national efforts in socioeconomic activities over the space of imagination as a subspace of the space of potential, and over the space of reality as a subspace of actuals [87]. In all my previous monographs, questions arise as to the relationships among variety ignorance, knowing, knowledge, understanding, identification and transformation leading to several questions as to what information is and what does it represent; and similarly, what knowledge is and what does it represent, and what are the similarities and commonness between information and knowledge on one hand and between knowing and knowledge on the other hand and what are the corresponding answers?

These questions and answers to them, in this epistemic framework, find meanings and expositions in the theory of identification of what there was (the past), what there is (the present) and what there would be (the future) under Sankofa-anoma tradition of the past-present-future time configuration. This Sankofa-anoma tradition splits into three dualities of past-present dualities, present-future dualities and past-future dualities consisting with the structure of time trinity that must be related to information trinity and knowledge trinity. The stated questions are followed by equally important questions regarding the forces at play between duals of dualities such as hope, dispaire, intelligence, motivation, courage in relation to elements in the space of success-failure dualities in knowing and information-knowledge accumulation in relation to neuro-decision-choice activities over the space of input-output dualities as defining factor of varieties in the space of static-dynamic dualities, where mistakes and corrections exists over the journey from relativity to absoluteness, from imperfection to perfection and from unsatisfying conditions to increasing satisficing conditions under the principle of non-satiation in the spaces of production and consumption. In the process, we must deal with a system of *production of flow of knowing* and a system of *accumulation of the stocks of the known* in relations to the dynamics of cognitive instruments of intergenerational national efforts as generated by conditions of neuro-decision-choice dynamics relative to intergenerational needs and wants in the space of input-output systems. These conditions of neuro-decision-choice actions find expressions in the national philosophical consciencism that exists and set boundaries of the space of rejection-acceptance dualities as defining the space of necessity-freedom dualities over the variety space. The space of necessity is an offspring of the space of possibility while the space of freedom is an offspring of space of probability relative to information conditions. Generally, the elements of the variety space present options over the space of imagination for variety transformations and over the space of reality for variety needs and wants of cognitive agents.

The system of cognitive instruments of national efforts in production of knowing and the accumulation of the stocks of knowns includes elements of individual and collective creativity, hope, innovation, judgments, intelligence, wisdom, incentives, motivations, disincentives and many preponderating factors under dualistic polar conditions in interrelated complex relations over the *relational field* which is to be defined, explicate and established in the monograph. From the general framework, the theories of info-statics, info-dynamics and fuzzy-stochastic entropy were advanced [475, 670, 671, 832] using the *fuzzy paradigm of thought* under the *general principle of opposites* composed of duality, polarity and organicity in relational continua and unity to laid down scientific-economic foundations for the understanding of diversity and unity of knowing and science [87, 174]. The foundationally epistemic *thinking-reasoning* structure is the *nyansapo principle* (adinkra symbolism) that relates beginning to end and vice versa in the space of beginning-end dualities, nothingness to somethingness and vice versa in the space of nothingness-somethingness dualities, in the parent to offspring and versa in the space of the space of parent-offspring dualities, negative to positive and vice versa in the space of negative-positive dualities and many other under dualistic-polar dynamics, where every variety belongs to the space of primary-derived dualities in the sense that every variety is a parent to some variety or itself as well as an offspring of another variety in the space of dualities and polarities in a complete universal system of oneness-dividedness dualities. Conceptually, dividedness is in oneness and from oneness emerges dividedness, and where oneness is in dividedness and from dividedness emerges oneness. Each of the previous monographs dealt with a specific set of variety elements in the understanding of information, knowledge and neuro-decision-choice process. Each monograph, therefore, is structured to be an illumination on a diversity as well as totality of the knowing process, information-knowledge-production and neuro-decision-choice actions in our understanding and awareness of our universe and production and reproduction of real life.

In the theory of entropy, it was indicated that there exists a system of the unity of knowing, the unity of science, the unity of engineering sciences, the unity of population science the unity of production-consumption sciences, the unity of input-output sciences and the unity of science of institutions and organization of institutions in dealing with information-knowledge constraints that generate the space of certainty-uncertainty dualities in the space of human neuro-decision-choice actions, the results of which define the historic paths from past through to the present and to the future in Sankofa-anoman traditions with self-correction systems. All these unities belong to the structured relational unity of knowing, the explanatory science, the prescriptive science, the technological science, and the engineering science, the science of production-reproduction of life, and the planning science as viewed from interactive information-decision-choice processes with respect to language formations, problem-solution dualities and problem-solution polarities of the family of systems of variety identifications, variety transformations, variety information-knowledge certainties and categorial formations of socio-natural varieties into categories of knowing, and the categories of information-knowledge structures. In the discussions, information was considered as a property of all varieties directly or indirectly abstracted from the matter-energy structure and defined as the characteristic-signal disposition, where the characteristic disposition is the content of the information and the signal disposition is the message of the characteristic disposition from sources to destinations and vice versa. In other words, information has two components comprising of content dispositions and the message dispositions over the source-destination dualities. Knowing is considered as an intellectual investment and as a flow while the information-knowledge systems are considered as intellectual capital accumulations and as stocks, the intellectual investment and capital accumulation constitute intellectual stock-flow processes. The intellectual stock-flow processes reside in the space of disequilibrium dynamics with multiplicity of socio-natural processes that induce info-statics for variety identifications and info-dynamics for variety transformations. As presented, knowing is about identifications of varieties and transformations of varieties as the primary categories of knowing, and where processes and technologies are derived categories of knowing for actions on variety transformability and non-transformability.

At the level of socio-natural problem-solution discussions, the emphasis is on information-decision-choice interactive processes where information is an input into decision-choice actions, and decision-choice actions are processing modules of information to generate knowledge and variety dispositions as new information outcomes. Every subject area of knowing finds a residence in either explanatory science composed of knowledge by acquaintance and knowledge by description, or prescriptive science, composed of knowledge of planning and knowledge of engineering which together constitute variety transformations in the system of applications. It is important to note that in the practice of variety information-knowledge search by neuro-decision-choice processes in the epistemological space, the explanatory science shows itself in terms of scientific and mathematical foundations of variety technologies while the prescriptive science shows itself in terms of technological and mathematical foundations of engineering and socio-natural planning for creations of new varieties or innovations in the existing varieties towards goal-objective accomplishments. The operating forces of human existence in neuro-decision-choice actions over the space of varieties with analysis and actions take place over the space of relativity and not over the space of absoluteness in relation to the social preference system of human desires, wants and needs under the general principle of non-satiation which is always expressed over the space of cost-benefit dualities that define an incentive system of decisions and actions. The cognitive operation of the principle of non-satiation is complex and under the general dualistic-polar conditions, where each dual in duality and each pole in polarity wants increasing benefits for themselves and increasing costs for the opposites in never-ending conditions. This is the game of dualistic-polar dominance over the space of mutuality of existence-destruction processes, where stability and peace require judicious neuro-decision-choice actions between the duals of dualities and between the poles of polarities.

Among all other varieties of knowing, the goals and objectives of both explanatory and prescriptive sciences are defined over the space of problem-solution dualities as seen in terms of characteristic dispositions of elements in the space of fundamental-applied dualities of knowing and information-knowledge accumulation. The relational structure of the information-knowledge conditions of the explanatory science and information-knowledge conditions of the prescriptive science as decision-choice activities provides the justification in support of organizing educational programs with combined science, technology, engineering and mathematics (STEM) as an integrated system for decision-choice activities over the space of problem-solution dualities, where cognitive activities mediate the understandings among dualistic-polar conflicts for variety identifications and transformations [177, 180]. Here, the ultimate goal-objective element in knowing, whether in science or non-science is to provide a quality information-knowledge system as inputs into decision-choice actions, where explanation, prediction and prescription within the space of static-dynamic dualities are elements with equal standing in the information knowledge systems. The problem with the analytical concentration on STEM is the failure to understand that its creation and successes depend on the successes of the socio-scientific effects on organization of neuro-decision-choice actions and development of institutions for the execution of neuro-decision-choice activities over the space of problem-solution dualities in relation to elements in the goal-objective space, where the conditions of STEM are always under the conditions of knowing and techno-scientific progress, and where the social costs and benefits of its results are under organized creations of relevant social institutions with judicious controls and management of social institutions of society. The organized creation of social institutions will be driven by conditions over the space of problem-solution dualities under the guidance of wisdom reflecting sagacity and discernment and cultural values as seen from the social philosophical consciencism of then generational state and stage.

The philosophical debates on the definition of science, structure of non-science and the unity of science are analytically interesting and require a general understanding from all areas of knowing and people who are concerned with neuro-decision-choice processes as the driving forces of knowing, knowledge accumulation, social transformations and socioeconomic developments requiring information-knowledge inputs. They also require the understanding of the meanings of information and knowledge relative to decision-choice processes. The argument in support of the unity concept in knowing, and science in particular, cannot simply proceed from the classical unity of science argument without clearly defining the information input of the argument [135, 925–961, 965, 974, 977, 983]. The connecting forces of the general unity find expressions in the theory of knowing through information-decision-choice-interactive processes in their complete space of variety static-dynamic problem-solution dualities around variety identifications and transformations. Here, in relation to problem-solution dualities, the emphasis rests on the interactive structure of information and the neuro-decision-choice structure, where information is seen as a property of matter-energy direct and indirect structures, varieties and their interactions. Any neuro-decision-choice process is a production process in the general *problem-solution space* that makes knowing an input-output process, where information is an input into the neuro-decision-choice modules and knowledge is an output under quality control to generate more information. Each identified new variety problem is new information under a neuro-decision-choice action and each corresponding identified variety solution is also new information under the same neuro-decision-choice action. Each position of viewing the unity problem in socio-scientific knowing is an illumination of information-knowledge problem-solution duality. The common elements of all areas of knowing are information defined in terms of characteristic dispositions. The elements of differences are the nature of the problem-solution dualities that establish categorial distinctions of varieties through the space of characteristic dispositions with an established distribution of characteristic dispositions supported by the corresponding distribution of signal dispositions where every variety is established and distinguished from others by its characteristic disposition [671]. The discussions of unity of science are also discussions on the unity of knowing and information-knowledge production that require critical familiarities with some important epistemic questions. Let us state some of these questions which are set in the space of primary-derived dualities.

1.2 Some Epistemic Questions on Knowing and Knowledge

A set of epistemic questions tends to arise in understanding the definitions, the unity and the relational connectivity of characteristic dispositions of the concepts of knowing, information, data, evidence, fact, knowledge, and evidential thing which are taken as known and understood in traditional discussions on theories and practices as they relate to human cognition and behavior over the spaces neuro-decision-choice actions and production-consumption dualities. Generally, and within the space of existing traditions in human behaviors, information is related to neuro-decision-choice actions and communications over the space of source–destination dualities without explicit definition of what information is. Information as is used here in theory, application and language, is about variety identifications and identity transformations in the space of static-dynamic dualities in questions surrounding communication, understanding and neuro-decision-choice actions. Some of these questions are primary and basic while some are derivative and non-basic and may be stated as:

(1) In what space can one find the activities of knowing and decision-choice actions and by what means?
(2) Similarly, in what space can one find the activities of information-knowledge production and neuro-decision-choice actions and by what means?

(3) What are the similarities and differences between knowing and knowledge, between the theory of knowing and the theory of knowledge and their relationship to information as may be defined; and what are the differences and their connectivity to ideas and thoughts?
(4) Who are the subjects of the activities of knowing and knowledge production?
(5) Who and what are the objects under search, knowing and knowledge production, and what roles do problems and solutions play in the unity of knowing and information-knowledge accumulation?
(6) In what space can one find problems and solutions and what are the relationships between problems, solutions, information and information?
(7) What are the relational structures among neuro-decision-choice actions, information-knowledge states, problem-solution elements and socio-natural varieties?
(8) How are these questions related to the unity of knowing, the unity of science and the fundamental-applied dualities in knowing and science?
(9) What are the identities, differences and similarities among the concepts of information, knowledge, data, evidence, and ideas, thought, concepts, imagination, potential, reality and actual, possibility probability, necessity and freedom in cognitive actions?

These are some fundamental questions in the cognitive search for answers to problems of meaning, variety existence and variety changes as they relate to understanding, environmental sustainability, human survivability, human capacity to produce and reproduce forms of new ideas and varieties in relation to nature and society, construction and destruction, war and peace, conflict and resolution, tension and change, retrogression and progress, stability and chaos and many such situations with dualistic-polar conditions. These questions find meanings in the space of information and knowledge from matter-energy varieties. Questions (1) and (2) define a problem-solution duality in the space of cognitive activities. Question (3) defines a problem-solution duality for categorial differences and similarities. Question (4) and (5) define a problem-solution duality of subject-object conditions. Question (6) defines a problem-solution duality for ontological-epistemological conditions. Questions (7) and (8) define a problem-solution duality of science-non-science relation and theory-application conditions in the space knowing and information-knowledge stock-flow relations over the space of production-consumption dualities. Question 9 defines a problem-solution duality of variety identification-transformation conditions over the static-dynamic processes relative to production-consumption and input-output processes. Question 9 divides into three important areas of cognition. One is variety identification which has been dealt with in the theory of info-statics [671]. The second is variety transformation which is dealt with in the theory of info-dynamics [672]. The third is acceptance-rejection system for control and management of elements in the input-out and production-consumption processes which is dealt with in the theory of fuzzy-stochastic entropy [831, 832, 835, 841].

The search for answers to all these questions takes place through continual epistemic actions under neuro-decision-choice activities of cognitive agents over the space of problem-solution dualistic-polar conditions relative to the possible answers to question 9 [87, 177, 671, 672, 832]. Nothing is known and nothing is produced without neuro-decision-choice action in the spaces of variety identifications and identity transformations. There are no variety identifications and transformation without knowledge; there is no knowledge without information, and there is no information without characteristic dispositions; there are no characteristic dispositions without varieties which are genetic offspring to matter-energy parentage. There is no problem or a solution in social space without a goal, an objective or a vision, and there is no goal-objective element without varieties, and satisfied and unsatisfied preferences in cognition, and there are no preferences without cognitive neural networks and varieties under the principle of non-satiation in the spaces of potential and imagination with elements of fantasy. The above stated questions find meaning and understanding in the space of question-answer dualities mapped into the space of problem-solution dualities, the identities of which require us to define and explicate knowing and knowledge and the corresponding theories to which we turn our attention to the basic outlines, where these phenomena and theories must be seen as establishing the understanding of the central foundation of life, production and reproduction of life which are anchored in the meaning and essence of human life that find definitions over the space of problem-solution dualities and polarities and over the space of imagination-reality dualities.

1.3 Knowing and Knowledge in Theory and Practice

In terms of establishing the unity of science, the approach to the unity of knowing is undertaken to bring into focus an alternative way of looking at the relational structures of knowledge and knowing, the conditions of fundamentality and the applied systems, the conditions of primary and derived varieties, the conditions of the nature of problems and solutions, the costs and the benefits, and incentives and motivations of neuro-decision-choice actions on varieties, intellectual investments and intellectual capital accumulations under decision-choice environment and cognitive actions. At various discussions towards knowing, knowledge is established over the epistemological space as a derivative from ontological *signal dispositions* of ontological *characteristic dispositions* of varieties, categorial varieties and conditions of relational connectivity of variety identities and their transformations. Here, knowing and information-knowledge production are considered as economic activities that take time and resources over the space of production-consumption dualities relative to the spaces of cost-benefit dualities and input-output processes, where the nature the input-output processes is driven by cost-benefit dualities that constitute the incentive structure for inspiration and motivation to action. This way of looking at knowledge may be made explicit by defining the *theory of knowing*, the *theory of knowledge* and *epistemic search* as working definitions in cognition over the space of variety fundamental-applied dualities. Knowing and knowledge are different varieties but interrelated, interdependent and constitute a system of organic intellectual production-consumption dualities as well as input-out dualities.

Social production of all forms is the product of neuro-decision-choice activities that relate variety solutions to variety problems under construction-destruction processes in relation to needs and wants over the spaces of real cost-benefit dualities and goal-objective preferences within the conditions of matter-energy equilibrium. Solutions to problems of individual and social production are cost-benefit transformations which generate unintended byproducts as either real benefit or real cost information in the space of variety destruction-substitution processes under knowing actions, where problems are identified and destroyed, and where solutions are created as replacements for continual identification-transformation dynamics which ensure the historic continuity of the past, present and future under the *Sankofa-anoma* tradition and quality control actions over the spaces of certainty-uncertainty and acceptance-rejection dualities and polarities. In fact, the acquisition of an information-knowledge item is a neuro-decision-choice activity under an epistemic search over the space of problem-solution dualities. It is also a transformation of forms with the same material over the space of destruction-construction dualities. Variety problem is an information; variety solution is also an information, where neuro-decision-choice actions are variety identifications, variety transformations or both in relation to knowing and knowledge within the space of social production-construction dualities. Knowing is a continual intellectual investment in cognition and as a flow in the social production-consumption dualities, while knowledge is a continual information-knowledge intellectual capital accumulation in cognition as stocks, also in the social production-consumption dualities. When it comes to the spaces of production-consumption dualities and input-output dualities an analytical care must be exercised to distinguish between the social space and the natural space in the space of static-dynamic dualities where the spaces of social production-consumption and input-output dualities are contained in the spaces of natural production-consumption and input-output dualities. Over the space of social production-consumption dualities, knowledge is accumulated stocks that generate differential intellectual capital input services to the production of knowing, while the outputs of knowing generate intellectual investment flows to update the stocks of intellectual capital accumulation which must be related to conditions of neuro-decision-choice actions teaching and learning. Both the phenomena of knowing and knowledge are derived from information as the primary category over the general spaces of production-consumption dualities and input-output dualities. To construct a theory of knowing and the theory of knowledge, we must specify the phenomena of knowing and knowledge.

Definition 1.2.1: Knowing Knowing is a neuro-decision-choice process for recognizing distinct and distinguished characteristic dispositions of varieties, and identifying their identities for their similarities and differences by their characteristic dispositions from their signal dispositions at the space of acquaintance to establish their identities as experiential information or knowledge by acquaintance over the space of *ignorance-knowledge dualities* relative to the space of imagination-reality dualities within the epistemological space. It is neuro-decision-choice input-output intellectual investment flows in the space of general production-consumption dualities, where variety ignorance is destroyed and in its place comes a replacement of variety knowledge in destruction-replacement dynamics, and where every knowing of a characteristic of the variety is an illumination of its identity in the identification process as well as an illumination of its transformation process in the space of static-dynamic dualities. The characteristic-signal disposition constitutes the information, where the characteristic disposition is the set of the variety contents while the signal disposition is the set of the instrumental carriers of the contents over the space of source-destination duality.

Definition 1.2.2: Knowledge Knowledge is an accumulation of neuro-decision-choice acceptances and rejections of outputs of knowing variety characteristic dispositions through some method and technique of variety verification of claims of variety characteristic dispositions over the space of *doubt-surety dualities* relative to the space of uncertainty-certainty dualities to establish knowledge by description from the knowledge by acquaintance as the experiential information. Knowledge is input-output stocks and hence an accumulation of the results of variety knowing, where such accumulation represents non-depletable intellectual inventories, the consumption of which leads to an increased knowing of variety identities and transformations. As a neuro-decision-choice input-output intellectual accumulation, the system of the knowledge stock is always in the space of error-correction dualities. where every knowledge stock of a characteristic of any variety is an intellectual capital, the services of which is an increasing knowing for the identification process as well as an increasing illumination of variety transformation process in the space of static-dynamic dualities, where old forms are augmented in the dualistic-polar spaces while retaining the old knowledge as well as adding new knowledge in the space of acquaintance-description dualities in the stock-flow disequilibrium dynamics.

Note 1.2.1: Differences and Similarities of Knowing and Knowledge It is noted that both knowing and knowledge are neuro-decision-choice intellectual actions, and so also every activity involving human action and life, where knowing is an intellectual flow from the space of variety ignorance-knowledge dualities as a primary mediation between the ignorance dual and the knowledge dual in the space of relativity, and where knowledge in an accumulated intellectual stocks from the space of doubt-surety dualities. Knowing is a neuro-decision-choice process of cognitively discovering the variety characteristics where the discovery of every characteristic of the variety is a reduction of ignorance of cognitive agents as well as an illumination of the variety identities in the identification and transformation processes. It is an intellectual investment flow as an experiential information or knowledge by acquaintance from the space of acquaintances where such experiential information becomes an input for intellectual refinement for neuro-decision-choice acceptance. Knowledge is a paradigmatically processed experiential information which becomes an accumulation of neuro-decision choice actions of cognitively accepted and rejected variety characteristics, where the acceptance and rejection of every characteristic of variety is a reduction of doubt of cognitive agents as well as an illumination of the surety of the variety identities for the intellectual inventories as knowledge by discription. The whole system of Information-knowledge intellectual stock-flow process passing over the spaces of production-consumption dualities and input-output dualities is self-exited and error-correction feedback dynamics, where the feedback processes connects knowing as an intellectual flow to knowledge as an intellectual stock, and where knowing and knowledge are intellectual stock-flow dynamics with knowing increasing stocks as outputs and knowledge increasing flows as inputs. Let us now specify the contents of the theories of knowing and knowledge with definitions.

Definition 1.2.3: The Theory of Knowing The theory of knowing is the study of ignorance reduction through the *epistemic search* for conditions of solutions to *ignorance-reduction problems* of variety identifications and variety transformations. It is also the study of input-output processes where the variety characteristic-signal dispositions are indestructible inputs and the outputs are the indestructible information-knowledge conditions about varieties and categorial varieties, where the characteristic dispositions are the contents of the information and signal dispositions are the messages in the information about varieties over the space of source-destination dualities. The theory of knowing is, thus, the study of an information-knowledge input-output of infinite flow processes of variety destruction-substitution enterprise for variety info-statics and info-dynamics over the space of problem-solution dualities. It is simply, the study of tactics and strategies to generate information-knowledge flows through the discovery of the variety characteristic dispositions and to reduce the variety cognitive ignorance over the space of actual-potential dualities. The information about variety ignorance reduction is an indestructible input and information about variety knowledge is the indestructible output, where knowing is a continual flow and an intellectual investment process to increase the information-knowledge system as an *intellectual capital stock accumulation.* Essentially, the theory of knowing is about the study of ignorance reductions as one moves from imagination dual to reality dual as well as moves from the variety ignorance dual to the variety knowledge dual within the space of ignorance-knowledge dualities as a subspace of the space of variety imagination-reality dualities. It is useful to keep in mind that the conditions of the variety ignorance are information and the conditions of the variety knowledge are also knowledge. The theory of knowing is simply the theory of intellectual investment in the space of general production-consumption dualities.

Definition 1.2.4: The Theory of Knowledge The theory of knowledge may be viewed as the study of tactics and strategies of degrees of *doubt reduction* surrounding the knowledge of varieties as an output of the theory of knowing through the *epistemic search* for neuro-decision-choice actions over the space of problem-solution dualities relative to *doubt-surety conditions* involving the outputs of variety knowing. It is also the study of input-output processes, where the degrees of uncertainty are destructible variety inputs of the process and the degrees of certainty are the indestructible variety outputs of the process to arrive at decidability conditions of information-knowledge acceptance. It is the study of the conditions of information-knowledge systems as an intellectual capital stocks, the services of which are to be used as inputs into the neuro-decision-choice activities over the space of production-consumption dualities. Essentially, the theory of knowledge is simply about the study of doubt reductions as one moves from the imagination dual to the reality dual as well as moves from the variety rejection to the variety acceptance within the space of variety imagination-reality dualities. The information-knowledge system is a system of input-output resources for neuro-decision-choice actions over the space of production-consumption dualities, where information-knowledge intellectual stocks are produced to serve simultaneously as productions and consumptions. The theory of knowledge is simply the theory of intellectual capital accumulation in the space of general production-consumption dualities.

Definition 1.2.5: Epistemic Search An *epistemic search* is a set of cognitive activities through the neuro-decision-choice activities over the space of problem-solution dualities to identify problem duals and the corresponding solution duals as well as methods, techniques and pathways for destruction-replacement processes in the information-knowledge stock-flow disequilibrium dynamics over variety goals, variety objectives and variety visions of cognitive agents to develop clarity in the knowing flows and the information-knowledge stocks of varieties over the space of actual-potential dualities that contains the space of imagination-reality dualities.

Note 1.2.2 A: Similarities and Differences between the Theories of Knowing and Knowledge To define and explicate the spaces of problem-solution dualities and polarities, it is useful to state and explain the similarities and differences of the theories of knowing and knowledge, and the concepts of knowing and knowledge in cognitive understanding and human decision-choice activities. The similarities of knowing and knowledge have meanings in the space of information as a characteristic-signal disposition to allow the comparative analytics of real and nominal concepts in cognition around spaces of theories and practices. The theory of knowing is a product of *cognitive ignorance* about the universal system of varieties and categorial varieties in the state of categorial dividedness and oneness over the space of existential actual-potential dualities, while the theory of knowledge is a product of *cognitive doubt* of the outputs of the process of knowing in the space of universal unity-disunity dualities. In the space of knowing and knowledge production, the space of actual-potential dualities constitutes the ontological space, while the space of imagination-reality dualities constitutes the epistemological space. In the process of increasing variety awareness, knowing is an intellectually information-knowledge investment flow. Knowing is a system of individual and social intellectual investment activities to enlarge the space of human awareness of the socio-natural environment and its contents of matter-energy varieties in the space of diversities through intellectual capital accumulation from the space of imagination-reality dualities. The space of imagination-reality dualities constitutes the epistemological space as a derived space from the ontological space. Knowledge is an accumulated information-knowledge capital stock that is expanded by an increasing knowing through cognitive intellectual investments by revealing the variety characteristic dispositions from the variety signal dispositions. Both ignorance and doubt are inputs from the space of uncertainty-certainty dualities through the information-knowledge processes in the reduction of the *epistemic distances* between ontological characteristic dispositions and epistemological characteristic dispositions of either identifications or transformations of varieties and categorial varieties. The inputs into the ignorance-reduction processes are the variety signal dispositions, and the outputs are the epistemological characteristic dispositions of the same varieties.

Let us keep in mind that information is defined as a characteristic-signal disposition, where the characteristic disposition a set of the contents and the signal disposition is the set of are the message channels in terms of codes from the variety characteristic dispositions that are transmitted by some energy through some matter medium over the space of source-destination dualities. Similarly, both knowledge and certainty are outputs from the space of uncertainty-certainty dualities through the information-knowledge processes in the reduction of the *epistemic distances* between the ontological characteristic dispositions and the epistemological characteristic dispositions of either identifications or transformations of varieties and categorial varieties to establish variety differences, commonness and similarities over the space of imagination-reality dualities as a subspace of the actual-potential dualities.

The inputs into ignorance-reduction processes and knowledge improvement processes are the ontological variety signal dispositions, and the outputs are the claimed epistemological characteristic dispositions of the same varieties. Similarly, the inputs into doubt-reduction neuro-decision-choice processes are the claimed variety epistemological characteristic dispositions abstracted from the ontological signal dispositions, and the outputs are the paradigmatically accepted epistemological characteristic dispositions that carry with them epistemological signal dispositions as information transformation within the space of source-destination dualities. Let us keep in mind that information is defined as characteristic-signal disposition, where the characteristic dispositions are the contents and the signal dispositions are the message carriers in terms of codes from the variety characteristic dispositions that are transmitted by some energy through some matter medium over the space of source-destination dualities. Simply stated, the theory of knowing is a theory of intellectual investment as input-output flows to update intellectual capital stocks, while the theory of knowledge is a theory of intellectual information-knowledge capital accumulation as stocks and as inputs into neuro-decision-choice systems to strengthen the efficiencies of the intellectual investment flows over the space of production-consumption dualities. In the epistemic process, we have ontological information composed of ontological characteristic dispositions and ontological signal disposition on one hand and epistemological information also composed of epistemological characteristic dispositions and epistemological signal dispositions on the other hand, and where the ontological conditions are the primary and epistemological conditions are derivatives.

Note 1.2.2 B: On the Theory of Knowing The theory of knowing is about the ascertaining or finding out about *what there was*, *what there is* and *what there would be* in the universal existence to provide some information-knowledge structures about the universal matter-energy varieties in the process of time connectivity, where such information-knowledge structures become input-output structures in the general neuro-decision-choice processes over the space of imagination-reality dualities as a subspace of actual-potential dualities. The foundationally conceptual analytical blocks in the development of diversity and unity of knowing and knowledge about science, non-science, their relationship and utility are variety, identification, categories and transformation, given the oneness and dividedness of matter, energy and time. The central epistemic core of *the theory of knowing* is that every knowing is about variety, variety identification and variety transformation, where the static-dynamic universal information-knowledge system is a set of stock-flow disequilibrium conditions of the acts of knowing and information-knowledge accumulation in terms of intellectual sock-flow dynamics.

The analytical structure of knowing is defined within a system of problem-solution dualities under the principle of opposites in relation to *what there was* (past), *what there is* (present) and what *there would be* (future) thus providing the information time trinity of past-present-future connectivity of Sankofa-anoma principle of interrelated time dualistic structures of past-present duality, present-future duality and past-future duality. The universal knowledge system is simply the interactive relational structure of the global system of success-failure, problem-solution and certainty-uncertainty dualities under information conditionality and cognitive capacity limitations of cognitive agents over the epistemological space as a derivative of the ontological space. A theory of knowing, viewed in the space of either production-consumption-dualities or the space of input-output dualities, is a system of structured concepts, reasoning in the space of relationality with a defined logic to produce individual variety identities and the corresponding categories to establish differences, commonness and similarities of variety characteristic dispositions from their signal dispositions.

The theory of knowing is the unity of the *theory of info-statics* directed to the neuro-decision-choice activities over the space of *identification problem-solution dualities,* and the *theory of info-dynamics* directed to the neuro-decision-choice activities over the space of *transformation problem-solution dualities,* both in relation to *what there was* (past), *what there is* (the present) and *what would be* (the future) in the *Sankofa-anoma* time tradition of the time trinity as illustrated by an epistemic geometry of Fig. 1.1, where the present in the journey of knowing defines the conditions for the telescopic past and the telescopic future such that the conditions of the past is the foundations of the present history, while the conditions of the present neuro-decision-choice actions create the historic structure of the future and the forecasting from the past information and the discounting from the future anticipated information form the information-knowledge input into the present neuro-decision-choice actions.

These complex interrelated structural relations in diversity and unity of knowing, knowledge and science will become clear as we continue in the discussion with the support of a system of paremiological geometries in the cognitive space. The paremiological geometries belong to a system of *anansimatical* geometries from *anansimatics* (mathematics of the Anansi complex web and design in line with the

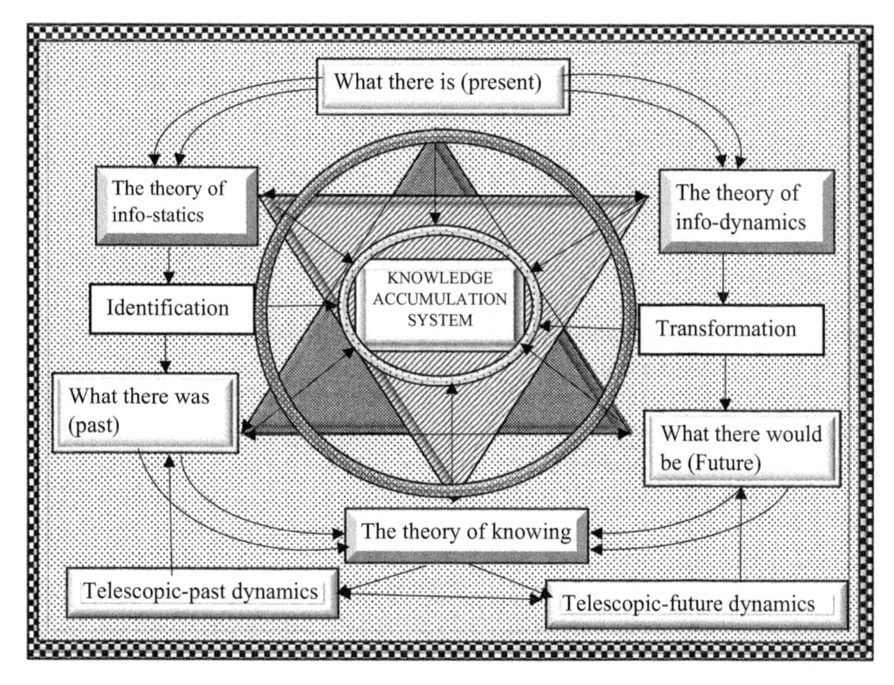

Fig. 1.1 A complex relational space for the development of the theory of knowing composed of the theories of info-statics and info-dynamics as stock-flow disequilibrium dynamics of the information-knowledge system

string theory) and interpreted in the *anansiological* conditions from *anansiology* (the philosophy of anansi storytelling in an epistemic setting) with more geometric symmetries in the adinkra symbols and *adinkramatics* with dualistic-polar interpretations from *adinkralogy*. Here, an attention should be paid to development of string theory, the algebraic geometry of spider (Anansi) web and the natural philosophy that the relational complexities present knowing processes of natural symmetry and stability, and their relationships to fractals of Africentric conceptual antiquity as seen under the principle of opposites and captured in adinkra traditions of information representations, and the philosophical contents of the Anansi story telling.

The theory of information-knowledge production, rather than the theory of knowledge, may also be called the theory of knowing where the latter is intellectual investment and the former is intellectual capital accumulation. Both the theory of knowing and the theory of knowledge production are cognitive activities in the space of problem-solution dualities as seen in all areas of knowing. The space of problem-solution dualities is an infinitely permanent structure and an inseparable part of the spaces of info-statics and info-dynamics in relation to variety identifications and transformations in the space of varieties as they are related to the space of production-consumption dualities. The space of the problem-solution dualities links all sciences into unity through knowing, where knowing is about cognitive activities over the space of problem-solution dualities generating intellectual investment flows to update intellectual capital stocks for developing variety transformation processes to change variety characteristic dispositions in the space of old-new dualities. The techniques and methods of solving knowing problems of knowledge varieties may as cognitive tools may, however, vary from one variety to another and one category to another category depending on their characteristic decompositions and methods of groupings. Some methods and techniques of a variety knowing may be transferable to other varieties and categories but not to all. In this respect, there is no unity of methods and techniques in the process of knowing. Some categories of knowing may have methods and techniques in common with some variations through innovation and enhancements depending on the similarities, commonness and differences of their characteristic dispositions. It is, thus, not the methods of solving the identification problems or the transformation problems of varieties that define the disunity or unity of science through knowing.

Through knowing, the unity of science in addition to the unity of non-science and unity of both, find expressions in the *space of problem-solution dualities and polarities*. The space of problem-solution dualities and polarities links all aspects of science and non-science into unity and yet maintains the scientific and non-scientific individualities through variety knowing, where each individual area of knowing brings specific illuminations to the collective understanding of *what there was* (the past), *what there is* (the present) and *what there would be* (the future) under the *Sankofa-anoma* (the time trinity) rationality. The Sankofa-anoma rationality (the time trinity) relates to acquisition of information-knowledge input into decision-choice actions by forecasting from the past to the present, and acquisition of information-knowledge input into decision-choice actions by discounting from the future to the present.

It is also through the impact of illuminations of individual varieties on other knowing activities that science acquires an increasing collective understanding of the universal existence in unity and the expansion of the understanding of the nature of the universal existence. In fact, the importance and utility of science and non-science in the general space of knowing and information-knowledge accumulation find meaning in conditions of social set-ups governed by different analytical systems of social sciences. Organization of knowledge of science is also the science of organization of social varieties into dividedness and unity within the space of production-consumption dualities. Let us keep in focus that knowing is an *intellectual investment process* and knowledge is an *intellectual capital-accumulation process* in the information-knowledge system for varieties and categorial awareness in the space of input-output dualities. The information-knowledge system is an input into the neuro-decision-choice systems and an output of the same decision-choice system broadly defined and specified, where the meaning and essence of human life and existence are governed by information-decision-knowledge-interactive processes over the space of problem-solution dualities to establish cultures, social progresses, social decays, social evils and social cruelty, social construction and destruction, environmental destruction, remediation and engineering, and civilizations from generation to generation under changing conditions of intergenerational preferences over systemic conditions within the space of good-evil dualities under the changing philosophical consciencism in time, over time, within each generation and over generations.

Note 1.2.2 C: Reflections on the Traditional Theory of Knowledge A reflective note on the traditional epistemology or theory of knowledge will be useful at this epistemic juncture relative to the approach of the economic theory of production over the space of input-output dualities, where the inputs of knowing and knowledge production is information which is different from knowledge. The ordinary understanding and use of the concept of knowledge resides in a complex confusion with linguistic interchangeability of facts, knowledge, skills, awareness and many others. Each of these interrelated and interdependent terms must be viewed as linguistic varieties requiring definitions and explications. The traditional philosophically theoretic definition and understanding of knowledge is even not better and introdes extra abstract complexities in the process of explication, where knowledge is defined and explicated as the *justified true belief* (JTB) which is with us as well as part of teaching, learning, discussions and research in departments of philosophy on the subject matter of knowledge. Our main interest is on the definition and explication of knowledge as *justified true belief* where such such a concept of *justified true belief* finds expressions and meaning in the space of imagination-reality dualities that generates hope and creativity between duals of dualities and between poles of polarities in the space of possibility-probability dualities within the space of ignorance-knowledge dualities with continuum and unity under the general principle of opposites.

There are several epistemic difficulties with the phrase of JTB as it relates to the concept of knowledge. All three terms of justified, true and belief are qualitative in nature and defined in the space of fuzziness with degrees of belonging to the respective fuzzy set. For the concept of knowledge to be understood, one must relate justified to the space of justified-unjustified dualities, true to the space of false-truth dualities and belief to the space of belief- non-belief dualities. These dualistic-polar concepts must be related to something if they are to be analytically useful and provide a framework of measurability. This something is information that must help in establishing verbal identities for distinction in a language. What is belief and how do we know it? What is true and how do we know it? Similarly, what is justified and how do we know it? Additionally, how is the "justified true belief" related to concepts of data, fact and evidence? When one defines knowledge as a *justified true belief*, one has in mind the idea of belief as the central focus with justified and true as qualifying distinction. Since belief is a linguistic variable it becomes necessary to restrict the domain of meaning and understanding. This restriction finds residence in *true* which is also a linguistic variable with dualistic conditions as a fuzzy variable requiring a dualistic conditionality. The dualistic conditionality on the linguistic variable *true* finds expression in the *justified.* The justification finds residence in the spaces of decidability-undesirability dualities and doubt-surety dualities with corresponding information as inputs into neuro-decision-choice actions. The definitional concept of *justified true belief* belongs to the space of quality control which involves verification, falsification, corroboration or all of them over the spaces of possibility-probability dualities, necessity-freedom dualities, anticipation-expectation dualities and risk-benefit dualities for any variety to be accepted as knowledge to be added to intellectual capital stocks, where knowledge is an input into the neuro-decision-choice actions for further knowing regarding variety processes, technologies and transformations.

Here, increasing complexities and confusions arise in the theory of knowledge since the classical paradigm of thought operates with exact information over the space of exactness under the principle of opposites with the excluded middle to artificially create a space of absoluteness in thinking and neuro-decision-choice activities to the exclusion of the elements of the space of relativity. In this classical condition, the definition of knowledge is a definition from the path of reductionism within the space of constructionism-reductionism dualities as simply a *quality control* without regard to its production. The thinking algorithm for the reasoning process is to provide the needed properties on the claimed knowledge items as whether they qualify to be in the classical space of *justified true belief* relative to the space of absoluteness in reasoning and neuro-decision-choice actions for variety decidability over the space of acceptance-rejection dualities. Viewed the same conditions of *justified true belief* as three interrelated linguistic variables over the space of relativity under the principles of opposites with continuum and unity, the thinking algorithms for reasoning require the resolutions of the conflicts in the spaces of belief-unbelief dualities, justified-unjustified dualities, doubt-surety dualities and true-false dualities for deciding elements in the space of *justified true belief* with corresponding membership characteristic functions of the duals. The importance of dualistic-polar structures in cognition and decision-choice processes cannot be underestimated and must be understood. The whole of human actions centers indirectly or directly is a system of mediations between duals of dualities and between poles of polarities specified in the space of relativity but not in the space of absoluteness.

The belief $\mathcal{B}$ as a fuzzy set of degrees of belief with membership characteristic function $\mu_{\mathcal{B}}(\cdot)$ is constrained by a fuzzy set of degrees of unbelief $\mathcal{U}$ with a membership characteristic function $\mu_{\mathcal{U}}(\cdot)$ with a fuzzy decision set $\mu_{\mathcal{D}}(\cdot) = (\mu_{\mathcal{B}}(\cdot) \wedge \mu_{\mathcal{U}}(\cdot))$ with an optimal solution defined as $\underset{x \in \mathcal{D} = \mathcal{B} \cap \mathcal{U}}{\text{Opt}} \mu_{\mathcal{D}}(x) = \alpha*$, justified as a fuzzy set of degrees of justification $\mathcal{Q}$ with a fuzzy membership $\mu_{\mathcal{Q}}(\cdot)$ which is constrained by a degree of -justification, $\mathcal{P}$ with a fuzzy membership function $\mu_{\mathcal{P}}(\cdot)$, leading to a fuzzy decision set $\mu_{\mathcal{D}}(\cdot) = (\mu_{\mathcal{Q}}(\cdot) \wedge \mu_{\mathcal{P}}(\cdot))$ with an optimal solution defined as $\underset{x \in \mathcal{D} = \mathcal{Q} \cap \mathcal{P}}{\text{Opt}} \mu_{\mathcal{D}}(x) = \beta*$. True is a fuzzy set of degrees of truth $\mathcal{M}$ with a fuzzy membership function $\mu_{\mathcal{M}}(\cdot)$ and constrained by false as a fuzzy set of degrees of falsity $\mathcal{N}$ with fuzzy membership characteristic function $\mu_{\mathcal{N}}(\cdot)$ leading to a fuzzy decision problem of the form $\mu_{\mathcal{D}}(\cdot) = (\mu_{\mathcal{M}}(\cdot) \wedge \mu_{\mathcal{N}}(\cdot))$ with an optimal solution defined as $\underset{x \in \mathcal{D} = \mathcal{M} \cap \mathcal{N}}{\text{Opt}} \mu_{\mathcal{D}}(x) = \omega*$ in the spaces of degrees of belief-unbelief, justified-unjustified dualities and true-false dualities where $(\alpha*, \beta*, \omega* \in (0, 1))$ are the corresponding fuzzy-stochastic conditionality in that $(\alpha * +\varepsilon) \in (0, 1)$ is the optimal degree of belief. Similarly, $(\beta * +\varepsilon) \in (0, 1)$ is an optimal degree of justification and $(\omega * +\varepsilon) \in (0, 1)$ is an optimal degree of true such that knowledge is an $(\alpha * +\varepsilon)$-degree of belief with an $(\beta * +\varepsilon)$-degree of justification and $(\omega * +\varepsilon)$-degree of truth where (ε) is extra-confidence of subjective judgmental correction factor. The general epistemic understanding of this classical knowledge definition and supporting conditionalities require a connectivity to information which must be defined and explicated. The whole meaningfulness of the classical theoretical process in knowledge production must be connected to the space of input-output dualities with an explicit information definition and explication over the space of primary-derived dualities as well as connected to the theory of languages in terms of the theory of codes for storage and communication over the space of source-destination dualities where every information from the source duality to destination duality contains contents as characteristic disposition and message channel as the signal disposition. The knowing is about deciphering of the signal disposition by the destination agents to abstract the variety characteristic dispositions from the space of acquaintance or the space of sources to reveal the identity of the variety in question. The follow up is the neuro-decision-choice actions to evaluate the decoded knowledge contents to be added to the intellectual capital accumulation as inputs for increasing production and consumption.

In my approach to the theory of knowing and information-knowledge accumulation over the space of primary-derived dualities, information is taken as the third dimension of existence that is connected to all existential varieties in the space of actual-potential dualities for presenting variety problems and finding variety solutions to the variety identification and transformation problems over the space of input-output dualities mapped into the space of production-consumption dualities. This third dimension is materially the characteristic disposition that projects signal disposition for knowing as intellectual investment flows and knowledge as accumulated intellectual capital stocks through the quadrilogy of the epistemic process of journeys from the potential to imagination, from imagination to the possible, from the possible to the probable and from the probable to the reality and then to the actual which is consistent with the rationality of knowing and knowledge production over the space of constructionism-reductionism dualities under conditions of fuzzy-stochastic conditionality. Every variety knowing is a journey on the path from variety ignorance to variety knowledge under information and paradigmatic constraints and their usage efficiencies. The path of the cognitive quadrilogy is due to cognitive capacity limitations and analytical short sidedness of cognitive agents over the space of acquaintance-description dualities.

Knowledge is about two important processes of variety-identification process and variety-transformation process defined over the space of identification-transformation dualities, where knowledge about the variety characteristic dispositions initializes the conditions for variety transformations from the variety identifications. The initial conditions present the external necessity for variety change, while the internal conditions present the internal freedom and sufficiency of methods and techniques for effecting the variety change over the space of inner-outer dualities. The necessity for change is generated by conditions of categorial conversion, while the freedom and sufficiency for change are generated by social philosophical consciencism that exists at the relevant period to create the sufficient conditions for change. As explained, the justified true belief finds meaning, acceptance and belief for each variety within the space of characteristic-signal dualities, where the degree of belief is defined by an epistemic distance between ontological disposition and epistemological disposition of any variety in some measurable domain over the space of quality-quantity dualities. Every knowledge is a derivative from information by using a paradigm of thought for abstracting contents of descriptions from the experiential information under fuzzy-stochastic conditionality. The details of this line of argument is provided in [670, 671] where all claims of information, data, knowledge, fact, evidence, knowing, and others are about variety characteristic dispositions; and awareness, familiarity and many others are about variety signal dispositions where the signal dispositions are coded in forms called languages and knowing is a process of decoding the variety signal dispositions. To understand the processes of knowing as production of intellectual investment flows and knowledge as intellectual capital accumulation stock, the concepts of information, data, fact, knowledge and evidence must be defined, explicated, distinguished and shown their interconnectedness in input-output processes for them to be usefully applicable to all areas of human experiences, language developments and communications. The basic structure is provided in this section. The three dimensions of matter, energy and information imply that every universal existence over the space of actual-potential dualities is inseparably composed of matter, energy and information, defined as characteristic-signal disposition, and where the different characteristic dispositions allows one to identify the varieties through the decoding of the signal dispositions, where both knowing and knowledge are connected to the knowledge square in a quadrilogical journey over the space of ignorance-knowledge dualities [177, 671, 672].

Chapter 2
The Knowledge Square, Knowing and Knowledge Production: Conceptual Pathway Analytics

Abstract This chapter presents conceptual pathway analysis of the theory of knowing and the theory of knowledge as neuro-decision-choice activities in the space of ontology-epistemology duality, where the knowing process and information-knowledge production proceed from the space of imagination as a sub-space of the potential to the space of possibility to the space of probability and to the space of reality as a sub-space of the space of actuals in a continual search for information and knowledge through the neuro-decision-choice activities in support of the general neuro-decision-choice actions over problem–solution dualities relative to the space variety identifications and variety transformations. The problem–solution dualities establish the principle of diversity and the unity of knowing, the principle of diversity and the unity of information-knowledge accumulation as well as the principle of diversity and the unity of science. The concepts of ontology and epistemology and the corresponding spaces are defined and explicated to be useful to economic-theoretic approach for the understanding of the knowing process as the production of intellectual investment flows and the production of the information-knowledge system as the accumulation of intellectual capital stocks.

2.1 Pathway Analytics and Logical Connectivity

Conceptual pathway analytics are compressed visual directional pathways to simplify the complexity of cognitive approach to ascertain knowledge about variety identity at static states and variety transformation at dynamic states conditional on cognitive capacity limitations, limitativeness and limitationality with respect to information and paradigm of thought. The conceptual pathway analytics is thus, to show the analytical connectivity that our information-knowledge system viewed in terms economic production has four basic blocks of with connecting pathways and unbreakable continuity to decipher the complexities in the theory of knowing as intellectual investment flows and the theory information-knowledge accumulation as an intellectual stock. These blocks and connecting pathways also form the basis of social variety production that allow movements from the space of imagination as a subspace

K. K. Dompere, *The Theory of Problem-Solution Dualities and Polarities*,
Studies in Systems, Decision and Control 405,
https://doi.org/10.1007/978-3-030-90279-7_2

of the space of potential to the space of reality. The cognitive activities in knowing take place in the space of imagination which is connected to the space of reality through possibility and probability spaces.

The conceptual pathways allow the development of the relationship between the principle of opposites and cognitive actions as problem–solution processes with careful balances between duals of dualities and between poles of polarities. The pathway analytics help to establish the primary dualities and polarities on one hand and derived dualities and polarities on the other under the parent-offspring dynamics at each round of cognition. The initial dualistic-polar structure is the space of the actual-potential dualities which define the space of ontological-epistemological duality from which the space of imagination-reality dualities is a derivative. Between the space of imagination-reality we have the connected subspace of possibility-impossibility dualities and probability-improbability dualities in the cognitive journey between the imagination duals to the reality duals and between the imagination poles to reality poles where all knowing and information-knowledge accumulation take place in the space of static-dynamic dualities relative to the space of identification-transformation dualities.

2.1.1 The Pathway Analytics and Paremiology of Geometries

The pathway analytics will help to generate a system of epistemic geometries describing the epistemic dependencies of stages in the space of ignorance-knowledge dualities and the pathways from ignorance duals to knowledge duals to establish a known element to be update the information-knowledge stocks. Such geometric representation will provide the rise of the theory of possible worlds from the space of possibility-impossibility dualities and how this space is defined and connected to the space of probability-improbability dualities and where the space of possibility-impossibility dualities is connected to the spaces of true–false dualities, correct-incorrect dualities and many relevant ones and then connected to the space of acceptance-rejection dualities as well as connected spaces of facts-alternative facts dualities, information-misinformation dualities, information-disinformation dualities and to input–output dualities in the space of decidability-undecidability dualities which is then connected to the spaces of error-correction dualities and cost–benefit dualities.

Under the dualistic-polar structures, ever space is defined in terms of dualistic-polar dominance in terms of characteristics of the characteristics such that every negative or cost duality is negative or cost duality becourse the negative or cost characteristic disposition dominates the positive or the benefit characteristic disposition and similarly, negative or cost poles are defined by the dominance of negative dualities and vice versa, while the negative or the cost polarities are defined by the dominance of the negative or cost poles and vice versa in the paremiological geometries which must find relational connectivity to the *spidermetry* (the geometry of the spider web or *anansiology*) as linked to *adinramatics* with interpretations from anansiology (the wisdom teachings from the anansi stories). The epistemic pathway analytics is connected to the dualistic-polar space of constructionism-reductionism dualities where constructionism links space of acquaintance to the space of description by paradigms of thought for the outputs and where reductionism links the space of description back to the space of acquaintance through the same paradigm of thought for the output acceptance through verification, corroboration and falsification to generate continual feedback process in the space of cognition. The basic cognitive geometry is the knowledge square to which we turn our attention.

2.2 Introduction to Knowledge Square as the Pathway Analytics

The knowledge of the universe is obtained through sequential know-hows of what there was, what there is and what would be about the elements in the universe, where these elements are varieties identified by differential characteristic dispositions in differences, commonness and similarities. In this respect, we specify the potential space of knowing as $\left(\mathfrak{U}, \mathbb{I}_{\mathfrak{U}}^{\infty}\right)$, where $\mathfrak{U}$ is the space of potential varieties with an infinite index set $\mathbb{I}_{\mathfrak{U}}^{\infty}$, the possibility space of acquaintance as $\left(\mathfrak{P}, \mathbb{I}_{\mathfrak{P}}^{\infty}\right)$, where $\mathfrak{P}$ is the space of possible varieties from awareness, with anticipation ~~under~~ necessity, and $\mathbb{I}_{\mathfrak{P}}^{\infty}$ is the index set of the possible varieties from the elements of acquaintance, the probability space of doubt as $\left(\mathfrak{B}, \mathbb{I}_{\mathfrak{B}}^{\infty}\right)$, where $\mathfrak{B}$ is the space of knowledge with doubt under freedom, and $\mathbb{I}_{\mathfrak{B}}^{\infty}$ is the index set of probable varieties, different from potential varieties, and the space of the actual of variety existence as $\left(\mathfrak{A}, \mathbb{I}_{\mathfrak{A}}^{\infty}\right)$ where $\mathbb{I}_{\mathfrak{A}}^{\infty}$ is the space of actual varieties under risk-benefit conditions and $\mathfrak{A}$ as its index set with the sequential paths of knowing and knowledge development as shown in Fig. 2.1 which represents the diagram of the knowledge square, where every variety of knowing process from the potential space, through the possibility space, the probability space and then to the space of the actuals has been extended to incorporate the spaces of imaginations, realities and the space of imagination-reality dualities as a sub-space of the actual-potential dualities for knowing, information-knowledge accumulation to illustrate the legitimate areas of social-natural operations.

The advantage of this extension is to provide the means and framework to examine important concepts as surprise, anticipation, hope, vision, goals, objectives expectations, perseverance, curiosity, subjectivity, ideas, art, music, theory and practice, design, construction and destruction of social varieties over the space of fundamental-applied dualities relative to the elements in the space of ignorance-knowledge dualities and its interactions with the space of qualitative-quantitative dualities. All human activities and behavior take place in the imagination-reality dualities as sub-space of the space of actual-potential structure in dualistic-polar conditions, where conditions of hesitancy of neuro-decision-choice actions are related to risk-anticipations from the space of imaginations and the benefit-losses from the space of realities. The whole system of knowing and information-knowledge production and cognitive actions is

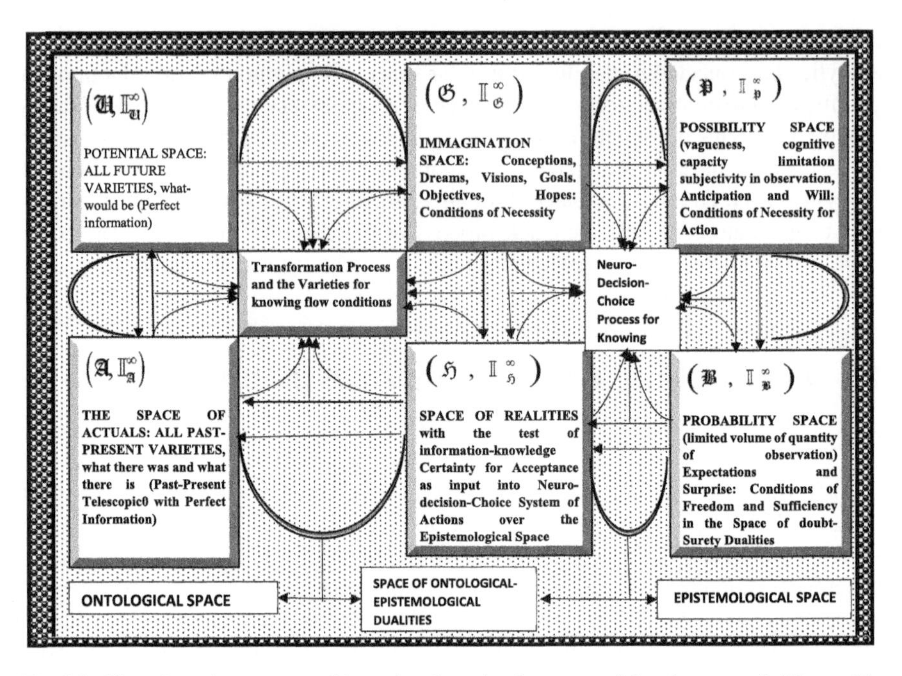

Fig. 2.1 The epistemic geometry of the variety knowing from potential as the space of all knowable varieties to the possible as the space of information acquaintance to probable as the space of information doubt to the actual as the space of degree of information-knowledge uncertainty and back to the potential in a never-ending process in the information-knowledge search about what *there was, what there is and what would be* of the universal objective set

initialized as the space of ontological-epistemological dualities containing the space of actual-potential dualities which contains the space of imagination-reality dualities under the general principle of opposites with relational continuum and unity to connect ontic conditions as a primary category of knowing to epistemic conditions as derived category of knowing. The dualistic-polar structure is given in Fig. 2.1.

The epistemic geometry of Fig. 2.1 may also be viewed as paremiologically cognitive geometry requiring some logical connectivity for completeness in understanding, and how this geometric structure representing knowing and information-knowledge search leading to the global intellectual capital stock as well as to produce global intellectual citizenry. The structure is such that the space of the potentials contains the space of imagination $\left(\mathfrak{G}, \mathbb{I}_{\mathfrak{G}}^{\infty}\right)$ such that $\left(\mathfrak{G}, \mathbb{I}_{\mathfrak{G}}^{\infty}\right) \subset \left(\mathfrak{U}, \mathbb{I}_{\mathfrak{U}}^{\infty}\right)$ while the space of actual contains the space of reality $\left(\mathfrak{H}, \mathbb{I}_{\mathfrak{H}}^{\infty}\right)$ such that $\left(\mathfrak{H}, \mathbb{I}_{\mathfrak{H}}^{\infty}\right) \subset \left(\mathfrak{A}, \mathbb{I}_{\mathfrak{A}}^{\infty}\right)$, the space of actual-potential dualities contains the space of imagination-reality dualities $\left((\mathfrak{H} \cup \mathfrak{G}), \mathbb{I}_{(\mathfrak{H}\mathfrak{G})}^{\infty}\right)$ such that the space of imagination-reality dualities is a proper sub-space with a fuzzy closure and hence $\left((\mathfrak{H} \cup \mathfrak{G}), \mathbb{I}_{(\mathfrak{H}\mathfrak{G})}^{\infty}\right) \subset \left((\mathfrak{U} \cup \mathfrak{A}), \mathbb{I}_{(\mathfrak{U}\cup\mathfrak{A})}^{\infty}\right)$. This is the case for all the geometric illustrations in this monograph and all other monographs of mine. From the epistemic geometry of the path of knowing, the relational structure of the theory of knowing, the theory of knowledge and the various components are presented in Fig. 2.2 to show their connections with the problem–solution dualities, where every problem is a solution and every solution is a problem as specified in terms of varieties and dualistic-polar processes relative to the elements in the goal-objective space.

The relational structure of the theory of knowing, the knowledge by acquaintance and the knowledge by description may be represented in an epistemic relational geometry as in Fig. 2.3 in terms of stock-flow conditions. The epistemic geometry shows how the theory of knowing and the theory of knowledge are connected to the theory the fuzzy-stochastic entropy for information-knowledge certainty as well

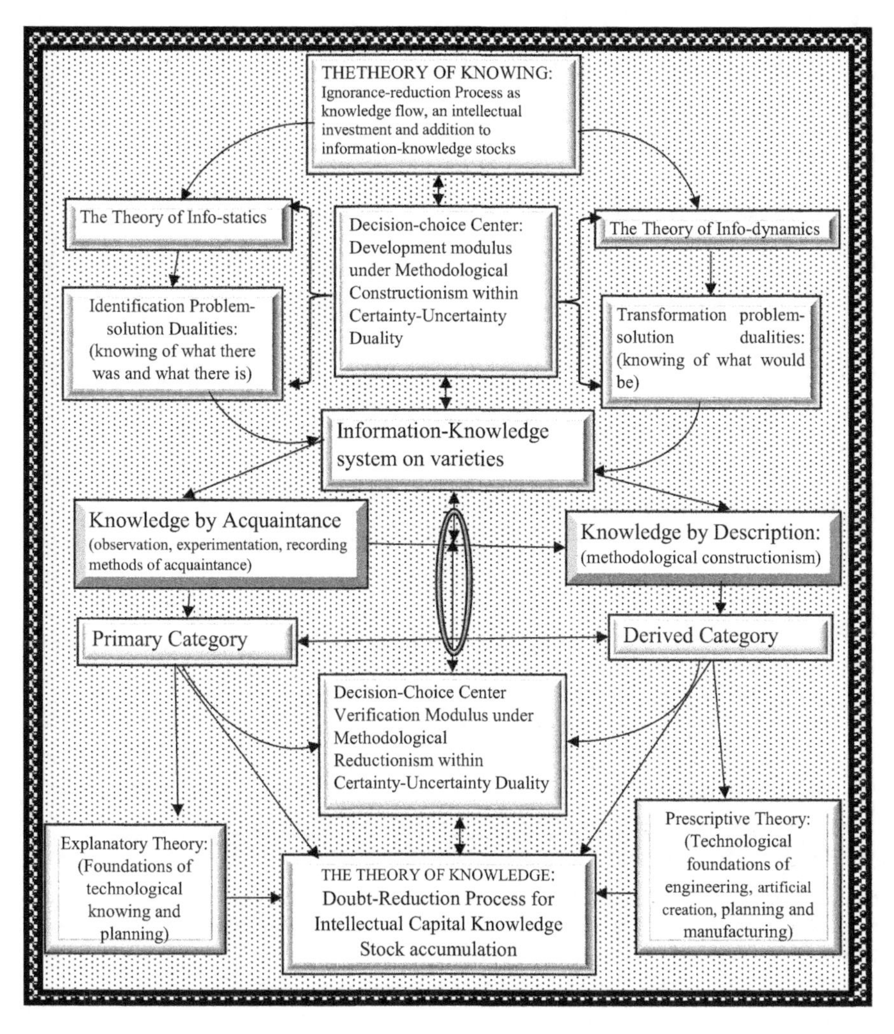

Fig. 2.2 A cognitive geometry of the relational structure of information-knowledge analytics, info-static analytics, info-dynamic analytics and the system of problem–solution dualities in the theory of knowing and the theory of knowledge

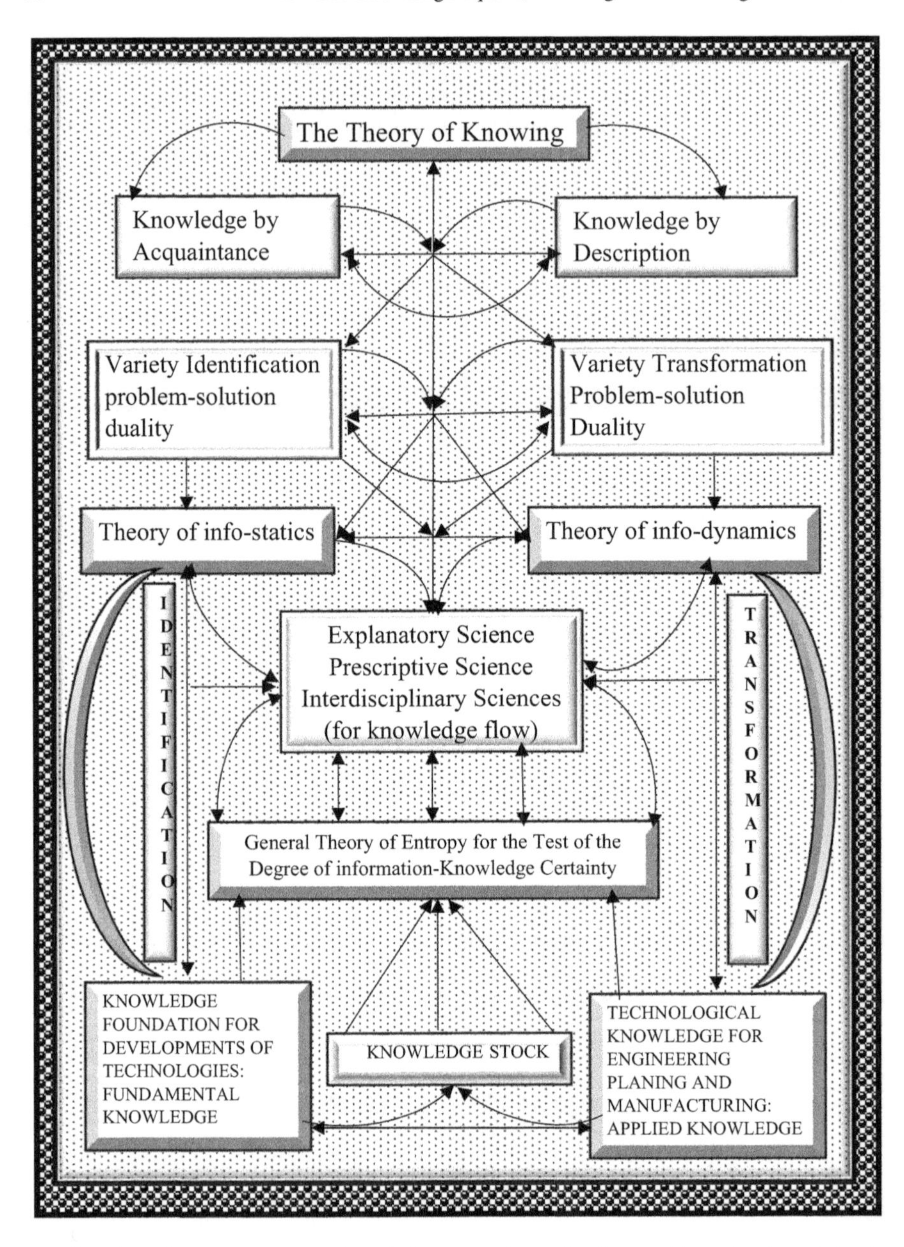

Fig. 2.3 An epistemic geometry of the relational structure of the theory of knowing, knowledge by acquaintance, knowledge by description, applied and engineering knowledge, explanatory science, prescriptive science and entropy

as showing the basic knowledge and applied (engineering) knowledge and where information is always an input into knowing and knowledge-stock processes over the space of primary-derived dualities. The theory of the fuzzy-stochastic entropy is an epistemic system of quality control methods in a backward telescopic process for methodological reductionism. The knowing is a production process with methodological constructionism under the parent-offspring processes which must be related to the space of input–output dualities. It must be kept in mind that the theory of knowing examines the epistemic conditions of ignorance around varieties over the space of static-dynamic dualities, while the theory of knowledge examines the doubt conditions over the space of certainty-uncertainty dualities. The theory of knowing is to find an answer by explanation to the question of how intellectual investments take place in our societies. The theory of knowledge is to find an answer by explanation to the question of how intellectual capital accumulations take place in societies. The theories of knowing and knowledge should explain their relationships and their utilities to the social setup.

Generally, the theory of knowing is the *primary category* of the theory of the information-knowledge process. The theory of knowledge by acquaintance and the theory of knowledge by description are integrated parts of the theory of knowing. The knowledge by acquaintance is the primary input category of knowing while the knowledge by description is a derived category of knowing using a paradigm of thought. The activities of knowing and knowledge production are general to all areas of knowing, including the sciences, and non-sciences, social sciences and non-social sciences and the exact and inexact with reference to all varieties in the space of fundamental-applied dualities. Knowing, knowledge production and engineering are all problem-solving activities under neuro-decision-choice processes with different ontological signal dispositions as inputs to generate outputs as epistemological characteristic dispositions. The phenomenon of problem–solution dualities represents the essential analytical construct and development of information-knowledge structure, disequilibrium dynamics of varieties and categorial varieties with further development of variety forms to facilitate the socio-natural existence of the continual transformation of forms from matter-energy equilibrium to sustain preferences and the dynamics of personal, interpersonal, generational intergenerational, national and international preferences over the space static-dynamic dualities. These transformation processes are defined over the space of cost–benefit dualities requiring judicious decision-choice-choices over the space of input–output dualities.

All areas of knowing find expressions with varieties over the space of problem–solution dualities. It is within the phenomena of the variety problem–solution duality and the space of variety problem–solution dualities that one finds a unity and diversities, oneness and dividedness among knowing, science, non-science, engineering sciences and non-engineering sciences to deal with the organic concepts of cognitive search and decision-choice actions to accumulate intellectual capital stocks. As presented, knowing relates to a flow of knowns and know-hows, and knowledge relates to the stock of knowns and know-hows. The similarities and differences of knowing and knowledge are very important to understand the principles of diversity and unity and categorial diversity of techniques and methods in the journey of epistemic search, and the oneness and dividedness of the general problem–solution dualities of information-knowledge systems. Here, it must be observed that institutional departments of teaching, research and learning within the system of the world educational organizations are not set up in accordance with methods and techniques of knowledge search, but rather they are set up on the basis of categories of problem–solution dualities that define disciplinaries of formal epistemic activities to improve individual and collective curiosities of knowing, skill acquisition for employment and decision-choice actions over the space of problem–solution dualities given the development of paradigmatic toolboxes for thinking algorithms. The thinking algorithms are developed within paradigms of thought for reasoning strategies over the spaces of doubt-surety dualities, uncertainty-certainty dualities and decidability-undecidability dualities by methods and techniques of knowledge search. These departments are, rather, set up on the basis of categories of problem–solution dualities that define the disciplinaries of formal epistemic activities to improve curiosity of area knowing, neuro-decision-choice actions over the space of problem–solution dualities and the development of paradigmatic toolboxes for thinking algorithms over the space of ignorance-knowledge dualities, where cognitive search in knowing and knowledge is a continual intellectual stock-flow process.

2.3 The Concept of Cognitive Search Activities in an Epistemic Abstract

Conceptually, there are two spaces of importance to understand information, neuro-decision-choice activities, knowing and knowledge and conditional environment of general existence. They are the ontological space and the epistemological space. For the analytical convenience of their connectivity, we will view them in the space of ontological-epistemological duality with continuum and unity. There are differences and similarities between the concepts of ontology and the ontological space and between the concepts of epistemology and the epistemological space as they are employed in the epistemic framework of this monograph. There are also similarities and differences between the concepts of ontology and epistemology, and between the concepts of ontological and epistemological spaces. Ontology and epistemology are taken as areas of study with reference to philosophical inquiries and how they relate to socio-natural existence, change and progress. The ontological space is taken as housing ontological varieties and natural actions from the past to present and the future, while the epistemological space is taken as a derived abstraction of the ontological space for social actions from past to present and to future regarding knowing and neuro-decision-choice actions that connect the two spaces. The differences and similarities of these concepts are established by the nature of their *characteristic dispositions* that define their identities in the language of communicational source-destination dualities with relational continua and unity for interactive understanding and further development. In the general process of information-knowledge development, language resides in a critical position in the enterprise of knowing, information-knowledge development and communication of knowing as well as performing the role of an instrument for monitoring and retaining the information-knowledge accumulation and its disequilibrium dynamics, interpersonal and intergenerational transmissions. Let us reflect on the concepts of ontology and epistemology.

2.3.1 The Concepts of Epistemology and Ontology in a Cognitive Search Over the Spaces of the Problem–Solution Dualities and Polarities

Let us conceptually provide working definitions for ontology and epistemology relevant to the development of the argument in support of the *unity of science* and the *unity of engineering sciences* through the *unity of general knowing* and how this unity is the foundation of social change and changes in intergenerational preferences. The definitions must be such as to maintain the identities and the relational structures of all entities in continuum and unity with a specific understanding of what information is. The concern in this monograph is primarily focused on the unity of science through the unity of knowing. The differences and similarities of varieties of knowing are established by the nature of variety characteristic dispositions as the content of information that define the internal essences and external natures of the varieties. The conceptual and working definitions of the ontological and epistemological spaces will be provided after the definitions of ontology and epistemology that are in use in this monograph.

Definition 2.3.1.1: Epistemology Epistemology is a created area of activities of cognitive agents concerning information-knowledge production, decision-choice actions over the space of problem–solution dualities and the creation of artificial forms, methods and techniques through engineering science, applied science, planning, their certainties and the degrees of justified belief attached to claims of information-knowledge items from the socio-natural environment. It is important part of epistemic organization to deal with conditions of necessary-sufficient and problem–solution dualities of knowledge in relation to source structures and uses in decision-choice systems for social transformations, environmental coping and the limits of knowledge, where the limits of knowledge find expressions in the space of certainty-uncertainty dualities which is created by cognitive capacity limitations of agents of knowledge regarding vagueness, incompleteness of observations and imperfections of laws of thought in thinking and reasoning.

Note 2.3.1.1: On the Concept of Epistemology The definition of epistemology includes terms and concepts that require clarifications to bring explications to the concepts of epistemology and epistemological space under an epistemic usage in knowing and information-knowledge stock-flow system. Among these defining concepts are knowledge, certainty, belief, necessary-sufficient duality and problem–solution duality, which are related to varieties of ideas, thoughts, notions and concepts, the understanding of which will be required for clarity of the definition. The cognitive activities involve analytics of *systems of question–answer dualities* which are translated into analytics of *systems of problem–solution dualities*, where human history is essentially history of resolving conflicts over the space of problem–solution dualities in search of dualistic-polar elements within the space of question–answer dualities with defining images of history of ideas and essence of life that gives meaning to existence within the dualistic-polar framework of imagination and reality, where thought and reasoning are the connecting forces to map the elements in the space of imagination-reality dualities onto the space of production-consumption dualities. Every question–answer duality has a corresponding problem–solution duality, and every problem–solution duality has a corresponding question–answer duality. From the definition of epistemology, the essential questions are: What is knowledge? What is the morphology of knowledge? What is knowledge production? What is the mechanism of knowledge production? What is knowledge production about? What are the inputs of knowledge production? What is the certainty of knowledge? What is justified belief? What are the paradigmatic methods of knowing? What are the scopes and utility of knowledge? The answers to these questions must be abstracted from the analytics of either the space of problem–solution dualities or the space of question–answer dualities as they relate to the meaning and essence of human life, existence, maintenance, threats and survivability.

The analytics of the system of problem–solution dualities are the decision-choice activities of cognitive agents, where every question–answer duality and every problem–solution duality are in relational continua and unity. Knowledge production, therefore, is the analysis of the system of question–answer dualities projected into the space of problem–solution dualities such that an answer to a question generates a new question requiring a search for a new answer over the space of question–answer dualities, and where a solution to a problem brings in a new problem requiring a new solution in the system of success-failure dualities at both static and dynamic states for universal and particular existences over the space of quality-quantity dualities. In this monograph, the concepts of universal and particular are explicated to be different from their traditional usage but are in relational diversity and unity within the space of real-nominal dualities where the uniting force of knowing and knowledge is information defined in terms of characteristic signal disposition under a paradigm of thought for conditions of epistemology, where neuro-decision-choice activities connect the space of acquaintance to the space of descriptions through a paradigm of thought which is also a creation by neuro-decision-choice action from the cognitive space.

The definition of epistemology relative to the analytics of the systems of question–answer dualities and problem-solving dualities moves the concept of epistemology to the space of decision-choice activities under cognitive control, and links it to cognitive problem solving and conditions of the construct of artificial intelligence and machine learning over the problem–solution space where the information-knowledge system is self-exiting and self-correcting cognitive production-consumption system. In this respect, the understanding of the conditions of epistemology may be analytically weaponized into the mechanism of non-human intelligence into the problem–solution space over the space of source–destination dualities, where machines are designed to mimic the cognitive capacity of human neuro-decision-choice action, learning and storage of outcomes from problem–solution acquaintances, and where the machine intelligence acquires an automatic capacity of the input–output process over the space of problem–solution dualities with learning and error-correction capacity in terms of feedback processes. The analysis of input–output conditions in the space of quality-quantity dualities remains as an analytic on the space of problem–solution dualities and the space of question–answer dualities. Let us provide a working definition for ontology to distinguish it from epistemology. Let us also keep in mind that definitions are processes to assign characteristic-sign dispositions to words, concepts, phenomena, things and ideas to establish varieties, differences, commonness and similarities in both ordinary and abstract languages to distinguish variety imagination from variety reality and to establish variety information from variety knowledge [80, 87, 177, 451, 467, 490–492, 504, 669, 671].

Definition 2.3.1.2: Ontology Ontology is a cognitive construct of the conceptual domain representing *what there was*, *what there is* and *what would be* in an infinite universe within the time trinity, of the infinite past, infinite present and infinite future in terms of actual and potential existence and non-existence relative to individual and collective varieties, entities and identities about things, states, processes and relations among them which ensure their being and non-being in a manner that presents challenges to knowing, knowledge accumulation and understanding in the cognitive space of human actions over the social organization of economic structure, political structure and legal structure in relational continuum and unity.

Note 2.3.1.2: On the Concept of Ontology The definition of ontology, not simply as a branch of philosophical studies, is important to allow the establishment of analytically epistemic connecting paths between ontology and epistemology. The connecting paths are decision-choice processes of thinking, reasoning, action and production of cognitive agents with activities within the space of ontological-epistemological dualities. Epistemology, as an area of study of activities of cognitive agents concerned with knowing and knowledge production, must also include knowledge of being, existence and transformation as reflected in states and processes regarding variety identification and transformation and the system of actions that makes them possible. Ontology, as an area study of activities of nature in the most abstract form, presents information that becomes inputs into epistemological studies by creating analytically epistemic structures for linking epistemology to ontology through neuro-decision-choice activities that must create epistemic paths that allow the revelations of what there was, what there is and what there would be. *What there was* presents information on the past varieties in the space actuals. *What there is* presents information on the current varieties in the actual space. *What there would be* presents information of future varieties in the space of the potentials.

The definition of ontology is to create conceptual system of the phenomena of socio-natural or ontological varieties which are linked to epistemological varieties as analytical derivatives with variety distinction and similarities which form a family of matter-energy categories. Simply stated, ontology is about existence of being and epistemology is about the knowing of being. The analytical space that houses ontological varieties and the epistemological varieties is the space of ontological-epistemological dualities. There are certain concepts and terms that have been used to create these variety concepts that allow the epistemic interactions within *ontology-epistemology duality* in relational continuum and unity, where cognitive agents epistemologically journey from the ontological characteristics to epistemological characteristics by decoding the ontological signal dispositions at space of acquaintance to produce experiential information of cognitive agents that becomes input for deriving knowledge by description.

The relational connectivity within the ontology-epistemology duality is important to the development of the theory of knowing and knowledge accumulation that reveals the awareness and understanding of the human internal and external conditions for management and control of the organization of social existence. It is also relevant to the development of the theory of knowledge and the theory of variety production and transformation to change and maintain variety values in time and over time.

The conditional results of the theory of knowing in relation to intellectual investment flows are important to the development of theory of knowledge in relation to intellectual capital-accumulation stocks and the uses to which they may be put. Among these definitional concepts are concepts of existence, being, identity, states, processes, relations and others within the space of varieties. These concepts establish the nature of the phenomenon of ontology as well as the subject matter of knowing at the level of epistemology to furnish us with the conditions to define ontological and epistemological spaces.

The defining concept of ontology is created at the level of epistemology through the linguistic process for analysis, understanding, communication and neuro-decision-choice activities of the conditions of *what there was* (the past), *what there is* (the present) and *what there would be* (the future). The study of the conditions of universal beginnings is about the understanding of *what there was.* The study of the conditions of the universal currents is about the understanding of *what there is and* how *what there was* became *what there is*. Meanwhile, the study of the ending of the universe is about *what would be* and how *what there is* will become *what would be*. The conditions contained in the definitions provide indications as to what the activities at the levels of epistemology and ontology are about. The definitions of ontology and epistemology help to answer the questions arising at the level of epistemology as to what are the objects, subjects and instruments of knowing and knowledge production as well as how is the epistemology conceptually connected to ontology. As structured, ontology is a *conceptual primary category* of existence while epistemology is a *derived conceptual category* of existence in terms of knowledge and neuro-decision-choice actions of cognitive agents in understanding their environment, meaning and essence of existence and its foundational support.

Seen in conceptually relative terms, ontology establishes the conditions of universal existence of the nature of matter, energy and the characteristics at both static and dynamic states, while epistemology seeks to establish the knowing and knowledge about the conditions of matter, energy and their characteristics at both static and dynamic states relative to ontology. Ontology refers to conditions of all the infinitely known and unknown varieties of the universe. It, thus, refers to the conditions of actual and potential existence. The space of the actual varieties is the collection of the epistemologically known and unknown varieties, where the space of actuals contains the space of realities, which is the space of the known varieties of the space of the actuals. The potential space is the collection of varieties in the active ontological variety transformation processes of the variety conditions between the actuals and the potentials where the space of potential contains the space of imaginations for cognitive activities under the development of new ideas, goals, objectives, vision and others.

Epistemology refers to conditions of a general state where knowledge is acquired through knowing of actual varieties as well as an intentional search for knowledge by cognitive agents on the infinitely unknown varieties and potential varieties based on neuro-decision-choice actions. It has been pointed out that the definition of epistemology, as a study of knowing and information-knowledge poses the questions regarding what the concepts of information and knowledge are, what are they about, and how do they relate to each other, cognition and human action. It has been argued that *knowledge* is the intellectual accumulation of known variety characteristic dispositions as stocks that allow variety identifications and transformations, and hence their identities and changing identities to be ascertained while *knowing* is an intellectual investment of variety characteristic dispositions as flows [573, 576, 577, 671, 672].

2.4 The Spaces of Relativity and Absoluteness in Knowing, Knowledge and Decision

The concept of information has been defined as characteristic-signal disposition that presents us with conditions of distinctions for establishing identities, diversities, commonness and analytical groups, sets and categories and unity. The theory to establish these nominal and real varieties is what has been discussed in the monograph entitled the theory of info-statics [671]. The dynamics of these varieties in terms of changes and transformations are presented as the theory of info-dynamics [672]. The conditions of the neuro-decision-choice actions over the spaces of doubt-surety dualities and acceptance-rejection dualities are presented as a general theory of entropy over the epistemic conditions of relativity and absoluteness in cognition and existence. In this respect, how does the characteristic-signal-based definition of information relate to the phenomena of absoluteness and relativity in the spaces of ontology and epistemology in terms of existence, knowing, paradigm of thought and neuro-decision-choice action?

Let us keep in mind that knowing is not knowledge even though knowledge relates as a derivative from knowing which is its primary category of an epistemic process, while knowing is a derivative from information as the primary category in the chain of the epistemic of parent-offspring process to give birth to data, fact, evidence and evidential things. In general and specific reflections over the space of human understanding and actions, how do the phenomena of relativity and absoluteness relate to and affect information-knowledge development and human neuro-decision-choice behavior over the space of decidability-undecidability dualities irrespective of how they are conceived and viewed in theory and practice ? Similarly, how do they affect the principles of diversity and the unity of knowing and science over the space of doubt-credulity dualities? The answers to these questions require us to establish their defining characteristics to give their identities for distinction and commonness. We, thus, seek definitions involving decision-choice analytics, information-knowledge analytics and input–output analytics. Such definitions must be general enough to relate to all varieties and their corresponding characteristic dispositions including nominal and real varieties.

Definition 2.4.1 Relativity and the Space of Relativity A relativity is a set of conditions that presents comparative analytics of two or more varieties in relational continuum and unity where each variety presents a frame of reference of their existence and motions relative to one another over the space of quantity-quality dualities presenting two motions of *qualitative motion* for changes in properties and *quantitative motion* for changes in relations and places in continuum; and where there is the *nominal relativity* in reference to linguistic systems of varieties and the *real relativity* in reference to matter-energy systems of varieties, and where the nominal relativity is a derivative from real relativity to give rise to nominal and real definitions. The collection of all concepts and ideas of relativity is the *space of relativity* which is also the space analog.

Note 2.4.1: Relativity and the Space of Relativity Some explanatory notes to the definitions of relativity and the space of relativity will be useful to the understanding of their uses in the analytical and epistemic works in this monograph and extensions that they may be connected in general space of human actions. The concept of relativity requires the existence of diversities which are referred to as varieties in this monograph and are extensively discussed in the theory of info-statics concerned with conditions of variety identification, identity and categorization [671]. It is also concerned with intertemporal conditions of variety transformation as discussed in [672]. The existence of varieties with relational continuum and unity provides frames and multiple frames of reference for comparative analytics and possibilities of ranking over the space of time trinity of past-present-future conditions and the establishment of order, based on some acceptable criterion. The theory of info-statics establishes the initial conditions for categorial conversion [669]. The theory of info-dynamics establishes the transformational conditions for categorial conversion through the space of primary-derived dualities [672]. The comparisons for distinction, differences and commonness take place over the space of relativity.

The varieties for comparative analytics include states, objects, processes, technologies, nominal, real and many others which are identified with the space of varieties. The varieties exist as diversities in relational continua and unity under dualistic-polar conditions in give-and-take sharing modes within the input–output and production-consumption unbreakable chains where internal and external comparative analyses are epistemically allowable for the understanding of diversity, preference ranking and knowing. Epistemically, the development of the theory of choice, whether in mathematics, economics, psychology or in general decision, is impossible if there are no interframes of variety references that allow ranking by some common criteria. Preferences, decision-choice actions and all motions imply the existence of relativity for variety comparative analytics. Similarly, the development of communications through different forms of languages and linguistic theory in either the family of ordinary languages (FOL) or the family of abstract languages (FAL) is impossible without verbal diversities in inter-relational modes for shades of meanings and understanding in the relativity of linguistic constructs. The nominal definition, real definition and the theory of definition are not possible if there is no space of relativity that allows shades of meaning and approximation; linguistic creativity, computational approximation and creative interpretation of events and representations.

It is the conditions of give-and-take relational modes that allow the interconnectivity of varieties in comparative analytics to access medicinal plants, active ingredients for the space of health-delivery system and many others. The space of relativity provides residence for varieties with differential give-and-take sharing modes and differential conditions of frames of reference over the epistemological space for the understanding of the behavior of ontological elements in varieties. The conditions of the universe and continual transformation from new to the old and vice versa, from life to death and vice versa, from forms to multiple forms under the principles of dividedness and the universal recycling of matter-energy systems have meaning and understandability only in terms of relativity of varieties and characteristic dispositions with continual frame and multiple frames of references. The understanding of the nature and behavior of varieties is always in a multiplicity of continual relativity with other varieties. In this respect, every area of knowing is defined in the space of relativity where comparative analysis of cognitive results may be undertaken to establish distinction, commonness and relational unity.

All decision-choice actions take place over the space of relativity that contains varieties and categorial varieties under some criterion of comparison and ranking over the space of decidability-undecidability dualities in relation to the spaces of acceptance-rejection dualities, doubt-surety dualities, correcting-noncorrecting dualities and many other dualities in the space of relativity that contains all dualities. The criterion of comparative analytics and ranking for the development of the theories of decision, choice and order over the space of quantity-quality dualities finds expressions, meanings and understanding over the space of relativity. Our number system and its legitimacy are established under the principle of relationality through comparative ranking and order over the space of relativity. There are many criteria for comparative analyses in different areas of neuro-decision-choice actions. Each of these criteria is reducible to a general important criterion that may be used to induce all variety ranking under comparative analytics. This criterion is general to all varieties and categorial varieties in neuro-decision-choice actions over the epistemological varieties and over the ontological varieties for human communication and understanding.

The general criterion is the set of cost–benefit conditions of varieties as the frame of reference in the space of preferences where varieties may be net cost or net benefit relative to a goal-objective element(s) of a decision-choice agent. The cost–benefit dualities are also input–output dualities which are also production-consumption dualities, all with relational continua and unity. In this respect, matter, energy, information and time appear simultaneously as internal cost dispositions and internal benefit dispositions within conditions of relativity that establish the ranking identities under neuro-decision-choice actions. The basic analytical structure is to provide conditions that will help to explain both the internal and external transformations of varieties and decision-choice actions as problem–solution processes. The cost–benefit conditions are generated by negative–positive characteristic dispositions to define the space of variety negative–positive dualities where every variety is generally defined by the conditions of negative and positive duals under various characteristic combinations in relative standings with relational continuum and unity.

An important analytical note, here, is that the identity of every variety is established by negative and positive characteristic dispositions in degrees of combination over the space of relativity, where the negative and positive dispositions appear as duals of the variety duality. Over the space of relativity, the negative and positive duals of any variety project themselves individually as internal cost–benefit conditions that are assessed by cognitive agents as cost dual and benefit dual to constitute the cost–benefit duality of the variety over the space of neuro-decision-choice actions. The *internal cost–benefit duality* is the *internal relativity* that may vary over cognitive agents establishing differences in variety preferences. Over the general space of varieties, there is the negative–positive polarity of comparative varieties with negative poles and positive poles that give rise to external cost–benefit conditions of each variety at the level of categories in the space of relativity, where the negative pole may be either a benefit or cost pole in relation to goals, objective and visions. Similarly, the positive pole may be either a cost or benefit pole to establish either a negative or positive polarity, where the internal cost–benefit conditions are translated into external cost–benefit conditions to give rise to cost–benefit polarity. The *external cost–benefit polarity* is the *external relativity*. The internal cost–benefit dualities provide the identities of the varieties in neuro-decision-choice value while the external cost–benefit dualities provide conditions for the relative standings of the varieties within an interframe comparability of preferences for ranking and order over the space of quantity-quality dualities. This epistemic frame provides the analytical foundations for the development of the theory of cost–benefit analysis as a general approach to neuro-decision-choice theory where the comparisons are made possible by information-knowledge conditions over the space of varieties.

The epistemic structure is that the negative–positive and real cost–benefit dispositions connect us to the analytical space of opposites in terms of duality with negative and positive duals for variety characteristic dispositions then to cost and benefit duals in terms of cognitive utility and are then projected to polarities with a cost pole and a benefit pole in relationally variety transformational dualistic-polar games over the space of relativity. The space of varieties of knowing and information-knowledge accumulation is thus the space of dualities and polarities in relational continua and unity for variety identities and transformations. The space of relativity, and the principle of opposites with relational continua and unity present the conditions of analog through which knowing as an intellectual flow is actionized to accumulate the information-knowledge system as intellectual stocks in disequilibrium processes. The set of cost–benefit conditions as a general criterion over the spaces of ranking, order and decidability-undecidability dualities means that every criterion characteristic disposition over these spaces is analytically reducible to the general real cost–benefit criterion without a remainder over the space of relativity. In the space of relativity, there exists primary variety and derived variety as parent-offspring duality. There is also the primary category and derived category of existence and knowing where the primary categories serve as identity to the derived categories in the space of construction-reduction dualities with relational connectivity and unity in the space of relativity. By constructionism, the derived is obtained from the primary just as from the parents the offspring are obtained. By reductionism, the primary is revealed from the derived, just as with the offspring the identities of the parents are made known under opportunity cost processes. By nominalism, the relativity is revealed by varieties with defined characteristic dispositions which may contain the characteristics of absoluteness through the concept of relativity-absoluteness duality.

Definition 2.4.2: Absoluteness and the Space of Absoluteness An absoluteness is a set of conditions that presents noncomparative analytics of varieties in the space of quality-quantity dualities with relational separations, diversities and disunity and no-give-and-take sharing mode, where each variety presents a non-interframe of reference of its existence and motions relative to one another over the space of quantity-quality dualities, thus, presenting independent *quantitative* motions for changes in relation and places and independent qualitative motions for transformations in relation to properties, where there is internal absoluteness and external absoluteness in identity and existence such that there is *nominal absoluteness* in reference to linguistic systems of varieties and *real absoluteness* in reference to matter-energy systems of varieties. The collection of all actual and potential concepts, real, nominal and ideas of absoluteness is the *space of absoluteness* (space of digital).

Note 2.4.2: On the Absoluteness and the Space of Absoluteness At the space of absoluteness, every variety exists as an identity onto itself with uniqueness. The space of absoluteness and the principle of diversities with relational separation and the excluded middle for disunity present the *conditions of digital structures* through which variety knowing is actionized as absolute for intellectual investment flows to accumulate the information-knowledge system in the disequilibrium processes as intellectual capital stocks. The elements in the space of absoluteness reside in internally and externally noncomparable modes since each one exists as an absolute variety in the space varieties with no relational give-and-take sharing mode. The concept of absoluteness requires the existence of non-interdepended diversities which are referred to as absolute varieties in this monograph and is discussed in the theory of info-statics which is concerned with conditions of variety identification, identity and categorization with the conditions of either belonging or non-belonging and not both to a category [671]. The existence of absolute varieties with relational separation and disunity provides no frames of reference with other absolute varieties for comparative analytics and no possibilities of ranking and the establishment of order based on some acceptable criterion for some level of general acceptance. The absolute varieties exist as diversities in uniqueness, where internal and external comparative analyses are analytically impossible and hence, indisputably critical understanding of relative diversities, ranking, unity and knowing is not available where each variety is either negative or positive and not both, and similarly it is either cost or benefit and not both. Epistemically, the development of the theory of choice action is difficult and even impossible if there are no interframes of variety references for ranking by some common criteria. Similarly, the critical understanding and the development of neuro-decision-choice actions and all variety motions over the space of quantity-quality dualities are difficult to abstract. Over the space of absoluteness, it is difficult to understand and explain the technology and process of decay, changes in variety characteristic dispositions and similar processes in a gradual process of decay with a partial decay or partial transformation from wellness to sickness and from sickness to wellness.

Similarly, the development of communications through different forms of languages and linguistic theory in either the family of ordinary languages (FOL) or the family of abstract languages (FAL) is impossible with absolute verbal diversities where every normal meaning and understanding must exist in absolute uniqueness with exactness and completeness. In this space of absoluteness, linguistic creativity will be zapped by verbal absoluteness; and pademiology and multiple meaning in communication within the space of source-destination dualities will be impossible. Similarly, the whole idea of error-correction process, that gives rise to mathematics of asymptotic methods, theory of approximation and the understanding and appreciation of the beauty of rainbow will be meaningless. The structure of rainbow is natural way of indicating the non-dominance of absoluteness which resides on the fringes of relativity. Any critical observation of diversities of natural forces will understand that the graduality of natural processes takes place over that space of relativity and not over the space of absoluteness on the average.

By the conditions of the state of absoluteness, the total internal characteristic dispositions of varieties exist as either negative or positive characteristic dispositions but not both, and hence there is no internal duality, since the internal duality admits of simultaneous existence of opposites in the same space, in the sense of contradiction where *nothing can both be and not be* and in the sense of the principle of the excluded middle where *everything must either be or not be*. By denying internal duality, the conditions of absoluteness also deny the existence of external duality and polarity, since the existence of either external duality or polarity requires comparative analysis based on relativity. The lack of a give-and-take sharing mode in the space of absoluteness is such that every area of knowing cannot be understood by the method of comparison where the epistemological characteristic dispositions must be exactly the same as the ontological characteristic disposition in order to claim knowledge where there is no partial knowledge, partial truth and partial existence. The lack of partial truth, partial knowledge and partial existence is not consistent with the world as we know it where there are continual balances between duals of dualities and poles of polarities.

The important analytical note, here, is that the identity of every variety over the space of absoluteness is established by either negative characteristic disposition or positive characteristic disposition but not both where there are no degrees of combination between the negative and positive internal dispositions creating internal absoluteness and external absoluteness where there are no internal duals conflicts and external logical conflicts. Over the space of absoluteness, the variety characteristic dispositions project themselves individually as either internal cost disposition or internal benefit disposition but not both, where for example, a gun is either bad (cost) or good (benefit) but not both as assessed by cognitive agents over the space of decidedness defined over the space of acceptance-rejection dualities. In this way, the are no internal dualistic conflicts to generate internal force, and hence the space of absoluteness denies internal self-change or self-transformation of the variety characteristic dispositions such as life-death duality with internal self-transformations from life to death and death to life or cognitive ignorance-knowledge duality with transformations from ignorance to knowledge by a process or peace-war duality with transformations from peace to war and war to peace by a process and many others. The concepts of primary category, derived category and parent-offspring process are not analytically useful over the space of absoluteness for the study of internal transformations of varieties where there are continual transformation of nature and society over the space of qualitative-quantitative dualities.

An important question arises as to how one can construct a theory of decision-choice action over the space of absoluteness without relational connectivity and unity under the principle of the excluded middle. How are ranking and order established, and under what criterion of comparison? There is also the explanatory problem with primary-derived processes, where from the primary category, one obtains the derived category of existence and knowing, where the each primary category serves as identity to itself as a derived category in the space of knowing. The epistemic frame of absoluteness does not provide us with useful analytical foundations for the development of the theory of cost–benefit analysis as a general approach to neuro-decision-choice theory in the action space over the space of problem–solution dualities requiring the existence of information-knowledge input–output process about internal cost–benefit characteristic balances and external cost–benefit balances, where every variety can serve as either benefit or cost relativity to a goal, objective or vision, under transformation-chains of input–output, problem–solution, production-consumption, dual-dual and ignorance-knowledge dynamics within the space of either primary-derived dualities or parent-offspring dualities. By nominalism, absoluteness is also revealed by varieties with defined characteristic dispositions which will not contain characteristics of relativity through the absence of the concept of relativity-absoluteness duality.

2.5 Useful Dualistic-Polar Spaces for Knowing and Knowledge Stock-Flow Dynamics

The development of the chapters of this monograph will take advantage of dualistic-polar analytical spaces for establishing conditions of knowing as intellectual investment flows and information-knowledge systems as accumulation of intellectual capital stocks. Let us keep in mind that the general theory of economic production is a theory of input–output process, where investment is made to produce capital as output which is then used in further production of capital to provide capital services as inputs for variety transformation. The theory of consumption is also a theory of input–output process, where investment is made to produce a variety capital as output to provide capital services as inputs to produce variety consumptions to generate intellectual and physical labor to provide labor service inputs over the space of production-consumption dualities. The understanding of human behavior takes place through the understanding of neuro-decision-choice actions which are expressed over deferent areas of existence. These neuro-decision-choice activities take place under the general principle of opposites that establishes a system of spaces of dualities with continua and unity, where the dualities have corresponding polarities to define a system of dualistic-polar games to resolve internal and external conflicts for transformations over the space of input–output dualities where every variety is simultaneously input and output relative to the elements of the space of imagination-reality dualities.

The number of spaces of dualities and polarities are many and continually changing relative to cognitions and human actions that are set to mediate the conditions between the duals of dualities, between the poles of polarities and between the dualities and the polarities through information-decision-choice-interactive processes in an accordance with a dynamic system of preferences guided by philosophical consciencism that has taken hold over a generation and historic period. Among the dualistic-polar spaces relevant for the development of the chapters that follow are ontology-epistemology duality, cost–benefit duality, actual-potential duality, nothingness-somethingness duality, imagination-reality duality, unity-disunity duality, possibility-probability duality, true–false duality, necessity-freedom duality, relativity-absoluteness duality, primary-derived duality, negative–positive duality, oneness-dividedness duality, ignorance-knowledge duality, rejection-acceptance duality, construction-destruction dualities and many others that may be introduced strategically.

Chapter 3
The Theory of Epistemic Spaces for the Diversity and the Unity of Knowing and Science

Abstract In this chapter, the concept of epistemic space is advanced, defined and explicated for the understanding of fuzzy spaces in the development of the theory of knowing as the theory of intellectual investment flows and the theory of information-knowledge system as the theory of intellectual capital stocks. In this framework, the theory of intellectual investment-capital process is seen as the theory of intellectual stock-flow dynamics, where the intellectual information-capital stocks provide capital-service input flows into the general neuro-decision-choice space and the intellectual investment flows provide continual updating of intellectual stocks in the space of quality-quantity dualities. The relevant spaces and sub-spaces are introduced leading to the definition and explication of the epistemic space and the development of theory of epistemic space for the understanding of both the theories of knowing and knowledge seen in terms of neuro-decision-choice activities over the space of problem–solution dualities.

3.1 An Introduction to the Concept of Epistemic Space

Chapter 2 was concluded with discussions on the concepts of relativity and absoluteness and then extended to the spaces of relativity and absoluteness. These concepts provide a framework to discuss the theory of epistemic spaces relevant to the understanding of the process of knowing, knowledge development and decision-choice processes. They will also become critically relevant to the development of variety information as defined by the abstraction of variety characteristic dispositions from the process of acquaintance. The abstraction of variety characteristic dispositions from the acquaintance will be affected by the adherence to the choice of either the relativity and the space of relativity or absoluteness and the space of absoluteness. From the concept of absoluteness and the space of absoluteness emerge the defining properties of exactness, completeness and perfection of the information system from the process of acquaintance by assuming no cognitive capacity limitations leading to the development of an exact information system and the classical paradigm of thought. Similarly, from the concept of relativity and the space of relativity emerge

K. K. Dompere, *The Theory of Problem-Solution Dualities and Polarities*,
Studies in Systems, Decision and Control 405,
https://doi.org/10.1007/978-3-030-90279-7_3

the defining properties of inexactness, incompleteness and an imperfection of information system from the process of acquaintance due to cognitive capacity limitations leading to the development of the fuzzy information system and the fuzzy paradigm of thought. Both the classical and fuzzy paradigms of thought will receive further discussion in this monograph (See also [174, 177, 669, 832]).These paradigms are the general guidance for the development of thinking algorithms to be used as reasoning framework for cognitive search activities to find information-knowledge items in relation to nominal and real decision-choice activities.

3.2 The Concept of Cognitive Search Activities in Analytical Spaces

A study is a knowing activity in an *action space* involving the characteristic-dispositions of subjects, objects and their relational interactions through processes under matter-energy conditions. The meaning and the nature of the concepts of characteristic disposition, the signal disposition and the characteristic-signal disposition have been discussed in the theory of info-statics where the characteristic disposition is the content of information and the signal disposition is the message of the information [669, 671]. What are the subjects, objects and their relational interactions in the action space? We shall first have some working definitions that will allow us to point to the answers that will lead us to create the needed cognitive spaces and their relational diversities and unity which will help in extending the concepts of diversity and unity to not only the space of knowing but also to the general space of production-consumption dualities as well as the space of input–output dualities. In this economic-theoretic approach to the understanding of the diversity and unity of knowing and science, we must keep in mind the relational connectivity of the space of production-consumption dualities and the space of input–output dualities and how they are connected to the space of parent–offspring dualities, with the conceptual understanding of bio-genetics, techno-genetics, process-genetics and the space of varieties and their interdependencies in the input–output chain, the concepts of food chain, supply chain, demand chain, power chain, and in general the neuro-decision-choice chain.

Definition 3.1.1: Action Space, ($\mathfrak{A}$) The action space, ($\mathfrak{A}$) with a generic element ($\mathfrak{a} \in \mathfrak{A}$) is the space of all cognitive actions as a sub-space contained in the union of the space of the realities as a sub-space of the space of actuals and the space of imaginations as the sub-space of the space of the potentials of neuro-decision-choice activities with inputs and outputs from the space of variety input–output dualities in relational continua and unity for each duality and among different dualities. The relational continua and unity of each duality ensure that there is a dual connectivity as well as multiple connectivity with the non-excluded middle among the varieties. The relational continua and unity among dualities also ensure that the knowledge on one duality has a potential to reveal aspects of knowledge on other varieties through conditions of relationality in such a way that knowledge on varieties are interdependent in give-and-take sharing modes such that there is a variety's internal relationality and external relationality in the form where the action space is neuro-decision-choice defined over the space imagination-reality dualities which is a subspace of actual-potential dualities, where the action space is composed of the union of the sub-space of actions on variety reality $\left(\mathfrak{A}^{\mathfrak{H}}\right)$ and sub-space of actions on variety imagination $\left(\mathfrak{A}^{\mathfrak{G}}\right)$ such that $\left(\mathfrak{A} = \left(\mathfrak{A}^{\mathfrak{H}} \cup \mathfrak{A}^{\mathfrak{G}}\right)\right)$.

Note 3.1.1: On Action Space Every element $(\mathfrak{a} \in \mathfrak{A})$ is a neuro-decision-choice action to change the relationship between a variety reality and a variety imagination in the place of the actual. In other words, we have $\mathfrak{A}(\boldsymbol{\mathfrak{A}}, \boldsymbol{\mathfrak{U}}) \leftrightarrow \mathfrak{A}(\boldsymbol{\mathfrak{U}}, \boldsymbol{\mathfrak{A}})$ which implies that $(\mathfrak{A}(\mathfrak{H}, \mathfrak{G}) \leftrightarrow \mathfrak{A}(\mathfrak{G}, \mathfrak{H})) \subset (\mathfrak{A}(\boldsymbol{\mathfrak{A}}, \boldsymbol{\mathfrak{U}}) \leftrightarrow \mathfrak{A}(\boldsymbol{\mathfrak{U}}, \boldsymbol{\mathfrak{A}}))$ in a never ending process of transformation to generate new varieties and new information for either qualitative, quantitative transformation or both changing the relationality of realities and imaginations in the space of imagination-reality dualities. Every transformation is either a motion that affects either a change in relation, a change in characteristics or both in the space of varieties through a process in the input–output space. What are the inputs, outputs and the input–output transformation modules in the space of input–output dualities within the action space? If the outputs are knowledge, then what are the inputs? What is the space of action and what is its morphology? The concepts of ontology and epistemology are general but with definitional distinctions from each other. The questions arising from the definitions of these concepts are also general and universal from which, particulars may be abstracted by epistemic processes. The conditions of interdependence may be solution-enhancing or problem-generating over the action space. In our search of diversity and unity of knowing and science we must establish the differences and similarities among information, data, knowledge, evidence and evidential things such that a clear distinction may be made among the nature of the theory of information, the theory of data and data analysis, data science, the theory of knowledge, the theory of evidence and the other derived concepts and theories such as possibility, the theory of possibility, probability, the theory of probability, statistics and the theory of statistics and how they are related to each other and the general theory of neuro-decision-choice actions and practice of ideas over the spaces of certainty-uncertainty and doubt-surety dualities.

3.3 The Concepts of Epistemological and Ontological Spaces in Cognitive Search for Knowing, Information and Knowledge

To link the concepts of ontology and epistemology to transformational technologies and decision-choice actions and action space, several other relevant spaces must be specified with the indication of their connecting cords for epistemic clarity. The immediate spaces to be specified and defined are the *ontological space* relative to ontology and the *epistemological space* relative to epistemology. Corresponding to both the ontological and epistemological spaces is the space of characteristic-signal dispositions which constitutes the ontological *information space* that connects all spaces and sub-spaces that may be created. The ontological information space is also the *ontic space* containing the actual characteristics required to abstract the characteristic dispositions for variety identities. The ontological space is equipped with ontological information and an infinite set of ontological technologies and processes, while the epistemological space is equipped with epistemological information and a set of epistemological technologies abstracted from the ontological signal dispositions the size of which depends on knowing and acquaintance space. The epistemological information space is also the *epistemic space* containing the derived actual characteristics from the ontological variety signal dispositions required to derive the epistemological variety characteristic dispositions for the establishment of epistemological varieties to be compared to the ontological varieties in terms of correctness of variety identification as variety knowledge under cognitive capacity limitation. The space of ontological technologies is the collection of all ontological actual and potential transformation processes which are generated by natural decision-choice modules for knowing and information-knowledge accumulation. The use of epistemic space is different from other uses in the literature. The relationship between the ontic and epistemic space will be discussed in understanding knowing and information-knowledge accumulation as neuro-decision-choice activities over the space of problem–solution dualities as it also relates to the space of production-consumption dualities.

The system of dualistic-polar structures is such that at the level staticity, we have ontological varieties established by ontic characteristics and characteristic dispositions as ontological variety identities on one hand and epistemological varieties established by epistemic characteristics and characteristic dispositions as epistemological variety identities on the other hand. At the level of dynamics ontological processes and transformation technologies of varieties, on one hand, and epistemological processes and transformation technologies as derivatives from those of ontological processes and transformation on the other hand. we have the ontological The space of ontological technologies and processes is independent from and not under the control of any cognitive agents even though it responds to the results of cognitive activities of neuro-decision-choice outcomes that have direct effects on the ontological varieties. The space of epistemological technologies and processes is the collection of all actual and potential epistemological transformation capacity processes which are generated by cognitive agents through individual and social neuro-decision-choice modules through the activities over the space of *imagination-reality dualities* within the epistemological space. The space of epistemological technologies and processes is the collection of all actual and potential epistemological transformation capacity processes which are generated by cognitive agents through individual and social neuro-decision-choice modules through the activities over the space of *imagination-reality dualities* within the epistemological space.

The nature of the concepts and phenomena of these spaces are to emphasize the neuro-decision-choice processes in a manner that views knowing as the economic production of intellectual investment flows and information-knowledge accumulation as the creation of economic intellectual factories that provide intellectual capital services as inputs into neuro-decision choice activities in the space of problem–solution dualities which incorporates the space of question–answer dualities. In this respect, the traditional theory of knowledge is about quality control and the methods of quality control as seen in the space of methodological constructionism-reductionism dualities. Our focus here is to view information-knowledge items as economic production in a general input–output space with a system of quality controls, where the outputs become information inputs into neuro-decision-choice activities in the action space, and where the outputs of the neuro-decision-choice actions become inputs into the action space. The useful starting point of the definitions of the relevant spaces in understanding the diversity and unity of knowing and science as economic productions at the level of epistemology is the *structure of the knowledge square*, which is composed of spaces of potential, possible, probable and actual varieties and categorial varieties where all spaces are linked by the conditions of the information space, where a number of dualistic-polar subspaces, such as the spaces of imagination-reality dualities and possibility-probability dualities arise within the system of variety knowing processes.

3.3.1 Defining the Characteristics of the Epistemological and Ontological Space

There are two types of definitions that will be provided for the epistemological and ontological spaces and how they unite as a duality containing the ontological dual and the epistemological dual. They are verbal and algebraic for each space in order to understand the neuro-decision-choice actions and their relations to variety identifications and transformations as they relate to the theory of knowing, the theory of knowledge and the theory of information and their relationality of mutual existence and destructions of varieties with the space of ontological-epistemological and ontic-epistemic dualities. We must always keep in mind that the role of definitions is to establish identities of varieties with distinction and diversities in the spaces of varieties, information and relations [87, 143, 451, 490–492]. The definitions of ontological and epistemological technologies, processes and their explanatory relations to transformations of varieties are provided in [669, 670, 672]. Here, the ontological processes, technologies and transformations are presented as primary categorial identities, while those of epistemology are presented as derived categorial identities.

In all these discussions, one thing stands out clear, and that is, there is no knowledge without the elements in the information space, and without this knowledge, there would be no input into neuro-decision-choice actions over the space of problem–solution dualities without which there would be no new social information and new problem–solution dualities. Neuro-decision-choice actions on the space of problem–solution dualities generate new information and new problem–solution dualities for new neuro-decision-choice actions in the space of problem–solution dualities. The definitions must connect to neuro-decision-choice actions over the infinite spaces of problem–solution dualities and source–destination dualities. Here, care must be taken to distinguish variety ideas from information-knowledge varieties which may also serve as inputs into the production of ideas. The information-knowledge varieties and variety problems belong to the space of reality as a subspace of the space of the actuals, while solutions, ideas, visions, goals and objectives belong to the space of imagination as a subspace of the space of potentials such that the space of imagination-reality dualities is contained in the space of actual-potential dualities and the space of problem–solution dualities is contained in the space of imagination-reality dualities.

Ideas, vision, goals and objectives may be actualized by processes into the space of reality as a subspace of the space of actual under the general principles of the *knowledge square*, where actualization is neuro-decision-choice process that moves from the space of imagination as a subspace of the space of the potentials to the possibility space under necessity and initial conditions of transformation under possibilistic uncertainty and then to the probability space under freedom and probabilistic uncertainty to the space of the reality as an information-knowledge element under the entropic principle over the spaces of decidability-undecidability dualities, doubt-surety dualities and acceptance-rejection dualities. The understanding of these neuro-decision-choice actions, the knowing process and the information-knowledge accumulation all require us to establish connectivity between the ontological space and the epistemological space which require distinctions through definitions. We have already defined the concepts of ontology and epistemology. These definitions will be extended to ontological and epistemological spaces. We shall offer verbal definitions and algebraic definitions as they apply in this monograph. As it will be discussed and explained in the monograph, social forces operate in the space of imagination-reality dualities as a subspace of actual-potential dualities in which natural forces operate. It is important to know the defining characteristics between reality and actual, and between the space of realities and the space of actuals. Similarly, it is important to know the characteristics between imagination and potential, and between the space of imaginations and the space of potentials. Reality, the space of realities, imagination and the space of imaginations are existentially dependent on cognitive agents, while actual, the space of actuals, potential and the space of potentials are existentially independent of cognitive agents.

Definition 3.2.1.1: The Ontological Space (Verbal) The ontological space is composed of matter, energy, varieties and information, where information is the organic characteristic dispositions of matter and energy, where matter and energy are decomposed into varieties of things, states and processes with their corresponding variety characteristic decompositions for identities, identifications and transformations. The ontological space is simply a collection of matter-energy actual and potential varieties of objects, states, technologies and processes of the past, present and future. The actual is made up of known and unknown varieties while the potential is made up of varieties in the making as well as actuals in continual processes of transformation to become potentials for the possible next actualizations.

The algebraic conceptualization and definition of the ontological space are a little more complex in the sense that the space and sub-spaces must account for an increasing number of sub-spaces in both static and dynamic conditions in the universal existence. Some of these static and dynamic conditions has been presented in monographs [174, 671, 672, 832]. There is, however, a need to pull together the connecting logical cords to present the case of the unity of knowing with the unity of non-engineering science and engineering science. The universe is organically viewed in terms of matter, energy and information where matter with defined characteristic disposition is the primary category of existence. Energy is an indispensable internal property of matter, while information is both the indispensable internal and external property of matter and energy. Matter, energy and information are related to processes, technologies and work to produce transformations of matter, energy and information into different forms of variety at both static and dynamic states of time. Diagrammatically, the ontological space may be presented in a pyramidal structure in relationality as in Fig. 3.1, where there is a matter-energy-information $(\mathbb{M} - \mathbb{E} - \mathbb{Z})$ pyramid and superimposed on it is a technology-process-transformation $\left(\widehat{\mathbb{T}} - \mathbb{P} - \mathbb{T}\right)$ pyramid. At the center is the universe that contains all things in static and dynamic

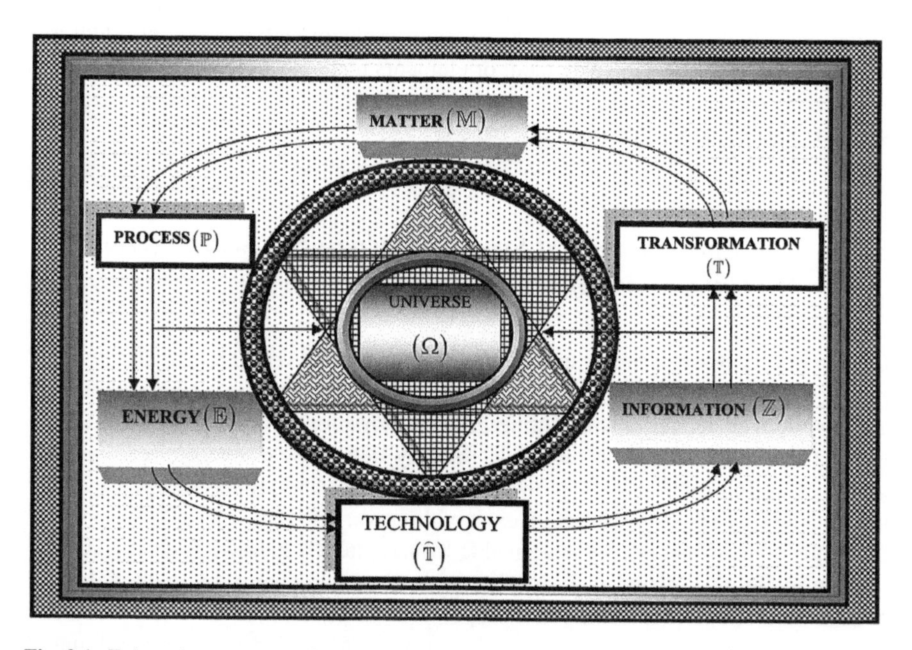

Fig. 3.1 Epistemic geometry of ontological space where the universe $(\Omega) = (\mathbb{M} \otimes \mathbb{E} \otimes \mathbb{Z}) \cup (\widehat{\mathbb{T}} \otimes \mathbb{P} \otimes \mathbb{T})(\mathbb{M} \otimes \mathbb{E} \otimes \mathbb{Z})$ is the primary pyramid and $(\widehat{\mathbb{T}} \otimes \mathbb{P} \otimes \mathbb{T})$ is the derived pyramid. It is a general framework in examining the definition of the ontological space and the corresponding sub-spaces

states, where the beginning-end polarity has a residing nothingness-somethingness duality in each pole.

The cognitive geometry of Fig. 3.1 requires some explanations. It is composed of interrelations of two pyramids in one, the universe. The universe at a static state is established by the *first pyramid* of matter, energy and information as three in one and one in three which constitute the essential challenges to knowing. On the first pyramid is imposed a second pyramid at a dynamic state of technology, process and transformation as three in one and one in three. Energy and information are central to matter at any state and matter is the primary category for the existence of energy and information which are matter-derived categories of universal existence at any static state. Process and transformation are central to technology which is the primary category of the existence of process and transformation, and hence process and transformation are technology-derived categories at the dynamic states. Every existence is defined in a combined polarity and duality, where each pole has a residing duality. The dualities function under the *general principle of opposites* to generate conflicts and force [651, 669, 670, 672, 673, 676, 713]. It is, here, that dualistic-polar games find expressions in internal and external transformations through decision-choice activities over the action space. The space of *static-dynamic duality* is intimately connected to the space of *primary-derived dualities* for the analytics of the destruction-replacement process of parent–offspring structures in variety transformations in an infinite system of problem–solution decision-choice processes where every problem constitutes a primary-derived duality and also every solution constitutes a primary-derived duality in the sense that a variety solution is a variety problem and a variety problem is a variety solution relative to an element or elements in a goal-objective chain towards a vision or visions where a vision or visions are ideas residing in the imagination space as a subspace of the potential space.

3.3.2 The Epistemological Space and Cognitive Search Over the Ontological Space

Certain questions arise in the spaces of *real primary-derived dualities* and *nominal primary-derived dualities* in the information space and in the space of methodological constructionism-reductionism dualities in all information-knowledge systems within the space of static-dynamic dualities. Broadly defined, can energy exist without matter? Can information exist without matter, energy and matter-energy varieties? Can technology and transformation exist without varieties and energy? The point of these questions is that energy is derivable from matter by methodological constructionism while information is derivable from matter and energy by methodological constructionism when information is properly defined in relation to variety identities. By methodological reductionism, one may trace information to energy and matter and energy to matter by methodological reductionism, at least, at the level of our current cognitive limitations.

Similarly, can a process take place without technology and can transformation take place without a process and technology? Given a technology, can a process take place without energy and can knowledge about varieties and changes in varieties be known without information? Another point that is being advanced here is that transformation is derivable from a process by methodological constructionism while a process is derivable from technology by methodological constructionism. By methodological reductionism one can trace transformation to a process and a process to technology under the principle of variety characteristic-signal disposition, where the characteristic disposition establishes the *content of information* and the signal disposition establishes the *message of information* for any variety and categorial varieties among source–destination entities. Can any variety be known without acquaintance, cognition and decision-choice action?

Finally, what is the relationship between the matter-energy-information structure and the technology-process-transformation structure? How and where do we find and know this relationship, and is this relationship the work of information, decision or both? The answer to these questions in the space of knowing is that the matter-energy-information structure constitutes the primary pyramid of universal existence while the technology-process-transformation structure constitutes the derived pyramid of universal existence with decision-choice activities over the defined action space given the existence of cognitive agents. The *derived pyramid* (trinity) is obtained from the *primary pyramid* by methodological constructionism and the primary pyramid is obtained from the derived pyramid by methodological reductionism. Without energy, force and work are not available to initiate a process with a technology for a transformation and without matter, nothing is transformed. The conditions of the primary pyramid and the derived pyramid constitute the foundations of the *input–output dynamics* for knowing, the development of the theory of knowing as intellectual investment-flow process and the theory of knowledge as the intellectual capital-stock accumulation process in information-knowledge systems.

The real conditions of the primary pyramid and the derived pyramid constitute the foundations of the input–output dynamics for change and development of the theories of info-statics and info-dynamics with the inherent conditions of cost–benefit dynamics constituting the development of theories of decision-choice actions while the uncertainties surrounding the decision-choice actions on variety identifications and transformations constitute the foundations for the development of the theory of fuzzy-stochastic entropy with its toolbox of sub-theories of possibility and probability for the design of measures of conditionality in the spaces of true–false dualities, decidability-undecidability dualities, acceptance-rejection and doubt-surety dualities with relational continuum and unity without the excluded middle, where all decision-choice-actions in thought and practices are exercised in the space of relativity and not in the space of absoluteness.

The conceptual definitions of ontology and epistemology are designed to create varieties of thought in a language of relevance for the understanding of the past-present-future variety existence. They do not say anything about socio-natural action space, the conditions of production of ontological existence and how ontology and epistemology are relationally connected in continuum and unity. The connectivity of relational continuum and unity requires further decompositions of the aggregates defining the blocks with specifying de-compositional and aggregative functions for epistemic analytics. The decomposition requires an application of a paradigm of thought to create subspaces for neuro-decision-choice analytics in the input–output spaces, where the same paradigm of thought provides the logical channels of composition and aggregation. The concept of the input–output space must be specific but not necessarily exact and then related to the specification of the universal existence at static and dynamic states. The objective of all these discussions is to link the theory of knowing as developed to decision-choice activities over the *space of problem–solution dualities* in order to establish foundational conditions for the justification of the *unity of science* through the *unity of knowing* and knowledge production to create foundations for the development of the theory of economic development and social transformation as decision-choice actions over the space of problem–solution dualities [670–672, 832].

The initial analytical complexity behind the analytical simplicity may be developed by a symbolic specification that will allow an algebraic definition and specification of the ontological space. The ontological space, like any conceptual space, is algebraically represented by different information structure in terms of characteristic disposition. Every symbol represents either macro-information or micro-information. Here, mathematics is a language representing information where the logic of mathematical manipulations are information processing modules to establish meaning for communication and understanding in the area of knowing. The complete specification of the ontological space is in terms of existence of *what there was* (the past), *what there is* (the present) and *what would be* (the future) to indicate the past-present-future existential conditions of matter-energy varieties in knowing and the information-knowledge system. The past-present-future existence can only be revealed by the information-knowledge structure to abstract the ontological space. The present existence reveals static conditions while the past existence and the future existence show dynamic conditions relative to the present time. The past reveals what has happened, the information of which is an instruction for the present decision-choice action. The future reveals what may happen relative to the present without indicating what will happen.

With variety characteristic dispositions, one may create mental images of some elements of the telescopic left (past) showing the socio-natural history of past events as irreversible conditions of decision-choice inputs as well as create mental images of telescopic right (future) showing the possibilities of visions, imagination and creativity available to the path of socio-natural transformations relative to the present. The past-present-future structure shows a *time trinity* in the *sankofa-anoma* tradition, where the present resides in the past and the future resides in the present and the past resides in the future in terms of the information-knowledge process, variety transformations and existential tomorrow under socio-natural decision-choice actions in theory and practice. The past relative to the present and the future relative to the present show the *telescopic left* and *telescopic right* respectively of past-present-future existence, the structure of which is shown in Fig. 3.2.

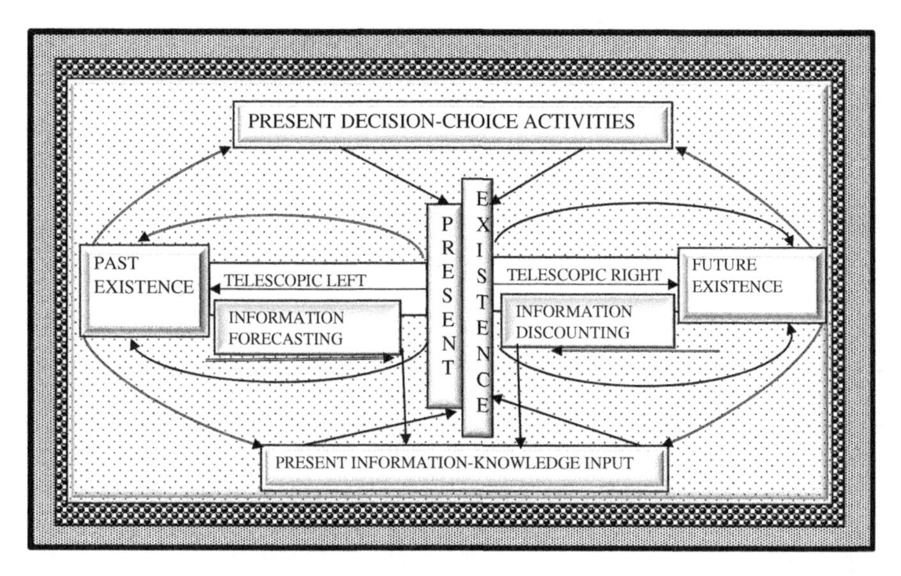

Fig. 3.2 The epistemic structure of the time trinity (Sankofa-anoma) in relation to information-knowledge input in decision-choice activities of cognitive agents

The telescopic left with past-present duality involves the knowing process to have an intellectual picture of the past from the position of the present to develop an information-knowledge structure as an understanding of the past and an input into current decisions for the actualization of future varieties. The telescopic left with past-present duality also involves an information-knowledge forecasting process for the input. The telescopic right with present-future duality involves the knowing process to have an intellectual picture of the future from the position of the present to develop an information-knowledge structure as an understanding of the future and as an input into current decisions for the actualization of future varieties. The telescopic right with present-future duality also involves an information-knowledge discounting process for the input. The combined telescopic left and telescopic right with past-future duality involves information-knowledge generation for cognitive activities with the elements in the action space relative to variety transformations.

It must be kept in mind that the time trinity presents an existence in three states of knowing for the definition of the ontological space with reference to the primary and derived pyramids which may be decomposed into three information-time dualities of *past-present duality*, *present-future duality* and *past-future duality* in knowing and information-knowledge stock-flow dynamics as inputs into neuro-decision-choice actions. In this way, one can speak of *ontological primary* and *ontological derived* pyramids in the past, present and future in terms of existence for knowing. The ontological space is the space of *ontological activities* in terms of *ontological decision-choice actions* which are decision-choices independent in identities but relationally interdependent and induced from the relational space of the system of varieties and transformations of ontological varieties as represented in Fig. 3.2. The primary category of existence initializes the transformation process for all derived existence over the space of ontology-epistemology dualities.

3.4 Algebraic Conceptions of the Ontological Space in Cognitive Search

The verbal definition of the ontological space is given in Definition 3.2.1.1. It is useful now to algebraically define the ontological and epistemological spaces. Let (Ω) represent the universe with a generic elemen ($\omega \in \Omega$), ($\mathfrak{U}$)the potential space with the generic element ($\mathfrak{u} \in \mathfrak{U}$), ($\mathfrak{A}$), the space of actuals with the generic element ($\mathfrak{a} \in \mathfrak{A}$), ($\mathbb{X}$), the general characteristic set with the generic element ($x \in \mathbb{X}$), ($\mathbb{S}$), the general signal set with the generic element ($s \in \mathbb{S}$), ($\mathbb{V}$) the complete set of varieties with the generic element ($v \in \mathbb{V}$), ($\mathbb{Z}$) information space with the generic element ($z \in \mathbb{Z}$), ($\mathfrak{T}$), technological space with the generic element ($\mathfrak{t} \in \mathfrak{T}$), ($\mathcal{P}$), the space of processes with the generic element $\left(p \in \mathcal{P}\right)$, ($\mathbb{T}$), space of transformation functions with the generic element ($\tau \in \mathbb{T}$), (Φ), space of phenomena with the generic element ($\varphi \in \Phi$), and ($\mathbb{I}$) is the index set, where the collection of all the set of the actual characteristics, constitute the ontic space ($\mathbf{O}$) such that ($x \in \mathbb{X} \subseteq \mathbf{O}$). These symbols will be used to define the ontological space as well as used in analytical reasoning. The concepts of variety and the variety space have been defined and specified in the theory of info-statics [669, 671] and the theory of info-dynamics [670, 672].

The essential structure is that, at the ontological space, every variety is composed of the characteristic set $\{(x \in \mathbb{X}_\upsilon) | \upsilon \in \mathbb{V}\}$ which is divided into a negative characteristic subset $\left(x \in \mathbb{X}_\upsilon^{\mathrm{N}}\right)$ and a positive characteristic subsets $\left(x \in \mathbb{X}_\upsilon^{\mathrm{P}}\right)$ in relational continuum and unity such that $\left(\mathbb{X}_\upsilon = \mathbb{X}_\upsilon^{\mathrm{N}} \cup \mathbb{X}_\upsilon^{\mathrm{P}}\right)$ to establish the identity of the variety ($\upsilon \in \mathbb{V}$) under dualistic-polar conditions, where the union is use to indicate the union of the duals to establish the duality. Let us examine the structural representation of the ontological space with a definition and then relate it to ontological information, variety and conditions of knowing in terms of potential and actual. The epistemic process is that all thinking and reasonings are based on the first principles of knowing in the problem–solution and question–answer processes under the defining principle of the characteristic-signal disposition that provides information for the problem–solution and question–answer identities. Here, the characteristic dispositions are the contents for identity verifications and the signal dispositions are the message communications of the nature of the identities for problem–solution and question–answer dualistic varieties.

Definition 3.3.1: Ontological Space (Algebraic) The ontological space is the universe with the generic element ($\omega \in \Omega$) which is the union of the space of the actuals ($\mathfrak{a} \in \mathfrak{A}$) and the potential space ($\mathfrak{u} \in \mathfrak{U}$), given the ontic space, where the potential space is the collection of all the potential varieties from the variety space

($v \in \mathbb{V}$) and the space of the actuals is the collection of all the actual varieties also from the variety space ($v \in \mathbb{V}$) equipped with the space of phenomena ($\varphi \in \Phi$). Each of the potential varieties and the actual varieties are identified by information structures from the information space ($z \in \mathbb{Z}$) which is composed of the characteristic set ($x \in \mathbb{X}$) and the signal set ($s \in \mathbb{S}$) with the characteristic set belonging to the ontic space, where the variety space is equipped with technological space ($t \in \mathfrak{T}$) that generates the space of processes $\left(p \in \mathcal{P}\right)$ to bring about the transformation space ($\tau \in \mathbb{T}$) that contains the transformation functions which bring about variety internal changes by setting the negative dual against the positive dual or the positive dual against the negative dual contained in each variety, and where the numerical counts of the elements of all spaces and sets are identified with the index set ($\mathbb{I}$) and where every variety has corresponding characteristic disposition for its identity. The algebraic structure is a system of interdependent spaces and sets at the *space of potentials* in the form:

$$\left.\begin{array}{l}
\Omega = \mathfrak{A} \otimes \mathfrak{U} = \left\{\omega = (\mathfrak{u}, \mathfrak{a}) \mid \mathfrak{u} \in \mathfrak{U} \text{ and } \mathfrak{a} \in \mathfrak{A}\right\}, \\
\text{where the space ofn potential is specified as :} \\
\mathfrak{U} = \left(\mathbb{V}_{\mathfrak{U}} \otimes \Phi_{\mathfrak{U}} \otimes \mathfrak{T}_{\mathfrak{U}} \otimes \mathbb{Z}_{\mathfrak{U}}\right) = \left\{(v, \varphi, \tau, z) \mid v \in \mathbb{V}_{\mathfrak{U}}, \varphi \in \Phi_{\mathfrak{U}}, \tau \in \mathfrak{T}_{\mathfrak{U}}, z \in \mathbb{Z}_{\mathfrak{U}}\right\} \\
\mathfrak{T}_{\mathfrak{U}} = \left(\mathcal{P}_{\mathfrak{U}} \otimes \mathbb{T}_{\mathfrak{U}} \otimes \mathrm{T}_{\mathfrak{U}}\right) = \left\{\mathrm{t} = (p, \tau, t) \mid p \in \mathcal{P}_{\mathfrak{U}}, \tau \in \mathbb{T}_{\mathfrak{U}}, t \in \mathrm{T}_{\mathfrak{U}}\right\} \\
\mathbb{Z}_{\mathfrak{U}} = \left(\mathbb{X}_{\mathfrak{U}} \otimes \mathbb{S}_{\mathfrak{U}}\right) = \left\{z = (x, s) \mid x \in \mathbb{X}_{\mathfrak{U}}, s \in \mathbb{S}_{\mathfrak{U}}\right\} \\
\mathrm{T} = \text{Time space}
\end{array}\right\} \quad (3.3.1)$$

Similarly, one can represent the *space of the actuals* with subscript ($\mathfrak{A}$) to complete the algebraic structure of the ontological space as:

$$\left.\begin{array}{l}
\Omega = \mathfrak{A} \otimes \mathfrak{U} = \left\{\omega = (\mathfrak{u}, \mathfrak{a}) \mid \mathfrak{u} \in \mathfrak{U} \text{ and } \mathfrak{a} \in \mathfrak{A}\right\}, \\
\text{where the space of actual is specified as :} \\
\mathfrak{A} = \left(\mathbb{V}_{\mathfrak{A}} \otimes \Phi_{\mathfrak{A}} \otimes \mathfrak{T}_{\mathfrak{A}} \otimes \mathbb{Z}_{\mathfrak{A}}\right) = \left\{(v, \varphi, \tau, z) \mid v \in \mathbb{V}_{\mathfrak{A}}, \varphi \in \Phi_{\mathfrak{A}}, \tau \in \mathfrak{T}_{\mathfrak{A}}, z \in \mathbb{Z}_{\mathfrak{A}}\right\} \\
\mathfrak{T}_{\mathfrak{A}} = \left(\mathcal{P}_{\mathfrak{A}} \otimes \mathbb{T}_{\mathfrak{A}} \otimes \mathrm{T}_{\mathfrak{A}}\right) = \left\{\mathrm{t} = (p, \tau, t) \mid p \in \mathcal{P}_{\mathfrak{A}}, \tau \in \mathbb{T}_{\mathfrak{A}}, t \in \mathrm{T}_{\mathfrak{A}}\right\} \\
\mathbb{Z}_{\mathfrak{A}} = \left(\mathbb{X}_{\mathfrak{A}} \otimes \mathbb{S}_{\mathfrak{A}}\right) = \left\{z = (x, s) \mid x \in \mathbb{X}_{\mathfrak{A}}, s \in \mathbb{S}_{\mathfrak{A}}\right\} \\
\mathrm{T} = \text{Time space}
\end{array}\right\} \quad (3.3.2)$$

The general ontological space for the epistemic studies and analyses of the unity of knowing may be specified as:

$$\begin{array}{c}
\Omega = (\mathbb{V}_{\mathfrak{U}} \otimes \Phi_{\mathfrak{U}} \otimes \mathfrak{T}_{\mathfrak{U}} \otimes \mathbb{Z}_{\mathfrak{U}}) \cup (\mathbb{V}_{\mathfrak{A}} \otimes \Phi_{\mathfrak{A}} \otimes \mathfrak{T}_{\mathfrak{A}} \otimes \mathbb{Z}_{\mathfrak{A}}) \\
\{\text{Potential varieties}\} \{ \text{Actual varieties}\}
\end{array} \quad (3.3.3)$$

A discussion on the algebraic definition of the ontological space will be useful for the development of the arguments on the epistemic diversity and unity of knowing and science and the general understanding of cognition, human actions and their relationship to the ontological space in this monograph. The ontological space is composed of a collection of varieties of matter, energy and technology which provide conditions of variety identities in the identification space and generate processes for variety changes in the transformation space where all the identities find expressions over the space of characteristics. The ontological space, at any time, is the union of the sets of all *that was*, all *that is* and all *that would be.* All that *was, is and would be* are coded in *ontological information*, the contents of which are specified in the space of the $(\mathbb{X}_{\mathfrak{U}} \cup \mathbb{X}_{\mathfrak{A}})$ and then then transmitted through the space of *onto-logical* $(\mathbb{S}_{\mathfrak{U}} \cup \mathbb{S}_{\mathfrak{A}})$ to all the elements in the ontological space. Thus, every element in the ontological space is simultaneously a source entity and a destination entity with continual give-and-take multiplicity of variety relations for signal dispositions constrained by natural limitations in the space of input–output dualities where every variety is simultaneously an input for some production processes and an output from some production processes.

3.4.1 Properties of the Ontological Space and Ontological Information

The ontological space is dynamic where each time point is a static state with established varieties, the identities of which are specified in terms of characteristic-signal dispositions that represent the *ontological information* of varieties and categorial varieties at any decision point of time. The set of ontological varieties is dynamic varying over different time points within the dynamic ontological space. The ontological space represents all that was, is and would be. It is the space that holds all that is to be known in time and over time It is the totality of the universe that is under the past-present-future (Sankofa-anoma) structure without a remainder. The ontological information is a property of matter that is extended to energy and all matter-energy derivatives such as processes, technology and others. The super-essential property of this conception is that the ontological information is perfect in the sense of completeness and exactness and is exactly equal to the *ontological knowledge* in the space of transformative ontological actions and activities of all forms. All information, defined as the characteristic disposition of varieties, is housed in the ontological library and filled under categorial shelves without information lost based on the Sankofa-anoma principle of past-present-future variety disequilibrium dynamics. The Sankofa-anoma principle simply affirms the combination of past-future information as input in present neuro-decision-choice actions.

The perfect ontological information and its equality with the ontological knowledge simply affirms that there are no uncertainties and risks in the ontological space, and hence *what there was*, will always be *what there was, what there is*, will always be *what there is*, and *what would be* will always be *what would be.* The epistemic implication in this respect is that the ontological space is an identity to itself and is defined in the space of absoluteness without comparative analytics, where conditions of diverse relationality are fixed by give-and-take modes of mutual existence in the space of construction-destruction dualities of transformations. There are no internal decoding and encoding mistakes, no disinformation, no misinformation, no fake news and no propaganda in the ontological messaging system in the space of ontological source–destination dualities. There is always perfect coding and sending from the source and perfect interpretation and response to codes in response to ontological dynamics as generated by ontological decision-choice actions. Each object is a variety with information for establishing its identity for universal identification or the category to which it belongs and the mechanics of its internal transformation. The technologies for static and dynamic behaviors are ontologically internal and self-creational. The ontological space is a *natural creation*, self-maintaining, self-destructing, self-exiting, self-evolving, self-creating and self-containing at all states and processes under natural input–output processes, where the inputs are the real costs in terms of transformation of *what there is* and the outputs are the real benefits in terms of transformation. The ontological real cost–benefit dualities are ontologically meaningless but are epistemologically meaningful under neuro-decision-choice system of actions and cognitive interpretations of the input–output results of natural processes. Let us see why by first defining the epistemological space which will analytically help to explain the relationships among the risk-non-risk dualities, cost–benefit dualities and certainty-uncertainty dualities of the cognitive worlds of operations under the dualistic-polar principle of the universal existence with continuum and unity in thought and practice.

3.5 The Epistemological Space, Cognitive Search and Connectivity to the Ontological Space

Let us turn our attention to the epistemological space, its relationship to the ontological space and its role in cognitive connectivity in terms of decision-choice actions, knowing and information-knowledge formation. The ontological space contains all the universal objects, states, technologies and processes as ontological varieties that can be experienced and known by all elements in the ontological space through the *space of acquaintance* with epistemic instruments acting on the ontological signal dispositions to decipher the codes in the ontological signal disposition and develop an individual or collective *epistemological information structure* from the ontological variety characteristic-signal dispositions. The ontological characteristic dispositions constitute the contents of ontological information structures which become inputs into the development of the epistemological information structure which is knowledge by acquaintance or experiential information which enters the epistemic space containing the epistemic variety characteristic dispositions as its contents. The quality of the knowledge by acquaintance of experiential information depends on a number of things such as cognitive capacity limitations, geomorphological limitations, geographical limitations and a number of things that create information-knowledge disparities to limit the interpretive view of the ontological signal dispositions, therefore creating a defective epistemological information structure, where this defectiveness can be reduced but not eliminated in the spaces of exactness-inexactness and certainty-uncertainty and doubt-surety dualities.

The epistemological information structure with its quality becomes an input into the development of languages, communication, methods of information processing and other relevant activities, the results of which become inputs into other epistemic actions such as neuro-decision-choice activities which tend to mimic the ontological variety's static and dynamic behaviors. All cognitive activities over the epistemological space find meaning as *problem-solving activities* with problem-generating activities and solution-generating activities which will require further problem–solution dualistic-polar actions. The epistemic processes, here, follow the conditions of the *first principles* in relation to variety, characteristic disposition, opposites and information.

From the first principle, the ontological space presents three fundamental problems in knowing that must be solved by cognitive agents with decision-choice actions in the epistemological space. The three fundamental problems are defined in the space of *identification problem–solution dualities*, the space of *transformation problem–solution dualities* and the space of *information-knowledge certainty problem–solution dualities*. The whole of the epistemological space is the union of the space of identification problem–solution dualities, the space of transformation problem–solution dualities and the space of *degrees of information-knowledge certainty* problem–solution dualities. All the actions to know by cognitive agents are about decision-choice actions over the space of variety identification problem–solution dualities and the space of transformation problem–solution dualities in terms of variety relations and variety places. At the center of the epistemological space is the set of cognitive agents with neuro-decision-choice activities to learn and mimic the ontological statics of *what there is* and the dynamics of ontological variety-transformation in the past and present, where the results of knowing and learning show themselves as *basic knowledge* and the results of mimicry show themselves as *applied knowledge* in terms of engineering sciences, prescriptive sciences, technological science, management science and other cognitive actions to organize and induce variety transformations through decision-choice activities over the space of problem–solution dualities relative to goals, objectives and visions. Let us first give a verbal definition of the epistemological space from which an algebraic definition will be offered to distinguish it from the ontological space and other spaces.

Definition 3.4.1: The Epistemological Space (Verbal) The epistemological space is a conceptual creation for the analysis and understanding of how cognitive agents relate to the elements in the ontological space and the varieties within it that have actual and potential variety dispositions in their existence and their socio-natural environment in terms of knowing, information-knowledge accumulation and utilization of the results of knowing as intellectual investment flows and the information-knowledge accumulation as intellectual capital stocks for the maintenance of survival in relation to continuity of the race of the cognitive agents and production of epistemological varieties in support of needs, wants and comfort. From the cognitive relational continuity with aspects of ontological information, the *experiential information* is developed by neuro-decision-choice actions through the interpretation of signal dispositions of ontological varieties to establish the epistemological information which then acts as an input into other forms of neuro-decision-choice actions. The epistemological space is composed of the *space of acquaintance*, the *space of problem–solution dualities* and *the space of decision-choice modules* where the space of acquaintance is the relational connectivity between ontological characteristic-signal dispositions and the cognitive agents providing inputs to develop the general epistemic space, the characteristic dispositions as misinformation, disinformation, propaganda or knowledge become integrated into the evolving social philosophical consciencism.

3.5.1 Symbolic Representation of the Epistemological Space

The epistemological space is the union of two pyramidal structures of acquaintance-variety information $(\mathbb{A} \otimes \mathbb{V} \otimes \mathbb{Z}_{\mathfrak{C}})$ as the *primary epistemological pyramid* and the pyramidal structure of decision-problem–solution-duality-uncertainty $(\mathfrak{D} \otimes \mathfrak{R} \otimes \mathbb{U})$ as the derived epistemological pyramid, where the epistemological information $(\mathbb{Z}_{\mathfrak{C}})$ is the derived characteristic-signal dispositions from ontological matter $(\mathbb{M})$ and energy $(\mathbb{E})$, and where the ontological matter and energy are decomposed into varieties $(\mathbb{V})$ of states, technologies and processes with their corresponding variety characteristic-signal decompositions, $\mathbb{Z}_{\mathfrak{C}} = (\mathbb{X}_{\mathfrak{C}} \otimes \mathbb{S}_{\mathfrak{C}})$ for identities and identification, and where the neuro-decision-choice space is governed by the structure of information-goal-constraint conditions defined in positive–negative dualities in the space of quality-quantity dualities which are revealed as cost–benefit characteristic dispositions, or danger-safety characteristic dispositions to cognitive agents. The *epistemological space* is simply a collection of actual and potential problem–solution dualities which is composed of identification-transformation problem–solution dualities in elements, states and processes of past, present and future relative to goals, objectives and visions under the principle of non-satiation as guided by social philosophical consciencism at generational points. The actual is made up of known and unknown problem–solution dualities, while the potential is made up of problem–solution dualities in the making as existing problems are solved to open emerging problems which are the actual in continual processes of change. The elements in the potential space are either goals, objective or visions defined in the space of imagination-reality dualities to which cognitive agents gravitate towards. The epistemic geometry of the structure of the definition of the epistemological space is shown in Fig. 3.3 the morphology of which may be compared to Sec. 3.2.1 of ontological system of pyramids.

The cognitive geometry of Fig. 3.3 requires some explanation. It is composed of the interrelation of two pyramids in one to define and establish the epistemological space for static and dynamic analysis in distinction from the ontological space relative to epistemology and ontology. The epistemological space is a collection of all the information-knowledge-decision-choice-interactive systems which is

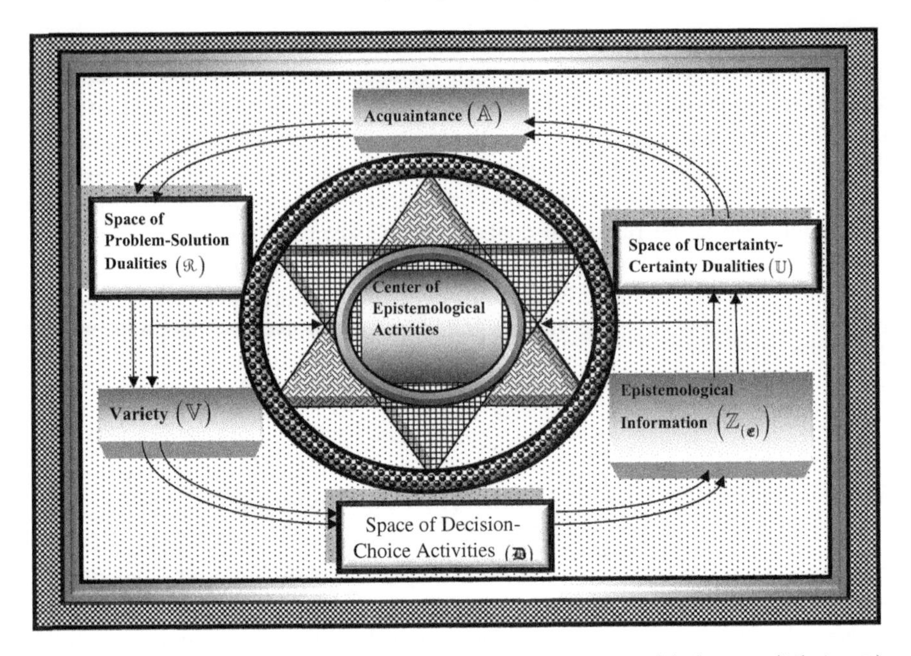

Fig. 3.3 An epistemic geometry of epistemological space where, $(\mathfrak{E}) = (\mathbb{A} \otimes \mathbb{V} \otimes \mathbb{Z}_{\mathfrak{E}}) \cup (\mathfrak{D} \otimes \mathfrak{R} \otimes \mathbb{U})$, $(\mathbb{A} \otimes \mathbb{V} \otimes \mathbb{Z}_{\mathfrak{E}})$ is the primary Epistemological pyramid and $(\mathfrak{D} \otimes \mathfrak{R} \otimes \mathbb{U})$ is the derived epistemological pyramid. It is a general framework in examining the definition of the Epistemological space and the corresponding sub-spaces that are linked to the ontological space

a union of the spaces of information-decision activities, problem–solution dualities, the space of uncertainty-certainty dualities, decidability-undecidability dualities, acceptance-rejection dualities, doubt-surety dualities and space of applied systems research, production, experiments and the general cognitive actions. The epistemological space is derived, self-maintaining, self-destructing, self-evolving, self-containing, self-exiting and self-transforming at all states and processes over the space of error-correction dualities. In this respect, the epistemological space is a cognitive representation of the ontological space under cognitive capacity limitations, the system of uncertainty-certainty dualities and risk-success dualities in decision-choice processes. Generally, therefore, the cognitive behaviors over the epistemological space are constrained by the automatic activities over the ontological space. It is through the decision-choice actions that cognitive agents become connected to ontological actions and develop awareness, understanding and knowledge through the interactive processes of problems and solutions. Furthermore, any state of social progress is an enveloping of the dynamics of neuro-decision-choice outcomes of problem–solution dualities over the epistemological space under the *asantrofi-anoma principle* of cost–benefit decision-choice processes and under the *funtummereku-denkyemmereku principle* (two-headed-crocodile in diversity and linked together by one stomach into relational continuity and unity with give-and-take sharing mode in unity and destruction) which is a dualistic-polar structure in relational continuum and unity with give-and-take sharing mode to move through the space of variety ignorance-knowledge (*onim-sua-ohu*) dualities. The asantrofi-anoma principle presents variety cost–benefit dualities in the same space violating the classical laws of thought. The *funtummereku-denkyemmereku principle* presents duality in relational continuum and unity, where the duals reside in each other creating a relational continuum and unity with give-and-take sharing modes of the external and internal mutual existence and destruction. The *onim-sua-ohu* principle presents the concept of the variety ignorance-knowledge duality where the duals are connected by "*sua*" learning to move the system from variety ignorance to variety knowledge through the reduction of the ignorance disposition and the increase in knowledge disposition.

It must be kept in mind that the *time trinity* presents an existence in three states of knowing for the definition of the ontological space with reference to the primary and derived pyramids which may be decomposed into three information-time dualities of *past-present duality*, *present-future duality* and *past-future duality* in knowing and information-knowledge stock-flow dynamics as inputs into neuro-decision-choice actions. In this way, one can speak of *ontological primary* and *ontological derived* pyramids in the past, present and future in terms of existence for knowing. The ontological space is the space of *ontological activities* in terms of *ontological decision-choice actions* which are decision-choices independent in identities but relationally interdependent and induced from the relational space of the system of varieties and transformations of ontological varieties as represented in Sec. 3.2.1.2. The primary category of existence establishes variety identities and initializes the conditions of the transformation processes for all derived existence over the space of ontology-epistemology dualities irrespective of the categorial classifications of the areas of knowing.

3.5.2 Algebraic Conceptions of the Epistemological Space in the Cognitive Search

We now turn our attention to the discussions on the conditions of the algebraic representations of the space and subspaces of the epistemological space. The idea is to distinguish it from the ontological space in terms of similarities, differences and elements of connectivity under neuro-decision-choice activities over the action space.

Definition 3.4.2.1: Epistemological Space (Algebraic) The epistemological space($\mathfrak{e} \in \mathfrak{E}$) is a collection of all the information-knowledge-decision-choice-interactive systems, which is a union of the spaces of information-decision activities ($\mathfrak{d} \in \mathfrak{D}$) and problem–solution dualities $\left(\rho \in \Re\right)$, where the space of decision-choice activities involves sequential interactions of the acquaintance space, information-knowledge processing space for *ignorant reduction* passing through the space of variety ignorance-knowledge dualities and information-knowledge-certainty space for *doubt reduction*. Algebraically, let ($\mathbb{A}$) represent the acquaintance space with a generic element ($\alpha \in \mathbb{A}$), ($\mathbb{V}$) the space of varieties with a generic element ($\upsilon \in \mathbb{V}$) , ($\mathfrak{P}$) the possibility space with a generic element ($\mathfrak{p} \in \mathfrak{P}$), ($\mathfrak{B}$) the probability space with a generic element ($\mathfrak{b} \in \mathfrak{B}$), ($\mathbb{Z}_{\Omega}$) the ontological information with a generic element ($\mathfrak{z}_{\omega} \in \mathbb{Z}_{\Omega}$), ($\mathbb{Z}_{\mathfrak{E}}$) the epistemological information with a generic element ($z_{\mathfrak{t}} \in \mathbb{Z}_{\mathfrak{E}}$), ($\mathfrak{K}$) the knowledge space with a generic element ($\mathfrak{k} \in \mathfrak{K}$), ($\mathbb{U}$) the space of certainty-uncertainty dualities with a generic element ($\upsilon \in \mathbb{U}$) and (Π) the space of paradigms of thought for decision-choice actions with a generic element ($\eta \in \Pi$) on information. Given these representations of spaces and subspaces,s the algebraic structure of the epistemological space is a Cartesian product of all the spaces of the form:

$$\left.\begin{aligned}
&\mathfrak{E}=(\mathbb{U}\cup\mathfrak{R})=\{\mathfrak{e}=(\upsilon,\rho)\mid\upsilon\in\mathbb{U},\rho\in\mathfrak{R}\},\\
&\text{where the space of certainty-uncertainty dualities is specified as}\\
&\mathbb{U}=(\mathbb{A}\otimes\mathbb{Z}_{\Omega}\otimes\mathfrak{P}\otimes\mathfrak{B}\otimes\mathfrak{D})=\{\upsilon=(\alpha,z_{\omega},\mathfrak{p},\mathfrak{b},\mathfrak{d})\mid\alpha\in\mathbb{A},z_{\omega}\in\mathbb{Z}_{\Omega},\mathfrak{p}\in\mathfrak{P},\mathfrak{b}\in\mathfrak{B},\mathfrak{d}\in\mathfrak{D}\}\\
&\mathbb{A}=(\mathbb{V}\otimes\mathbb{Z}_{\Omega}\otimes\Pi\otimes\mathfrak{D})=\{\alpha=(\nu,z_{\omega}\eta,\mathfrak{d})\mid\nu\in\mathbb{V},z_{\omega}\in\mathbb{Z}_{\Omega},\eta\in\Pi,\mathfrak{d}\in\mathfrak{D}\}\\
&\mathfrak{P}=(\mathbb{V}\otimes\mathbb{M}\otimes\mathfrak{D}\otimes\mathbb{Z}_{\mathfrak{e}})=\{\mathfrak{p}=(\nu,\mu,\mathfrak{d},z_{t})\mid\nu\in\mathbb{V},\mu\in\mathbb{M},\mathfrak{d}\in\mathfrak{D},z_{t}\in\mathbb{Z}_{\mathfrak{e}}\}\\
&\mathfrak{B}=(\mathbb{Z}_{\mathfrak{e}}\otimes\mathfrak{P}\otimes\mathfrak{D})=\{\mathfrak{b}=(z_{t},\mathfrak{p},\mathfrak{d})\mid z_{t}\in\mathbb{Z}_{\mathfrak{e}},\mathfrak{p}\in\mathfrak{P},\mathfrak{d}\in\mathfrak{D}\}\\
&\mathfrak{D}=(\mathbb{V}\otimes\Pi\otimes\mathbb{Z}_{\mathfrak{e}}\otimes\mathfrak{D}\otimes\mathfrak{K})=\{\mathfrak{d}=(\mathrm{v},\eta,z_{t},\mathfrak{d},\mathfrak{k})\mid\nu\in\mathbb{V},\eta\in\Pi\otimes\mathbb{Z}_{\mathfrak{e}},\mathfrak{d}\in\mathfrak{D},\mathfrak{k}\in\mathfrak{K}\}
\end{aligned}\right\}\quad(3.4.2.1)$$

Similarly, we have

$$\left.\begin{aligned}
&\mathfrak{E}=(\mathbb{U}\cup\mathfrak{R})=\{\mathfrak{e}=(\upsilon,\rho)\mid\upsilon\in\mathbb{U},\rho\in\mathfrak{R}\},\\
&\text{where the space of problem-solution dualities is specified as:}\\
&\mathfrak{R}=(\mathbb{Z}_{\mathfrak{e}}\otimes\mathbb{V}\otimes\Pi\otimes\mathfrak{D}\otimes\mathfrak{K})=\{\rho=(z_{t},\nu,\eta,\mathfrak{d},\mathfrak{k})\mid z_{t}\in\mathbb{Z}_{\mathfrak{e}},\nu\in\mathbb{V},\eta\in\Pi,\mathfrak{d}\in\mathfrak{D},\mathfrak{k}\in\mathfrak{K}\}\\
&\mathbb{Z}_{\mathfrak{e}}=(\mathbb{X}_{\mathfrak{e}}\otimes\mathbb{S}_{\mathfrak{e}})=\{z_{t}=(x_{t},s_{t})\mid x_{t}\in\mathbb{X}_{\mathfrak{e}},s_{t}\in\mathbb{S}_{\mathfrak{e}}\}\\
&\mathbb{Z}_{\Omega}=(\mathbb{X}_{\Omega}\otimes\mathbb{S}_{\Omega})=\{z_{\omega}=(x_{\omega},s_{\omega})\mid x_{\omega}\in\mathbb{X}_{\Omega},s_{\omega}\in\mathbb{S}_{\Omega}\}\\
&\mathfrak{K}=(\mathfrak{D}\otimes\mathbb{Z}_{\mathfrak{e}}\otimes\Pi)=\{\mathfrak{k}=(\mathfrak{d},z_{t},\eta)\mid\mathfrak{d}\in\mathfrak{D},z_{t}\in\mathbb{Z}_{\mathfrak{e}},\eta\in\Pi\}
\end{aligned}\right\}\quad(3.4.2.2)$$

We may unite the *space of certainty-uncertainty dualities* with the *space of problem–solution dualities* to define the structure of the epistemological space as:

$$\begin{aligned}
\mathfrak{E}=(\mathbb{U}\cup\mathfrak{R}) &= (\mathbb{A}\otimes\mathbb{Z}_{\Omega}\otimes\mathfrak{P}\otimes\mathfrak{B}\otimes\mathfrak{D})\cup(\mathbb{Z}_{\mathfrak{e}}\otimes\mathbb{V}\otimes\Pi\otimes\mathfrak{D}\otimes\mathfrak{K})\\
&\qquad\left\{\begin{matrix}\text{space of Certainty-Uncertainty}\\ \text{Dualities}\end{matrix}\right\}\cup\left\{\begin{matrix}\text{Space of problem-solution}\\ \text{Dualities}\end{matrix}\right\}
\end{aligned}\quad(3.4.2.3)$$

The space of the *actualized successful human decision-choice actions* is simply the the space of realities ($\mathfrak{H}$) which is a subspace of the space of imagination-reality (($\mathfrak{G} \cup \mathfrak{H}$) $\subset$ ($\mathfrak{A} \cup \mathfrak{U}$)) where ($\mathfrak{g} \in \mathfrak{G}$) is the space of imaginations, as well as the subspace of the space of the actuals as contained in the intersection of the ontological space and the epistemological space in the form:

$$\begin{aligned} \mathfrak{H} \subset \mathfrak{A} = \mathfrak{E} \cap \Omega = & \{(\mathbb{A} \otimes \mathbb{Z}_{\Omega} \otimes \mathfrak{P} \otimes \mathfrak{B} \otimes \mathfrak{D}) \cup (\mathbb{Z}_{\mathfrak{E}} \otimes \mathbb{V} \otimes \Pi \otimes \mathfrak{D} \otimes \mathfrak{K})\} \\ & \cap \{(\mathbb{V}_{\mathfrak{U}} \otimes \Phi_{\mathfrak{U}} \otimes \mathfrak{T}_{\mathfrak{U}} \otimes \mathbb{Z}_{\mathfrak{U}}) \cup (\mathbb{V}_{\mathfrak{A}} \otimes \Phi_{\mathfrak{A}} \otimes \mathfrak{T}_{\mathfrak{A}} \otimes \mathbb{Z}_{\mathfrak{A}})\} \end{aligned} \quad (3.4.2.4)$$

It must be noted that in dualistic-polar conditions $(\mathfrak{A} \cup \mathfrak{U}) = (\mathfrak{A} \otimes \mathfrak{U})$ and so with other spaces.

Note 3.4.2.1: On the Definition and Structure of the Epistemological Space The understanding of the decision-choice activities over the epistemological space is in terms of knowing, and in relation to *what there was* (past), *what there is* (present) and *what would be* (future), to indicate the past-present-future existential conditions in knowing into the beginning-end duality and the mysteries that it contains. Here, the understanding is induced by epistemic processes through the decision-choice analysis of the past-present-future existence which can only be revealed by the information-knowledge structure to abstract the epistemological space which will allow the understanding and knowledge about the ontological activities. The present existence reveals static conditions, while the past existence and the future existence show dynamic conditions relative to the present time. The past-present-future structure presents a *time trinity* in the *sankofa-anoma* tradition where the present resides in the past and the future resides in the present and thus future resides in in terms of information-knowledge processes about matter-energy varieties connecting information-knowledge inputs of forecasting and discounting into present neuro-decision-choice actions. The past relative to the present, and the future relative to the present show the *telescopic left* and *telescopic right* of past-present-future existence in the knowing process about socio-natural dualistic-polar structures. The left–right telescopic structure or the past-future telescopic structure around the present is shown as cognitive geometry in Fig. 3.4 It must be kept in mind that the time trinity presents existence in three interconnected stages of the unity of knowing for the definition of the epistemological space with reference to the pyramidal structure of the epistemological space. In the construct of the epistemological space, one can analytically speak of *epistemological primary* and *derived pyramids* presenting information on past, present and future in terms of variety existence. The epistemological space is the space of epistemological activities in terms of epistemological decision-choice actions and transformations to develop information-knowledge accumulation about ontological varieties as represented in Fig. 3.4.

Note 2.4.2.2: On the Epistemological Structure of Time Trinity There are some important observations about the time trinity or the Sankofa-anoma principle that are useful to explain in relation to contemporary theories on information-decision-choice interactive processes. Every decision-choice action at any current time requires information-knowledge input in relation to truth of telescopic past and in relation to telescopic future information-knowledge structure containing anticipations, surprises, expectations, visions, hopes and many other such varieties that reside in the space of false-truth dualities [468, 490–493, 497, 507]. An important question arises as to whether the inputs into neuro-decision-choice actions are either information, knowledge or both in the sense that information is not knowledge, but information

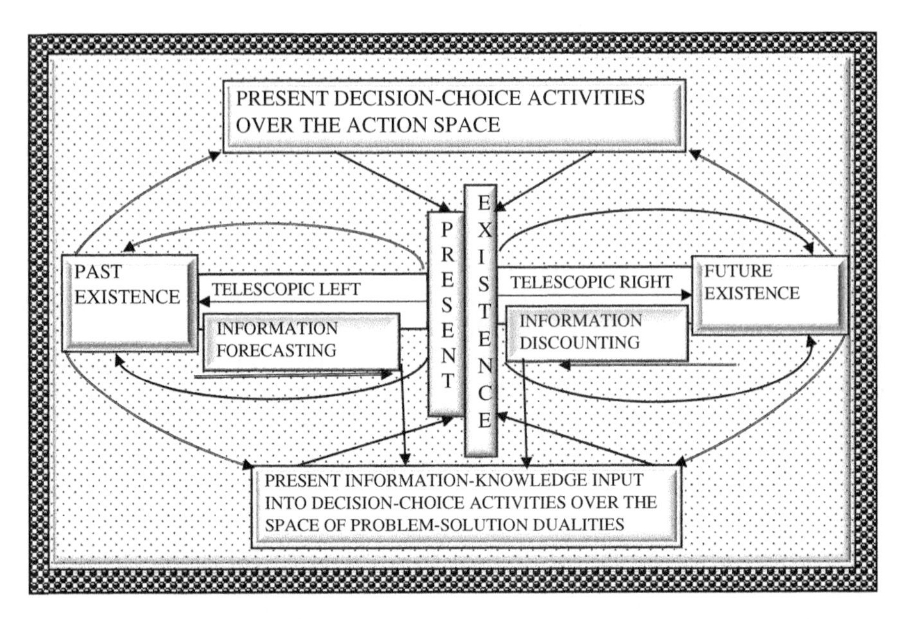

Fig. 3.4 The epistemological structure of the time trinity (Sankofa-Anoma) in relation to information-knowledge input in decision-choice activities of cognitive agents

derived. This important question has been discussed in the monographs [174, 176]. The information-knowledge inputs relate to the time trinity where the past is used to forecast the future to the present and future information outcome is used to discount the future information to the present, such that the current information-knowledge inputs into decision-choice actions retain the structure of the *asantrofi-anoma principle* of the time trinity in decision-choice actions to define the path to tomorrow destinations in order to retain the past-present-future information connectivity for neuro-decision-choice actions. A discussion on the verbal and algebraic definitions of the epistemological space, just like those of the ontological space will be helpful for the understanding of the general argument which is developed to establish the differences and similarities between the ontological and epistemological spaces in a manner that shows relational continua and unity of universal knowing in reference to cognitive ignorance, neuro-decision-choice uncertainties, doubts, risks and decidability associated with cognitive activities over the *space of problem–solution dualities* that will be useful for the development of the arguments in this monograph and how they connect with other ideas in my previous monographs [174, 177, 669, 670–672].

The epistemological space is the working space and workshop of the mind through the neuro-decision-choice processes. It is composed of a collection of problem–solution dualities with decision-choice outcomes defined over the space of certainty-uncertainty dualities relative to the ontological varieties of matter, energy and technology which generate the space of identification problem–solution dualities of *what there was* and *what there is* at static states as well as generate transformation problem–solution dualities of *what would be* in dynamic states and cognitive encounters over the acquaintance space. In a sense, the epistemological space involves activities of neuro-decision-choice actions by cognitive agents in dealing with the elements of space of the problem–solution dualities with information-knowledge stock-flow conditions, where such elements relate to ontological varieties in the success-failure-outcome space of decision-choice systems. Let us keep in mind that the cognitive agents are also elements in the ontological variety space, where the main activities of cognitive agents are decision-choice activities in the space of problem–solution dualities with cognitive capacity limitations in terms of behavior over the spaces of uncertainty-certainty dualities, doubt-surety dualities, decidability-undecidability dualities and acceptance-rejection dualities.

Just like every element in the ontological space acquires its identity and variety from the ontological information, so also every element in the epistemological space acquires its identity and variety from the epistemological information. The epistemological information is the creation of the decision-choice process from the acquaintance space that contains the ontological characteristic-signal dispositions of the matter-energy varieties, where the characteristic dispositions reveal the contents and the signal dispositions carry the massages of the ontological varieties under knowing and decision-choice activities over the space of uncertainty-certainty dualities which are generated by cognitive capacity limitations regarding vagueness and volume incompleteness. The contents of the epistemological information are the variety characteristic dispositions and are abstracted from the ontological signal disposition. The epistemological information from the ontological signal dispositions of the ontological varieties becomes the input into the *paradigmatic process* with outcomes of dynamic information-knowledge structures under a *fuzzy-stochastic conditionality* over the space of certainty-uncertainty dualities, where the information-knowledge structures of the varieties become inputs into the evolving decision-choice processes over the space of problem–solution dualities for the individual and social transformations of varieties.

The epistemological information is a derivative from the ontological information through the ontological signal disposition such that the epistemological characteristic dispositions may and usually do vary from the ontological characteristic dispositions of the varieties. The contents of epistemological information are specified in the space of the epistemological characteristic dispositions where these characteristic dispositions are abstracted from the ontological signal dispositions as encountered over the acquaintance space to reveal the epistemological characteristic-signal dispositions of varieties. In this respect, we have two characteristic dispositions which are the ontological characteristic disposition and the epistemological characteristic disposition for any variety. The cognitive agents must assess the closeness of the epistemological disposition to the ontological disposition for any variety of interest. We shall refer to the differences between the ontological characteristic disposition and the epistemological characteristic disposition, with some measure, as the variety *epistemic distance*.

The epistemological information is always defective and is never equal to the ontological information in the space of transformative decision-choice actions over the space of ignorance-knowledge dualities. The analytical explanation in this respect is that the epistemological space is a defective derivative from the ontological space, and it is only by chance that there will be an equality between the epistemological characteristic disposition and the ontological characteristic disposition irrespective of how they are measured. There are continual external and internal decoding and encoding of mistakes in the epistemological messaging system due to complicated cognitive capacity limitations with the space of acquaintance. The size of the space of acquaintance for individual and social collectivity is always limited by the space, means and technology of acquaintance. There are always imperfect interpretations called vagueness and poor, and inaccuracies with volume incompleteness relative to codes in response to epistemological dynamics as generated by epistemological decision-choice actions. Each epistemological object is an *epistemological variety* with information abstracted from the acquaintance space for establishing its *ontological identity* and universal identification or the ontological category to which it belongs as well as the understanding of the mechanics of its internal transformation in terms of relation, property or both. The neuro-decision-choice systems are the *epistemological technologies* for static and dynamic behaviors which are internal and self-creational. The epistemological space is self-correcting, self-exiting, self-maintaining, self-evolving and self-containing at all states and processes under the management and control of individual and collective decision-choice activities in the spaces of certainty-uncertainty dualities, doubt-surety dualities, decidability-undecidability dualities and acceptance-rejection dualities.

The algebraic specification of the epistemological space ($\mathfrak{e} \in \mathfrak{E}$) must be separated from the specification of the ontological space ($\omega \in \Omega$) and yet must be connected for relational continuum and unity for socio-natural existence and continual universal knowing through neuro-decision-choice activities over the space of problem–solution dualities. The activities of variety knowing are cognitive actions over the space of problem–solution dualities with abstracted epistemological information from the *space of acquaintance* ($\alpha \in \mathbb{A}$) passing through *the possibility space* ($\mathfrak{p} \in \mathfrak{P}$) which reflects necessities of neuro-decision-choice actions, to the *probability space*($\mathrm{b} \in \mathfrak{B}$) which reflects freedoms of neuro-decision-choice actions through the *space of uncertainty-certainty dualities* ($\upsilon \in \mathbb{U}$) due to cognitive capacity limitations of cognitive agents as linked to the spaces of doubt-surety, decidability-undesirability and acceptance-rejection dualities.

It must be epistemically understood that there is an inseparable link between possibility-probability dualities and necessity-freedom dualities in all cognitive individual and collective actions and neuro-decision-choice actions, the outcomes of which define the path of generational and intergeneration social history under the dynamics of intergenerational preferences. The epistemological space as a derived space is made up of a collection of derived ontological varieties of matter-energy structures under the space of neuro-decision-choice modules which induce processes for socio-natural variety knowing with relevant inputs as well as variety transformations in the social space relative to the natural space. It is, thus, the space of cognitive activities to know the ontological past, present and future and the translation of this knowing to neuro-decision-choice actions over the space of problem–solution dualities to set variety solutions against variety problems where variety solutions generate new variety problems in the parent–offspring disequilibrium dynamics connecting the epistemological space to the ontological space and the epistemological space to itself in the production and reproduction of varieties.

The epistemological space is also the space to know and mimic the *natural technologies* of transformations in the development of *engineering sciences* as well as *prescriptive sciences* and use them for social organization and social institutional transformations for social developments, where social developments are neuro-decision-choice transformations with differential velocities and accelerations over time points. Every social system at each point of time is, thus, a variety with a defined characteristic disposition that may be placed in a socio-economic category. In passing, it may be remarked that this is the best approach for the development of *the theory of economic development* and the theory of social change that consider neuro-decision-choice actions over the space of problem–solution dualities in the transformative process and categorial conversion within the space of quantity-quality dualities [668–672, 708, 778].

Any serious theory of economic development must indicate the structures of qualitative and quantitative motions for its transformations based on decision-choice activities and the corresponding information-knowledge structures over the space of problem–solution dualities. Every knowing and every cognitive action involving *variety identification* and *variety transformation* is neuro-decision-choice action in the space of problem–solution dualities. Transformations at the social space are conceptualized over the space of imaginations where the imagination varieties create problems that must be solved to bring the imagination varieties into the space of reality. It is here that the space of imagination-reality dualities plays determining role in the understanding of socioeconomic transformations and development. It is also here that the development of qualitative and quantitative equations of motion help to explain the directions of socioeconomic transformations and possible improvements in social improvements. The union of the space of neuro-decision-choice activities and the space of problem–solution dualities is the *action space*, ($\mathfrak{A}$). The question that arises is what is the morphology of the space of problem–solution dualities? The answer will be discussed in the next chapter. Let us now examine the relationship between the space of varieties and the principle of opposites.

3.6 The Principle of Opposites and the Variety Space

It is important to notice the complexity of the concept of the variety and the further complexity in relation to the *principle of opposites*, paradigms of thought, and principles of diversity and unity in knowing, science and neuro-decision-choice actions. The principle of opposites is central to all conflicts, decision-choice actions, internal and external transformation and generation of socio-natural forces. The understanding of the principle of opposite provides a framework to analyze conditions of variety problems, variety solutions and their relationships. The principle of opposites is composed of interdependent sub-principles of duality in relational continuum and unity, and sub-principles of polarity also in relational continuum and unity, where the duality and polarity are in relational continuum and unity with each other. The continuum and unity are defined by give-and-take mode in terms of existence over the space of construction-destruction dualities. Every variety in the universal system exists as a multiplicity of dualistic-polar objects. Every element in any given space is a variety such that the universe is a complete collection of varieties under identification and transformation, and hence equals to the *space of varieties*, which is an infinite collection of actual and potential varieties. The multiplicity of dualities and polarities includes *negative–positive dualities* and *polarities* in deferential relative combinations at the level of the ontological space. At the level of epistemological space where cognitive agents operate under neuro-decision-choice actions with preferences, each variety does not only appear as negative–positive duality and polarity, but the negative–positive duality and polarity are transformed into *cost–benefit dualities* and *polarities* which are then transformed into *problem–solution dualities* which in turn become identification-transformation dualities all in a complex system of dualities and polarities for decision-choice activities over the epistemological space in relation to the information-knowledge disequilibrium dynamics. The basic structure is shown as an epistemic geometry in Fig. 3.5.

The relational structures and interactive conditions of the principle of opposites and the concepts of duality, polarity and decision-choice actions over both the ontological space and the epistemological space will be extensively discussed in the chapter dealing with the theory of problem–solution dualities with a generous use of fuzzy information and paradigm of thought in the space of relativity. The roles that the concepts of duality and polarity play in understanding the variety identification problem–solution dualities have been discussed in the theory of info-statics [671] while the variety transformation problem–solution dualities have been discussed in the theory of info-dynamics [672]. It must be kept in mind that the theory of info-statics involves all the static conditions of qualitative and quantitative dispositions of varieties and categorial varieties, while the theory of info-dynamics involves the transformation conditions of qualitative and quantitative dispositions of varieties and categorial varieties in all areas of knowing including science and non-science. The presence of the space of quality-quantity dualities implies that variety transformations must deal with qualitative motion as well as quantitative motion in simultaneous processes involving changes in relations and changes in properties. Over the epistemological space, the knowing system of variety identifications and transformations are viewed in terms of neuro-decision-choice activities on the elements in the spaces of problem–solution dualities and uncertainty-certainty dualities. A general theory of entropy is then developed for dealing with the conditions within the space of doubt-surety dualities associated with cognitive limitations generated by information defectiveness and paradigmatic inefficiencies relative to the space of decidability-undesirability dualities.

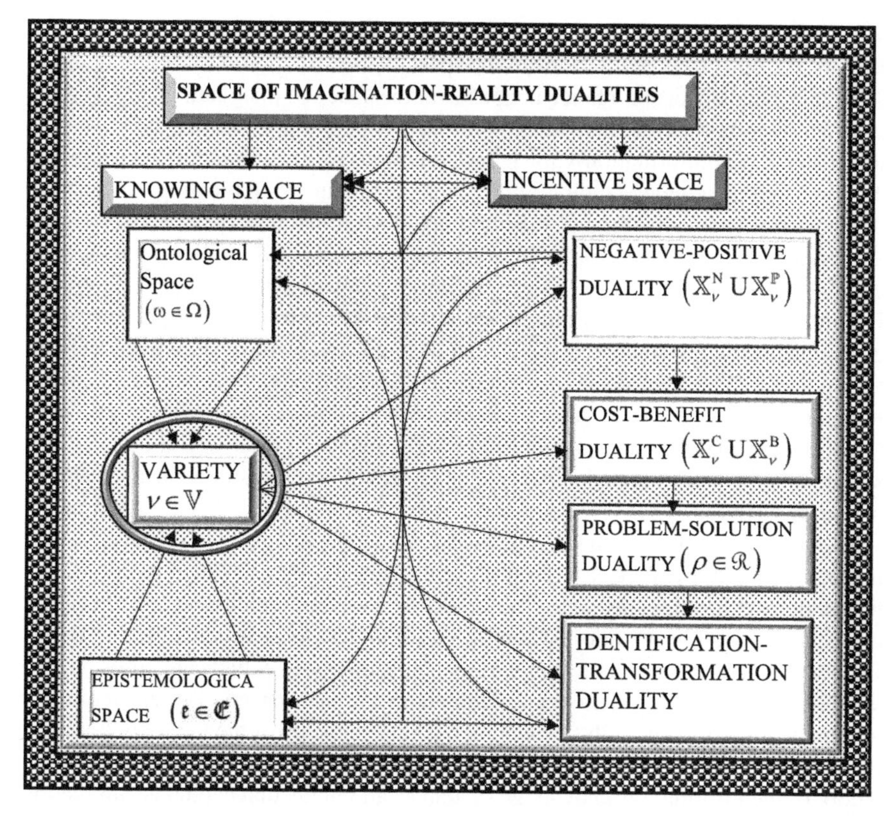

Fig. 3.5 An epistemic geometry of the relationships among the principle of opposites, ontological space, epistemological space and the space of varieties

3.6.1 The Principle of Opposites and Neuro-Decision-Choice Activities

The neuro-decision-choice activities include information-knowledge constructs and institutional creations for policy development and transmissions, as well as engineering of all forms including invention of new forms, innovations in some existing forms and destruction of some old forms. The social interest in STEM (science, technology, engineering and mathematics) is driven by social variety creations of, and from the understanding of the fundamental information-knowledge conditions of natural forces for the transformations of varieties to meet the dynamics of individual and social preferences for varieties. It is the understanding of the information-knowledge conditions of the natural forces that ensures the continual efficient understanding of the inter-supportive diversities, the unity of knowing and the efficient socioeconomic transformations of societies in intergenerational dynamics as new ideas are generated by the space of information-knowledge disequilibrium dynamics destroying some old elements and creating new elements in the social philosophical consciencism to restructure beliefs, hopes, preferences, creativities, goals, objectives and visions in the space of construction-destruction dualities relative to neuro-decision-choice actions over the space of imagination-reality dualities as the subspace of the space of the actual-potential dualities. It is also the understanding of information-knowledge conditions of social forces that ensures the continual efficient understanding of the diversity and unity of knowing, science and the indivisible understanding of the interplay of socio-natural forces of explanatory and prescriptive processes of the information-knowledge disequilibrium of knowledge stock-flow conditions, paradoxes, ill-posed problems and paradigms of thought. It is useful to note that the pyramidal system of the time trinity (past-present-future conditions) divides into a system of dualities and polarities of present-past dualities (polarities), present-future dualities(polarities) and past-future dualities (polarities) and corresponding to them there are (what-there-was)-(what-there-was-not), (what-there-is)-(what-there-is-not) and (what-would-be)-(what-would-be-not). The system of information-time dualities and polarities may be illustrated in an epistemic geometry as in Fig. 3.6.

Figure 3.6 presents a complex system of dualities and polarities as they relate to the understanding of the stock-flow disequilibrium dynamics of information-knowledge production with rationality under neuro-decision-choice processes over the space of problem–solution dualities, where information-knowledge items are the inputs into the neuro-decision-choice processes, the outputs of which are information-knowledge items which then become inputs in the next round. The epistemology within the investigation of this monograph is thus made up of the theory of knowing as the intellectual investment under the disequilibrium of flow conditions and the theory of the information-knowledge system as an intellectual capital accumulation under the disequilibrium of stock conditions. The theory of knowing and the theory of information-knowledge system constitute the theory of intellectual investment-capital production under the space of identification-transformation dualities as one moves over the space of ignorance-knowledge dualities embedded in the space of imagination-reality dualities which is a union of the spaces of imagination varieties and the space of reality varieties in the dualistic-polar structures. The space of imagination-reality dualities is a subspace of the space actual-potential dualities as a union of the space of the actuals and the space of the potential where the varieties reside in dualistic-polar structures with fuzzy relational conditionality and closure.

The space of incentive-disincentive dualities in support of the structure of neuro-decision-choice actions is embedded in the space of real cost–benefit dualities where the elements in the space are broadly defined under dualistic-polar conditions and fuzzy relational conditionality. The identification-transformation dualities are connected to the time trinity, the fuzzy paradigm of thought and the path connectivity of relationships between the duals of dualities, between the poles of poles and between the dualities and polarities in the space neuro-decision-choice actions where the diversity, equity, and inclusion (DEI) are integrated in the designs of institutions and organization of institutions into oneness to create a social environment that respects and values individual differences, commonness with give-and-take sharing modes along varying dimensions within the space of cost–benefit dualities in relation

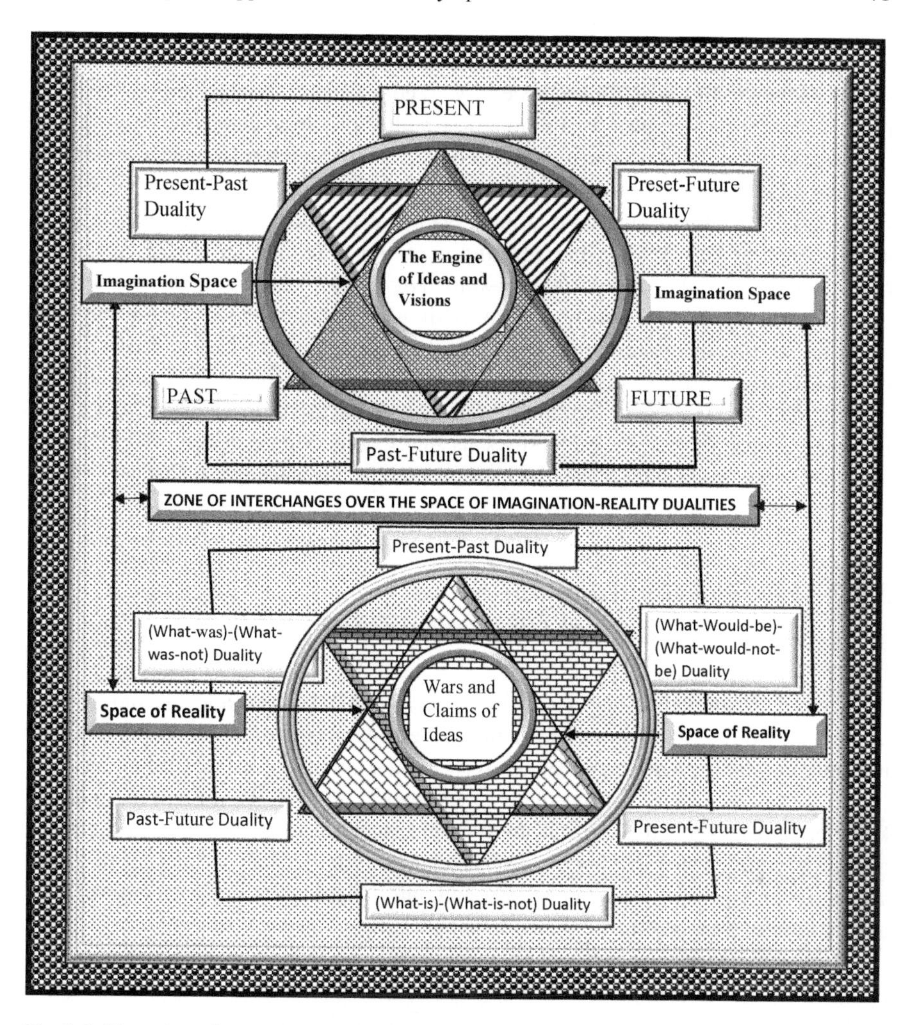

Fig. 3.6 The epistemic geometrical interaction and inter-change of information-time dualities and polarities in the space of imagination-reality dualities as a subspace of the space of actual-potential dualities that defines the space of ontological-epistemological dualities

to information-knowledge disequilibrium dynamics and the principle of individual, collective and generational non-satiation as encapsulated in the social philosophical consciencism that presents an evolving freedom as well as social categorial conversion that presents an evolving necessity for change with intrinsic forces, where the necessity points to external conditions for change and the freedom points to internal conditions to change.

The Fig. 3.6 gives us a complex system of epistemic journeys from the space of imagination to the space of reality that allows us to undertake path analytics for neuro-decision-choice actions, where these path analytics are in relation to the elements of the space of cost–benefit dualities, where every variety resides in the spaces of possibility-probability, necessity-freedom and certainty-uncertainty dualities contained in the space of imagination-reality dualities under the principle of information-knowledge disparity by transforming imagination to reality, where the space of imagination-knowledge or imagination-reality dualities is a proper subspace of the space of actual-potential dualities. Nature works directly within the space of actual-potential dualities by transforming potential to reality, and reality to potential under the priciple of *information-knowledge equality*. Cognitive agents work indirectly through the principle of information-knowledge inequalities through all activities weithin the imagination-reality dualities. The defining characteristic dispositions of variety imagination and variety reality have been specified and will be further explained in this monograph in relation to knowing as intellectual investment flow and information-knowledge stock as intellectual capital accumulation where the epistemological space is creatively connected to the ontological space.

3.6.2 Connectivity Within the Space of Epistemological-Ontological Dualities

The creative connectivity between the epistemological space and the ontological space leads to the discussions of the needed epistemic instruments on the paths of multiplicity of ontic-epistemic connectivity in the knowing processes. To understand the scientific foundation of the theory of knowing and the theory of information-knowledge stock-flow disequilibrium dynamic as an economic production central to all human actions, we must understand the connectivity between the epistemological space in which cognitive agents operate and ontological space in which nature operates and their relationship to neuro-decision-choice actions of cognitive agents on what are naturally dependent and socially non-controllable and what are naturally independent and socially controllable on human behavior, and what are independent and incontrollable on human behavior. The analytical connectivity requires a supporting geometric structure to Sec. 3.5.2 that shows the separation and unity. The principle of opposites is such that the potential space, the space of the actual and the space of actual-potential dualities are the domain of nature and independent of cognitive agents; however the space of imagination, the space of reality and the space of imagination-reality dualities are the domain of cognitive agents and constrained by natural forces over the space of actual-potential dualities and the cognitive capacity limitations of cognitive agents due to information-knowledge disparities in variety knowing and transformations. These conditions may be presented as a paremiological geometry in Fig. 3.7.

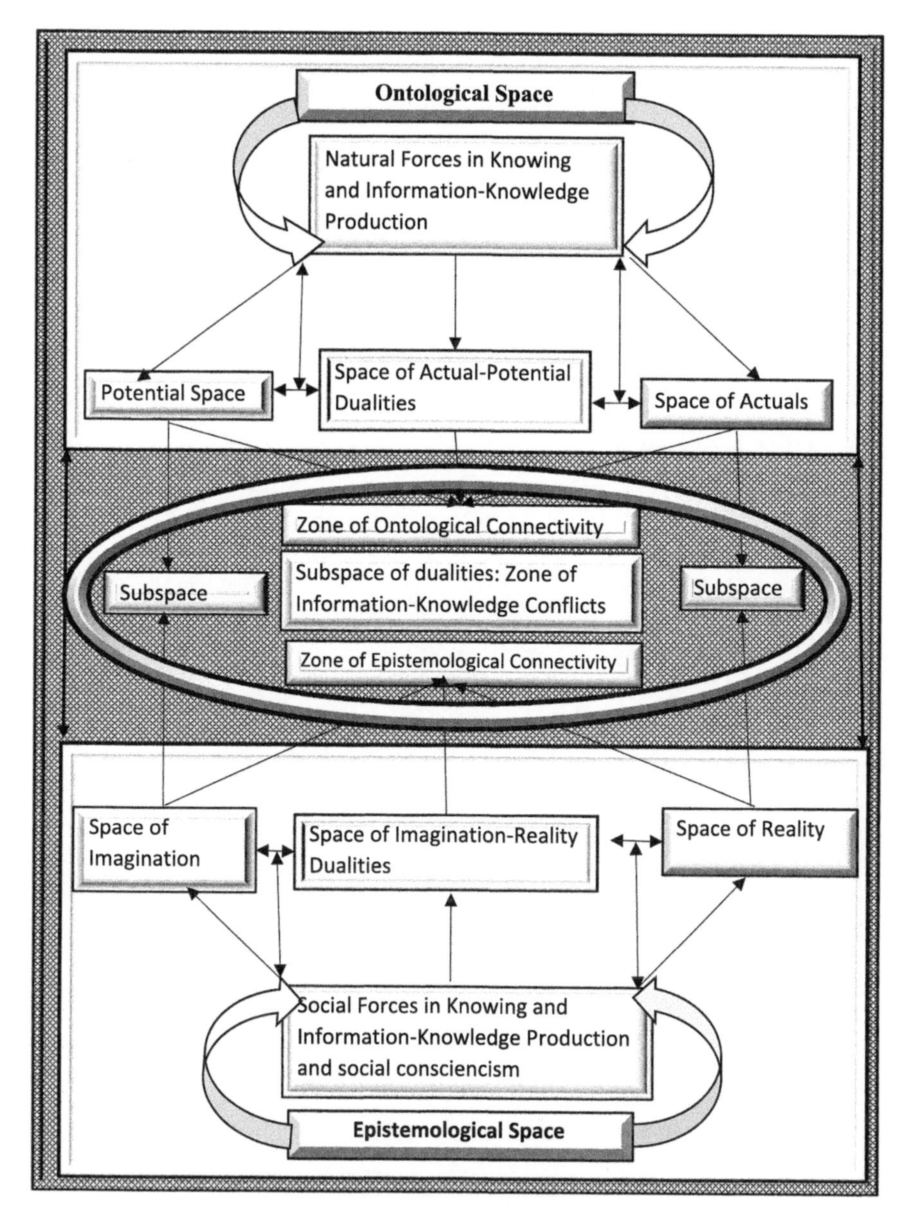

Fig. 3.7 Paremiological geometry of connectivity of actual-potential space to imagination-reality space over the space of ontological-epistemological dualities

Chapter 4
The Theory of Relational Problem-Solution Dualities, The Unity of Knowing, The Unity of Science and the Diversity in Knowledge

Abstract This chapter is used to present the theory of problem-solution dualities as the first part of the general theory of opposites where the second part is the theory of problem-solution dualities which will be taken up in the next chapter. It discusses the conditions of the nature of problems relative to solutions in dualistic-polar structures with relational continua, unity and give-and-take-sharing modes in the space of fundamental-applies dualities to provide conditional connectivity to the development of the theory of polarities. The concepts of ontological and epistemological spaces are introduced, defined, explicated and connected to the space of problem-solution dualities. and the connectivity of these concepts to the roles they play in the knowing process as intellectual investment flows and the production of information-knowledge system as intellectual capital accumulation stocks is then explained. The space of incentive-disincentive dualities is introduced and linked to the space of cost-benefit dualities in order to establish the concepts of primary-derived spaces and parent-offspring processes relative to the elements in the space of problem-solution dualities as transformational elements where problems are transformed into solutions to give rise to new problems over the multiplicity of cost-benefit or destruction-construction enveloping that establishes the space of problem-solution genetic paths of socio-natural variety dynamics. The space of problem-solution dualities are established over the space of imagination-reality dualities which is shown as a sub-space of actual-potential dualities that constitutes the space of ontological-epistemological dualities.

4.1 Introductory Reflections on Duality

All human actions are about neuro-decision-choice activities directed to a continual direct and indirect searching for solutions to socio-natural problems that constrain human conceptions of progress over the space of cost-benefit dualities. The results of the interactive problem-solution structures provide cognitive agents with awareness of socio-natural varieties environment. The direct and indirect awareness about the socio-natural varieties find meanings as knowing and information-knowledge

K. K. Dompere, *The Theory of Problem-Solution Dualities and Polarities*,
Studies in Systems, Decision and Control 405,
https://doi.org/10.1007/978-3-030-90279-7_4

intellectual stock-flow systems that constitute the elements in the space of the input-output dualities which is then connected to the space of decision-choice processes for knowing and information-knowledge systems that constitute the intellectual stock-flow disequilibrium dynamics. The various combinations of the variety problems and the variety solutions provide us with important channels of knowing and the development of information-knowledge about socio-natural general existence of varieties in the past, present and future over the space of ontological-epistemological dualities from which the ontological and epistemological spaces are conceptually defined and explicated for examining neuro-decision-choice actions over the space of problem-solution dualities. The results of neuro-decision-choice actions are outputs of knowing and information-knowledge items which then become inputs into the neuro-decision-choice space for further knowing and information-knowledge accumulation. The conditions of the space of variety problem-solution dualities as the primary factors in establishing the unity of knowing and the unity of science as seen in the spaces of ontological and epistemological varieties, where knowing is neuro-decision-choice actions to reconcile the relational conflicts between the set of characteristic dispositions of ontological varieties and the set of epistemic dispositions of epistemological varieties in the space of ignorance-knowledge dualities with relational continua and unity to set the knowledge duals against the ignorance duals to increase the intellectual investment flows and to update the information-capital accumulations as intellectual capital stocks. The study of the theory of knowing as intellectual investment flows and the theory of knowledge as accumulations of intellectual capital stocks in the primary-derived process is the study of neuro-decision-choice theory that unite all areas of human actions and the institutions and departments of sciences and non-sciences. Knowing is about uncovering of variety characteristics to reduce the degrees of variety ignorance while increasing the degree of knowledge. In this way, the variety identity is revealed by asymptotic processes of cognitive activities irrespective of the nature of the variety or the category of its belonging. Every neuro-decision-choice action is a mediation between opposite duals of a duality and between poles of a polarity, where the dualistic-polar process leads to setting one dual against the opposite dual of a duality or setting one pole against the opposite pole of a polarity in the space of internal-external dualities.

The relationships among concepts of problem, solution, question and answer relative to neuro-decision-choice actions on info-statics and info-dynamics in information-knowledge systems are related to the principle of opposites with the use of the fuzzy paradigm of thought where comparative analytics are undertaken with the conditions of the classical paradigm of thought. The concepts of epistemological and ontological spaces, as have been defined and explicated in the previous chapter, are utilized to create varieties of organic existence with relational connectivity within the spaces for the cognitive activities of knowing as intellectual investment flows and information-knowledge development as intellectual capital stocks to bring into focus the role that neuro-decision-choice activities play on the elements in the space of question-answer dualities involving our socio-natural universal existence and cognitive preferences leading to the emergence of the space of problem-solution dualities at both the static and dynamic states under the real cost-benefit principles of relativity and rationality.

This real cost-benefit principle is the *asantrofi-anoma* principle and rationality within the adinkra-symbolic traditions for the existence and the analytics of the opposites in the same space to understand intra-variety and inter-variety conflicts and tensions in dualistic-polar games that generate the never-ending dimensions of problem-solution (question-answer) dynamics, where solutions (answers) to problems (questions) generate new problems (questions) requiring a new solutions (answers) in continual dynamics supporting the general principle of unbreakable chain of intergenerational neuro-decision-choice actions. The conditions of the elements in the space of question-answer dualities as mapped onto the space of problem-solution dualities induce some interesting questions requiring answers relating to the universal beginning-end duality, present-future dualities, nothingness-somethingness duality, God-non-God duality, creation-non-creation duality, zero-infinity duality, primary-derived duality, existence-nonexistence duality of other intelligent being, materialism-spiritualism dualities with corresponding dualities and many dualistic-polar structures within the space of ontological-epistemological dualities.

The conditions of universal finite-infinite duality with all spaces of question-answer dualities and problem-solution dualities are the creations of cognitive agents operating over the epistemological space and searching for the understanding of the conditions on questions and answers surrounding their being and perplexing conditions of their problems and solutions over the space of cost-benefit dualities that translates into many conditions over the space of want-unwanted dualities of their past-present-future (*Sankofa-anoma*) time trinity of existence-nonexistence dualistic-polar structures over the ontological space. All conditions revealing questions, answers, problems, solutions and states of troubles find expressions in information and knowledge as they are presented in the space of variety characteristic dispositions. From the logic of the theory of economic production the intellectual investment and intellectual capital accumulation are placed in the space of input-output dualities and also over the space of supply-demand dualities relative to the input-output chain processes and destruction-construction enveloping paths under the epistemic conditions of primary-derived dualities with parent-offspring dualistic-polar structures, where every variety is both a parent and an offspring within the space of matter-energy equilibrium over the space of primary-derived dualities, where the primary varieties establish initial conditions of identities while the derived varieties establish conditions of the space of reality without revealing what varieties may be actualized from the space of imagination into the space of reality.

Alternatively perceived, the primary duality reveals the conditions of information-knowledge accumulation as intellectual stocks in the space of past-present dualities with variety identities, while the knowing reveals the path of intellectual investment flows without revealing what varieties may be known as updating of the intellectual capital stocks in the space of present-future dualities in the sense of revealing the intellectual light of the present but not the varieties in the past-future dualities under the universal sankofa-anoma conditions, where the present is known, while the cognitive agents work on telescopic-past knowing process to construct the information inputs for the current neuro-decision-choice inputs in the neuro-decision-choice actions on the telescopic-future process over the space of risk-benefit dualities with unknown varieties except the conditions of curiosity, courage, hope and perseverance over the space of imagination-reality dualities.

The analytical effort, here, is to establish an unbreakable relationship among neuro-decision-choice actions, problem-solution dualities, and question-answer dualities that helps to explain the important roles played by diversity and unity, oneness and dividedness, organization and disorganization in human conditions under the preferences of non-satiation as constrained by cognitive capacity limitations of different forms. The conditions of non-satiation reside in the space of imaginations as a subspace of the space of potential, where human actions are to transform the variety imagination into the space of realities as a subspace of the actuals through the space of cognition and practices as induced by neuro-decision-choice actions, where the transformations from the space of imagination to the space of the actuals are neuro-decision-choice actions to optimize the conditions of non-satiation constrained by the conditions of ignorance-knowledge dualities and polarities. The important focus of an epistemic understanding is that decision-choice theories as we know them do not preset frameworks that show journeys from the space of imaginations to the space of realities constrained by the conditions of ignorance-knowledge dualities. The whole human neuro-decision-choice action is the maximization of benefits subject to costs or minimization of costs subject to benefits as observed by Euler that "nothing happens in the universe that does not have a sense of either certain maximum or minimum". Every neuro-decision-choice problem either analytically or non-analytically presented is a reconciliation of a conflict in a dualistic-polar structure where the duals and the poles contain and constrain each other over the spaces of ignorance-knowledge, certainty-uncertainty, decidability-undecidability, doubt-surety dualities in arriving at information-knowledge output.

4.2 Primary and Derived Spaces and Information Structures

The ontological space is the *primary identity* and the primary category of elements in the space of the unknown-known dualities. These elements are characterized as ontological objects or varieties that represent things, states and processes around which ideas are formed to develop structures in the space of ignorance-knowledge dualistic-polar conditions. The knowing always takes place at the epistemological space. The epistemological space is the *derived identity* and the derived category of elements subject to knowing activities through neuro-decision-choice actions. These elements are characterized as epistemological objects or varieties that represent things, states and processes, the information-knowledge structures of which are to be verified against the characteristic dispositions of ontological elements. In this respect, there are always the epistemological varieties and corresponding to them are the epistemological characteristic dispositions, abstracted in the space of acquaintances, that must be compared with the corresponding ontological varieties, where each ontological variety is an identity to an epistemological variety as a derived identity. The knowing process is such that the ontological variety characteristic dispositions over the space of epistemic decidability-undecidability dualities involve the comparative analysis of ontological and epistemological variety characteristic dispositions regarding the values of their epistemic distances. The test of the degree of variety *epistemic distance* is undertaken with fuzzy-stochastic entropy through methodological reductionism in the space of constructionism-reductionism dualities. The epistemic distance is generated by information-knowledge inequality from the structure of cognitive capacity limitations of cognitive agents. In all the knowing processes, the ontological characteristic dispositions for establishing the ontological variety identities are complete and perfect with information-knowledge equality, while the epistemological characteristic dispositions for establishing the epistemological variety identities are vague, incomplete and imperfect due to cognitive capacity limitations of the knowing agents over the spaces of acquaintances and paradigms of thought. The imperfections of the conditions of the space of acquaintances and paradigms of thought create a system of disparities and conflicts in the space of ignorance-knowledge dualities leading to a system of *variety epistemic distances,* where the claims of knowledge items are qualified by *fuzzy-stochastic conditionality* with *fuzzy-stochastic entropy* as the quality-control modules [832].

The differences between ontological and epistemological information-knowledge structures are the *variety epistemic distances* which represent information-knowledge space of certainty-uncertainty dualities in the domain of quantitative and qualitative existence. Similarly, the concepts of epistemology and ontology have been defined as areas of study. The concepts of ontology and ontological space where defined and explicated in the previous chapter to point out the need for different analytical tools in the epistemic process, where the ontological space houses the natural decision-choice processes to induce natural transformations of natural varieties. Similarly, the concepts of epistemology and epistemological space were defined and explicated in the previous chapter to produce analytical tools useful for the understanding of the process of knowing and information-knowledge accumulation, where the epistemological space houses the social neuro-decision-choice processes to induce continual social variety transformations of epistemological varieties. The effects of the outcomes of ontological and epistemological decisions affect the neuro-decision-choice behaviors of cognitive agents as they respond to ontological and epistemological outcomes over the epistemological space with continual activities over the space of imagination-reality dualities. It will become clear that it is the cognitive behaviors over the space of *effect-response dualities* that sustain the space of problem-solution dualities where any response generates effects which then generate a corresponding response in the effect-response disequilibrium dynamics. Ontology and epistemology are areas study while the ontological space and epistemological space are areas of actions connected by neuro-decision-choice actions over the space of logical constructionism-reductionism dualities under the fuzzy-stochastic entropy and rationality as qualifications of cognitive capacity limitations over the space of acceptance-rejection dualities.

All these definitions have multiplicities of relationally individual and collective connectivity and unity that establish give-and-take sharing interdependencies for communications and understanding over the space of source-destination dualities. It has also been stated that the defining attributes of the unity of science, through the unity of knowing over the epistemological space, find meanings and logical tools over the *space of problem-solution dualities* $(\mathfrak{R})$ as they relates to identification and transformation relations of existence and non-existence of socio-natural varieties. In fact, the meaning and the essence of life at any time point are made up of nothing but neuro-decision-choice activities over the *space of problem-solution dualities* in the search of the meanings of elements in the *space of question-answer dualities*. It must be clear that the activities of knowing and the dynamics of stock-flow information-knowledge processes are about variety identification and transformations in the space of ontological-epistemological duality in relation to the space of the actual-potential dualities that contains the space of imagination-reality dualities as the dualistic-polar games for neuro-decision-choice activities setting variety solutions against variety problems in continual feedback dynamics with error-correction processes under the preference order of the principle of non-satiation that generates derived behavioral spaces such as justice-injustice dualities, kindness-wickedness dualities, fairness-unfairness dualities, greed-generosity dualities, selfless-selfishness dualities, love-hate dualities and many of such dualistic-polar human behaviors based on the distributional conditions over the space of cost-benefit dualities under individual and collective actions guided by the ruling social philosophical consciencism. It useful to understand the theories of languages and linguistic systems and how they are connected to the theory of knowing, the theory of information and the theory of knowledge. The synonym-antonym linguistic constructs reflect the dualistic-polar nature of language representations of the elements of knowing and information-knowledge accumulation, while real and nominal definitions represent the identities of the intellectual flows of the knowns and intellectual stocks of the flows, and explications establish the boundaries of application and interpretation in the space of source-destination dualities, where disinformation arises from the source and misinformation arises from the destination to amplify cognitive capacity limitations as well as influence the directions of the individual and collective neuro-decision-choice actions over the space of problem-solution dualities and polarities.

4.2.1 Reflections on Problems, Solutions, Problem-Solution Dualities in Knowing

The concepts of problems and solutions are central to the understanding of the complexity and simplicity of knowing as intellectual investment flows and information-knowledge accumulation as intellectual capital to all statics and dynamics of varieties and categorial varieties under human production and reproduction activities. The definitions of the concepts of problem and solution in the these epistemic analytics deviate from the general tradition where definitions are established and used in the space of absoluteness, by taking them from the space of absoluteness and placing them in the space of relativity with dualistic-polar definitions under fuzzy logical conditionality. In fact, there are no either negative or positive progress and change if there are no problems and solutions, and there are no problems and solutions if there are no either individual and collective preferences and there are no preferences if there are no varieties and many others in this line of reasoning. tensions, destructions, and transformations are foundations of internal and external variety changes over the space of *negative-positive dualities*, while *destruction* and *construction* are foundations of progress and retrogress over the space of variety negative-positive dualities as induced by socio-natural decision-choice processes. At the level of social set-ups, tension and progress are the results of conflicts in individual and collective preferences under the principle of non-satiation over the space of varieties which generates the spaces of question-answer and problem-solution dualities peace-war dualities love-hate dualities and many others relative to the elements of the space of goal-objective varieties as mapped onto the space of cost-benefit dualities.

It will become clear in this monograph that the existence of socio-natural varieties and preferences as induced by neuro-decision-choice processes is central to the development of the conditions of the spaces of problem-solution and question-answer dualities which demand the development of corresponding systems of the knowing processes and information-knowledge accumulations to define the meaning and essence of life in the sense that, the meaning of life is a system of continual problem-solving actions for the maintenance and the essence of life in the manner in which solutions are found over the space of problem-solution dualities. Life ceases to be for cognitive agents without neuro-decision-choice actions which also cease under the conditions of sameness as defined by characteristic disposition that deprives the existence of the space of problem-solution dualities, varieties and preferences. The essences of the neuro-decision-choice activities are the balancing of the conditions of outcomes of dualistic-polar games for greater and greater net benefit in the spaces of dualities and polarities.

The whole of socio-natural activities of neuro-decision-choice actions involves sets of problems and solutions at the epistemological space, where variety preferences are subjectively established by individual and collective conceptions of progress with increasing benefit as abstracted from the space of imagination, and neuro-decision-choice outcomes provide the individual, collective and social path of human history over the spaces of pain-pleasure dualities and success-failure dualities under the *Sankofa-anoma* tradition (past-present-future connectivity) as the socio-natural environment alters in relation to the universal existence of matter-energy varieties. The existence of problems and solutions is a characteristic of the epistemological space and not a characteristic of the ontological space. Problem-solution dualities exist because preferences exist within the space of imagination-reality dualities. Preferences exist because varieties exist. Varieties give rise to aspirations, hopes, goals objectives and visions under the principle of non-satiation in the space of imagination-reality dualities as a subspace of the potential space which gives rise to neuro-decision-choice actions to transform variety imaginations to variety realities of variety problems to variety solutions that subsequently give rise to the need for variety information-knowledge elements as inputs for neuro-decision-choice valuations of varieties and neuro-decision-choice actions over the space of problem-solution dualities.

The understanding of the concepts of problems and solutions is extremely complex and more difficult than we think as they are referenced over the space of neuro-decision-choice actions. In the nutshell, the totality of human existence is encapsulated in the space of neuro-decision-choice activities over multiplicity of spaces of dualities and polarities They exist in the space of relativity connected to asymptotic cognitive behavior rather than in the space of absoluteness connected to the space of absolute cognitive behavior. There have been many scholarly works on problem-solving activities where problems and solutions are seen in the absolute space, not fully defined and hence not well coordinated with the comparative analytics of cost-benefit information-knowledge conditions which exist in the space of relativity. The mathematical theory of comparative analytics is established over the space of relativity. Its development is not possible over the space of absoluteness, where variety relational continuum and unity are assumed away with the principle of excluded middle in the sense of non-contradiction and opposites cannot reside in the same space at the same time. Comparison and ranking of varieties require the existence of *diversities* as established over the space of quality-quantity dualities, as well as they require the existence of a system of *logical connectivity* through some form of relational continuum and unity. They also require the existence of paradigm of thought that allows the integration of diversities with relational continuum into unity with continual give-and-take-sharing system of variety existence, where the relational continuity allows us to understand the continuity of our food chain from the spaces of input-output dualities, destruction-construction dualities and production-consumption dualities.

The mathematical theory of category, group, sets, order, ranking, optimization and others have an implicit assumption of diversity and connectivity of varieties, where the diversities and connectivity are linked to the elements in the space of cost-benefit dualities for assessments and ordering. This implicit assumption on the diversity, connectivity, oneness, dividedness, organization, disorganization, simplicity and complexity must always be kept within the use of mathematics in theories of specific areas of knowing to provide contents of thought for actions to be guided by and to provide usefulness in developing theories of explanation of variety behavior and their applications over the space of problem-solution dualities, where the understanding of knowing and information-knowledge accumulation are taken from the first principle of necessity-freedom processes over the spaces of problem-answer dualities and question-answer dualities over the space imagination-reality dualities as constructed from the epistemological space which is a derived space from the ontological space as the general identity of knowing as intellectual investment of flows and information-knowledge as accumulation of intellectual capital stocks.

4.2.2 The First Principle, Information-Knowledge Structures and the Problem-Solution Processes

In general, therefore, our understanding or lack of understanding of human progress, conflicts, destruction, construction, war, peace, justice, injustice, governance, love, hate, general conditions and general decision-choice activities forces us to raise the following questions: What is a problem? What is a solution? What is a problem-solving process? What is trouble and how does it relate to problems and solutions? How are problems, solutions and troubles related to universally socio-natural information-knowledge conditions and neuro-decision-choice activities? How and when do we know that a problem exists, a solution exists, a problem has been solved or knowledge items has been acquired? What are the differences and similarities among the concepts and phenomena of information, data, knowledge and evidence and which of them enters the neuro-decision-choice modules as an input, and which of them exists as an output? Generally, the problem-solving activities which are related to question-answer activities must be conceptualized over both the ontological and epistemological spaces where our universe is seen by cognitive agents as composed of variety problem-solution dualities. However, all the scholarly works on these relevant questions in understanding human conditions relative to information and neuro-decision-choice actions have been restricted to the epistemological space without explicit links to the ontological space. The ontology as an area of philosophical studies must be linked to the ontological space as a place for the study of natural decision-choice processes over of ontological varieties in the relational past-present-future dualistic-polar time relative to the disappearance of the old and the emergence of the new.

The epistemology as an area of information-knowledge studies must be linked to the epistemological space as a place for the study of social neuro-decision-choice processes over the epistemological varieties in relational past-present-future dualistic-polar time relative to knowing, acquisition of information-knowledge accumulation, the practice of what is known from nature and mimicking of natural processes to create social varieties and manage the conflicts in the social opposites and the dualistic-polar game structures consistent with socio-economic preferences under the conditions of non-satiation, cost-benefit balances and resource constraints, where the resource constraints depend on transformation technologies and processes [669, 670, 672]. The development of this monograph will make explicit the relative roles of primary and derived spaces as we try to understand knowing as a production of intellectual investment flows and information-knowledge accumulation as the production of intellectual capital stocks. It will then become clear the conditions of the rise of intellectual properties, national security, surveillance system, and many other over the space of necessity-freedom dualities are the results of the claims of input-output conditions of efforts over the space of problem-solution dualities. Similarly, the existence of disinformation, misinformation propaganda are the perceived results of cost-benefit conditions over the space of imagination-reality dualities.

Proposition 4.1.2.1: The First Principle in Knowing Every seriously conscience process in knowing starts with a question or a problem, where the question resides in the problem to search for a solution that houses the answer to the question, where the solution and the answer must pass through test conditions for logical legitimacy to satisfy a collective acceptance over the space of individual and collective doubt-surety dualities, and where the problem-solution and question-answer dualities are conceptualized in the space of social philosophical consciencism that has taken hold of that generation in a specified period and operating over the space of imagination-reality dualities. The first principle involves the fulfilments of the statements of the questions under investigation in the space of question-answer dualities or the statement of problems of interest in the space of problem-solution dualities.

Note 4.1.2.1: On the First Principle in Knowing and the Information-Knowledge Processes The development of problem-solution dualities must always be related to the space of imagination-reality dualities where the identifications of problem space and solution space have residence in the social philosophical consciencism as it relate to cognitive activities over the space of acquaintance to generate experiential information under the guidance of generational social philosophical consciencism that prevails at that generation time within the Sankofa-anoma principle of time trinity that sprits into three dualistic-polar time structures, and information-knowledge trinity that that also splits into three dualistic-polar information-knowledge structures for variety identifications and transformations, where variety problems are identified in the space of reality and variety solutions are abstracted in the space of imagination to replace the variety problems in the space of realities from which new problems emerge with cognitive evaluations as the contents and signal interpretations of messages of the philosophical consciencism alter.

The evolving enveloping path of the generational social philosophical consciencism is such that every generation is defined by its philosophical consciencism that guides the knowing processes, the development of information-knowledge systems relative to the variety identifications and transformations. The existence of the corresponding general social philosophical consciencism is such that every generations confronted with changing elements in the space of problem-solution dualities, where what were unsolved problems in the space of the past may not be problem of the space of the present in the space of past-present dualities, while some unresolved present problems under the present generational philosophical consciencism may not exist in the future time within the space of present-future dualities and hence within the space of past-future dualities as related to the set of dualistic-polar time and the set of dualistic-polar information as seen over the spaces problem-solution dualities with the space of question-answer dualities under the generational social philosophical consciencism which mediates over the elements of the space of imagination-reality dualities and polarities. Present rules, intolerances, laws, does, don'ts, perceptions, behavioral acceptances, nonacceptances and many others over the neuro-decision-choice space may substantially differ from those of the past and those that may show themselves in the future time. These behavioral elements are input-output processes that define the path of social philosophical consciencism in interdepended mode where the social philosophical consciencism shapes the input-output processes which then shape the evolving path of the social philosophical consciencism affecting the dualistic-polar individual and collective interplays of dualistic-polar games over the spaces of problem-solution and question-answer dualities to find varieties with either increasing net benefits or decreasing net costs over the space of problem-solution dualities where cost-benefit evaluations are guided by the negative-positive dual contained in the social philosophical consciencism. In all the analytical processes, we must always keep the relational connectivity among the spaces of cost-benefit dualities, problem-solution dualities. question-answer dualities and the social philosophical consciencism.

In fact, it is usually forgotten that the existence of cost-benefit dualistic-polar conditions in the same variety allows us to examine the complex variety cost components of the natural environment of benefit productions and instruments of war productions from information-knowledge dynamics, where all human activities are seen as input-output relations over the space of variety cost-benefit dualities to justify the concept of dual use of varieties. The useful approaches to knowing, information-knowledge development and efficient neuro-decision-choice actions demand that we work with the *first principle* where we always ask: what are the questions and what are the answers? What are the problems and what are the solutions? The important epistemic approaches to knowing, information-knowledge production, learning, teaching and research is to make explicit from the beginning the problems to which solutions are sought, as well as the questions to which answers are to be found through theoretical and applied analytics. What the information-knowledge inputs are and hence what the developmental channels are for mapping the space of question-answer dualities onto the space of problem-solution dualities and then onto the space of information-knowledge inputs? This *first principle* is central to the development of each chapter of this monograph and others that will follow or precede it. The first principle in the development of any knowing process and information-knowledge accumulation begins by explicitly stating the questions and problems of interest, and then develops thinking to deal with the reasoning activities over the spaces of variety problem-solution dualities and question-answer dualities as they are encountered in the space of acquaintance within the epistemological space.

The first principle of knowing as intellectual investment flows and information-knowledge accumulation as intellectual capital stocks must be followed by a second principle of knowing also as intellectual investment flow and information-knowledge accumulation as intellectual capital stocks. The second principle is to provide us with the epistemic vehicle to move from problems to solutions and hence from questions to answers relative to variety identifications and transformations. The second principle is system of either explanations or prescriptions in theory or applications based on an acceptable paradigm of thought leading to the third principle of knowing for neuro-decision-choice actions over the spaces of decidability-undecidability, acceptance-rejection and doubt-surety dualities. These principles are cognitively developed based on the ruling philosophical consciencism as a guide to the development of thinking algorithms for reasoning in the space of ignorance-knowledge and identification-transformation dualities setting variety knowledge against variety knowledge through the neuro-decision-choice actions over the space of problem-solution dualities.

4.2.3 Philosophical Consciencism, National Personality and the Problem-Solution Processes

The theories of knowing and information-knowledge systems, whether in science and non-science or social science and non-social, are based on economic theory of human actions over the spaces of production-consumption and input-output dualities where cognitive agents, individually and collectively use neuro-decision-choice actions to balance the conflicts in production duals and consumption dual as well as balancing input duals and output dual to produce intellectual investment flows and intellectual capital stocks under general cost-benefit conditions within the space of real-nominal dualities. The efficiencies of the individual and collective neuro-decision-choice processes in knowing and information-knowledge system depend on the national personality that gives rise to individual and collective personalities as abstracted from the national personality as guided by a social philosophical consciencism of the social set-up of the generation. For the explanation clarity, let us provide an analytical definitions of philosophical consciencism and national personality from which other national personalities, such as American personality, European personality, African personality, Asian personality, Russian personality, Middle Eastern personality, Japanese personality and many others with sub-collective personalities within national personalities that may be identified with their corresponding characteristic dispositions.

Each national personality is continually evolving with specific exceptionalism that affects the individual and collective social neuro-decision-choice actions over the space of problem-solution dualities producing outcomes consistent with its personality. Each national personality is generated and supported by a corresponding social philosophical consciencism that has taken hold of the society and evolving over generations. To understand the relative commonness, similarities and differences of nations, societies within the space of global social dynamics, one must examine the relative social philosophical consciencism and the corresponding national and social personality. Similarly, the evolving same social philosophical consciencism and national personality with societal dynamics will help to understand generational differences and similarities of the same society in preferences and social adjustments to nature and the system of global societies as the contents of the evolving social philosophical consciencism alter by the evolving information-knowledge intellectual capital stocks.

Neuro-decision-choice actions are exercised over the spaces of problem-solution and question-answer dualities through institutions and organizations. The social institutions and organizations are the products of individual and national personalities which are in the last analysis products of the social philosophical consciencism that has taken hold of social collectivity and established the boundaries of neuro-decision-choice actions over the space of social acceptance-rejection dualities and polarities from which the individuals and the collectives are guided to react to dualistic-polar balances over the space of input-output dualities as socio-natural environments change to alter the evaluated conditions over the space of cost-benefit dualities. The driving forces of human behavioral dynamics and the enveloping of social history are neuro-decision-choice systems, individual and collective personalities, and their social philosophical consciencism on the basis of which empires, nations, kingdoms, individual and other social collectives rise and fall in the space of life-death dualities. To understand this unbreakable relational continuum and unity of knowing and information-knowledge development, let us provide working definitions to establish the essential characteristic dispositions of the concepts.

We have already established the characteristic dispositions of neuro-decision choice, information and knowledge, as such we will turn our attention to philosophical consciencism, national personality and then explain the relational give-and-take sharing modes within the evolving complex system of dualistic-polar structures which present evolving spaces of dualities and polarities. The evolving complexities find meaning in the organic system as infinitely closed socio-natural input-output and production-consumption self-evolving sub-systems, where a production is consumption and consumption is production, inputs are outputs and outputs are inputs, problems are solutions and vice versa depending on preferences, system of goals, objectives and visions as seen over the space of imagination-reality dualities. All spaces will be verbally and algebraically defined.

Definition 4.1.3.1: Philosophical Consciencism Philosophical consciencism is an intellectual enveloping super-matrix paths, socially logical framework and a system of connected and disconnected thoughts made up of relational diversities, continuum and unity of philosophy, culture, ideas, information-knowledge sub-systems and social consciousness, with supporting sub-systems of ideologies that are developed from the individual and collective experiential information structures in a given social set-up with its mode of thinking and reasoning for integrating competing social forces, and develop them in a manner that allows for the digestion and reconsideration of social contradictions within its system of dualistic-polar structures in such a way that these contradictions fit into a progressive social personality for social transformation of the given society in order to set the variety imaginations against variety realities or defend variety realities within the space of the variety actuals against emerging variety potentials as abstracted from variety imaginations within the space of imagination-reality dualities which constitutes the sub-space of the space of actual-potential dualities and polarities as defining the space of ontological-epistemological dualities.

Definition 4.1.3.2: National Personality National personality is defined as a dynamically organized characteristic disposition with a closter of neuro-decision-choice principles for domenstic and intenational social behaviors that harbor *social exceptionalism* and differs from other neuro-decision-choice behavior of other societies and nations, where the exceptionalism is encapsulate in the social culture uniquely different as the core of the national personality which is supported by a social philosophical consciencism over the epistemological space. It contains particular ways of responses over the space of incentive-disincentive dualities relative to the general cognition, emotions, feelings, thoughts thinking, reasoning, and attitudes, perceptions, tensions, national self-actualization and self-preservation, mood, extraversion, peace, war, hope and power acquisition, power distribution, internal conflicts, conception of society, conception of right and wrong, balancing duals of duality, poles of polarity over the space of violence-nonviolence dualities and many others.

Note 4.1.3.1: Relationality between Philosophical Consciencism and National Personality It is useful now to discuss the conceptual interdependencies between philosophical consciencism and national personality and how they connect to the national exceptionalism in telescopic past, present and future as mutually creating and destructing entities. There is an unbreakable conceptual feedback relation between social philosophical consciencism and the national personality. The national personality is the foundation for the development of the social philosophical consciencism. The social philosophical consciencism contains diverse instructions for individual and collective neuro-decision-choice activities. The national personality and the social philosophical consciencism help to establish the conditions for the understanding the path of national exceptionalism and history as well as provide the explanatory and prescriptive conditions of evolving path of philosophical consciencism over the space of static-dynamic dualities. The evolving social philosophical consciencism molds the path of national personality by molding individual personalities from which the national personality is obtained by fuzzy aggregation to affect the evolving character of the individual personalities that guide the individual neuro-decision-choice activities over the space of imagination-reality dualities regarding concept formations over the space of domestic-international dualities to shape preferences and responses to socio-natural varieties under the principles of non-satiation, where the national personality, national exceptionalism and the social philosophical consciencism are mutually interdependent, creating give-and-take sharing modes for their symbiotic existence and development. Similarly, individual and national personalities are mutually creating for existence and destruction.

All neuro-decision-choice actions in any social set-up reveal the complex nature of individual and national personalities over the space of problem-solution dualities producing knowing as intellectual investment flow, information-knowledge accumulation as intellectual capital stocks. The neuro-decision-choice actions over the space of imagination-reality dualities relate to the space of problem-solution dualities, where the nature of neuro-decision-choice activities are the work of national personality as fuzzy aggregations of individual personalities, where the characteristics of both the individual and national personalities are the works of the evolving social philosophical consciencism, the path of which is shaped by the national personality that also shapes the development of the national institutions as vehicles through which collective and individual neuro-decision-choice actions are exercised over the space of imagination-reality dualities containing varieties of goals, objectives, vision aspirations, greater net benefits as defined by dualistic-polar conflicts in the spaces of imagination-reality dualities and problem-solution dualities.

It is over the nature of relationality among the national personality and the evolving philosophical consciencism that we find characteristic dispositions which pertain to the consistent differences that exist among nations, in the sense of *national exceptionalism* defined over the interactive spaces of neuro-decision-choice actions and problem-solution dualities in knowing and information-knowledge production. It is also, here, that we find characteristic dispositions that reveal common essences among nations containing cognitive agents who are continually struggling over the space of problem-solution dualities in search of varieties with greater net benefits in replacement of varieties with greater net cost over the space of problem-solution dualities relative to the psychosocial principle of non-satiation and the national philosophical consciencism in relation to intellectual investment and intellectual capital accumulation.

The international differences over the space of problem-solution dualities lead us to the study and construct of *the theory of national exceptionalism* relative to international competition, rivalries, wars aggression and many such behaviors that make international cooperation difficult while the national commonness lead us to study and construct of *the theory of national similarities* relative to collaboration, cooperation, peace, nonviolence and many others that motivate diplomacy as the art and science of peaceful and nonviolent resolutions of international disagreements, conflicts and benefit claims and refusal of variety cost-ownerships over the global space of cost-benefit dualities. The socio-psychological foundation of the study and development of the theory of national exceptionalism is that nations at any generational point are equipped with personality dispositions that help to develop preferences and neuro-decision-choice actions different from other nations over the global spaces of problem-solution dualities, input-output dualities and cost-benefit dualities and that each national exceptionalism is the result of the contents of the corresponding national philosophical consciencism that has taken hold of that society at generation points and the manner in which they effect the national unity to affect the use of the general socio-economic laws that govern the enveloping paths of national histories of domestic and international relations over the space of violence-nonviolence dualities, where the preferred neuro-decision-choice of actions in settling disputes is variety violence-dispositions that dominate variety-nonviolence disposition.

Let us keep in mind that philosophical consciencism like any variety is defined by a dualistic-polar structure as a union of negative characteristic disposition that establishes the negative dual, and positive characteristic disposition which then establishes the positive dual The negative an the positive duals combine to define the conditions of the duality and polarity of the national philosophical consciencism. The negative and positive duals are themselves defined over the space of cost-benefit dualities with negative cost, positive cost, negative benefit and positive benefit. It is these complex dualistic-polar conditions of the national philosophical consciencism that split the national personality into competing groups with group personality difference in the nation to create continual conflicts and tensions for institutional formations for the development of social and personal policies leading to collective social behavior over the space of resistance-nonresistance duality to social change as may be established over the space of imagination-reality dualities.

The continual conflicts and tensions in societies are generated by deferential preferences and intense struggle for the acquisition of social decision-choice power to affect the nature of distributional advantage over the space of cost-benefit distribution dualities. In other words, the system of power distribution is intimately connected to the system of cost-benefit distribution. It is this struggle for cost-benefit distributional advantage in the social set-ups that assets preponderating effects on the directions of national history relative to freedom, oppression, empire development, colonialism, neocolonialism, resistance, national pride, international rivalries, cooperation, peace, animosities, war and many others within the set of both the domestic and global conditions. The outcomes of neuro-decision-choice behavior over the space of success-failure dualities relative to the space of problem-solution dualities depend on the national philosophical consciencism, the national personality, national exceptionalism and their give-and-take sharing structure in terms of their effects on the national cognitive capacity limitations over the space of acquaintance-description dualities of knowing and information-knowledge development contained in the epistemological space.

4.2.4 The Importance of Social Philosophical Consciencism and the First Principle of Knowing and Intellectual Stock-Flow Dynamics

Some reflections on the concept, phenomenon and the theory of philosophical consciencism and how they relate to the dynamics of the spaces of problem-solution dualities, and imagination-reality dualities in knowing as the production of intellectual investment flows and information-capital accumulation as the production of intellectual capital stocks is necessary and useful. The social philosophical consciencism is an organic social variety that affects all aspects of neuro-decision-choice action over the space of individual-collective dualities within any nation to effect its historic past, present and future telescopic processes through it effects on the national personalities as a fuzzy aggregation of individual personalities over the intergenerational enveloping path of the same nation regarding knowing and information-knowledge accumulation. Like every variety, the social philosophical consciencism is composed of the positive characteristic disposition that constitutes the positive dual which resides in its positive pole of its positive dualistic-polar structure, in addition to the negative characteristic disposition that constitutes the negative dual which resides in its negative pole of its dualistic-polar structure. The positive dual contains all the positive characteristics that help to define its identity. The negative dual contains all the negative characteristics which also help to define its identity. The positive and the negative duals of the duality define the identity of the national philosophical consciencism as establishing the existing conditions of intellectual capital stocks at any time within a generation to establish the initial conditions for change over the space of negative-positive dualities. The positive and negative poles of polarity of the philosophical consciencism define its continual dynamics to expand the size and replace the elements through knowing as intellectual investment flows which reside in its dualities.

The social philosophical consciencism, at any moment of time and generation point, contains rules, regulations, laws, procedures, costumes, beliefs, incentives, disincentives, knowledge, information, disinformation, misinformation, propaganda, and many others which are defined over the spaces social acceptance-rejection and incentive-disincentive dualities in the management and control of the individual and collective neuro-decision-choice processes over the spaces of the production-consumption dualities and input-output processes. The social philosophical consciencism is dynamic with evolving intergenerational differences and similarities over the enveloping path of social history where the social history includes the history of knowing and information-knowledge accumulation over the spaces of production-consumption dualities, input-output dualities and problem-solution dualities for the preservation and maintenance of the cognitive activities over the dualistic-polar game space of existence of varieties under the psycho-social principle of non-satiation.

At any moment of time for any defined generation of a social set-up, the social philosophical consciencism existing at that time point simplifies the complexities of neuro-decision-choice actions to create algorithmic short cuts for balances between the duals of the space of problem-solution dualities and the duals of the space of cost-benefit dualities under cognitive capacity limitations and the principle of non-satiation under the disequilibrium processes of knowing and information-knowledge production, where every knowing is about variety costs and variety benefits with their structural relativity, and every information-knowledge item is about the accumulation of intellectual capital stocks of the cost-benefit identification and transformation, their give-and-take sharing modes as inputs into neuro-decision-choice actions in managing the intra and inter dualistic conflicts and tensions within and between individual and collective preferences over the spaces of dualistic-polar varieties under the asantrofi-anoma (cost-benefit) principle of variety ranking and rationality.

The social philosophical consciencism is the creation of cognitive agents of the society of interest by the individual and the collective for the individual and collective usage to manage antagonistic social give-and-take sharing relations under the principle of individual and collective non-satiation over the socio-natural space of cost-benefit dualities for social stability in the input-output dynamics, where the social stability is linked to the nature of the neuro-decision-choice outcomes and input-output distribution over the the spaces of incentive-disincentive and cost-benefit dualities. The social philosophical consciencism of any nation contains all the cognitive limitations of cognitive agents where the cognitive limitations are amplified by the principle of non-satiation in the individual and collective preferences generating dualistic-polar solutions with either increasing injustices, war, wickedness, hate, individualism and with reducing justice, peace, kindness, love and collectivity for individual and collective net benefits. In other words, the positive dual and the negative dual are both defined over the space of cost-benefit dualities that demands the delicate balances of the individual and collective neuro-decision-choice actions over the space of problem-solution dualities to produce and use information-knowledge stocks which are continually updated by justified and unjustified conditions over the space of acquaintance-description dualities by the cognitive activities of knowing.

The important thing to keep in mind in the knowing as an economic production of intellectual investment flows and information-knowledge accumulation as an economic production of intellectual capital stocks is that both production of flows and stocks are the works and results of neuro-decision-choice actions under the guidance of philosophical consciencism over the space of problem-solution dualities by identifying problems and transforming then into solutions within the space of imagination-reality dualities as a subspace of the space of actual-potential dualities. The space of variety realities is the same as the space of epistemological variety actuals which is derived identity from the space of ontological variety actuals as the primary identity. The space of variety imaginations is the same as the space of epistemological variety potentials as a derived identity from the space of ontological potentials as the primary identity. The identification of variety problems takes place in the space of realities and the transformation into variety solutions takes place in the space of imagination and is guided by the contents of the philosophical consciencism which is also enriched by information-knowledge conditions of the problem-solution outcomes as intellectual investment flows and as additions to intellectual capital stocks in the spaces of production-consumption dualities and input-output dualities. In this respect, the direction of social history of each society is shaped by the contents of the evolving social philosophical consciencism that provides information-knowledge inputs into and as a guidance for individual and national neuro-decision-choice activities over the space of national-international dualities.

Viewed in this way, the comparative analytics of societies and nations are the comparative analytics of their social philosophical consciencism, the corresponding national personality and national exceptionalism, where differences and similarities reflect the differences and similarities of their contents of negative dispositions, positive dispositions and their dualistic-polar give-and-take intra-social and inter-social sharing relations over the space of individual-collective dualities to establish social behavioral conditions within the spaces of altruism-envy and selfless-selfish dualities. The social philosophical consciencism like information-knowledge system is always in disequilibrium dynamics over the space of internal destruction-replacement dualities and over the intergenerational enveloping paths. The social philosophical consciencism is the foundations of national social personality from which individual personalities are formed to be either consistent or opposed to the prevailing social personality that affects knowing and information-knowledge development over the spaces of production-consumption dualities and input-output dualities as projected onto the space of cost-benefit dualities. It is through the conditions of the evolving philosophical consciencism that we can explain intergenerational difference in preference shifting and changes in social actions.

The transformation behaviors of variety events over the ontological space are taken as natural decision-choice processes that cognitive agents have no control except to study, understand, learn and develop them into information-knowledge structures as inputs into technology, engineering and planning for social adaptation, progress, survival, needs, wants and comforts under the principle of non-satiation in preferences over the space of quality-quantity dualities. The ontological activities are nothing more than natural behaviors over the ontological space of problem-solution dualities under the forces of natural dualistic-polar conflicts for natural transformation of matter-energy varieties over the spaces of the new-old and life-death dualities. The main concern, here, is to define and explicate the concepts of *problem-solution duality*, the *space of problem-solution dualities* and their relationships to problem-solving activities, concepts of problems, solutions, questions, answers, question-answer duality, and neuro-decision-choice actions. The main concern will then be extended to making explicit the uniting forces in all areas of knowing, in *the diversity and unity principles of science* through knowing, to meet the conditions of the first, second and third principles in knowing and information-knowledge accumulation, where knowing and information-knowledge accumulation take place as intense interactions between neuro-decision-choice actions and the space of problem-solution dualities in a never-ending problem-solution and question-answer dynamics. It will become clear that all cognitions take place within the complex systems of interactions through production-consumption and input-output dualities under cognitive capacity limitations as neuro-decision-choice agents execute the games of life over the space of imagination-reality dualities and polarities as a sub-space of actual-potential dualities, where the rewards of the games are better variety cost-benefit dualities.

4.2.5 Social Philosophical Consciencism, Information Manipulation, and Neuro-Decision-Choice Actions

The contents of philosophical consciencism as input-output processes that are central to neuro-decision-choice actions provide pathways to the understanding of the effects of disinformation, misinformation, propaganda, conspiracy theory and fake news in the game space of the problem-solution dualities to set variety solutions against variety problems over the space of imagination-reality dualities. The characteristic dispositions of disinformation, misinformation, conspiracy theory, propaganda and many other elements in the social philosophical consciencism become part of input-output system of communicating factors over the source-destination dualities to control and shape the neuro-decision-choice directions and outcomes of domestic and international destination agents in order for the source agents to abstract greater net benefits while either reducing the net benefits or increasing the net costs of the destination agents. The process involves the activities of information-knowledge manipulations of the contents of the national philosophical consciencism from which neuro-decision-choice inputs are abstracted by the destination agents while at the same time taking directions from the dualistic-polar dispositions of the national philosophical consciencism which may affect the responses to the elements in the zones of surprise that is contained in the space of acquaintance.

The information-knowledge manipulations of the contents of the social philosophical consciencism include:

(1) Creating *fear system* over the space of incentive-disincentive dualities in response to reward-punishment elements over the space of cost-benefit dualities.
(2) Creating a *system of variety phantom problems* with lower net benefits in the space of imagination and offering variety solutions from the space of reality with net greater benefits through destruction-replacement processes to change the social cost-benefit distribution among neuro-decision-choice agents operating over the space of imagination-reality dualities.
(3) Creating a *system of distractions,* where intellectual foci are shifted from problems in the space of realities to phantom problems and solutions in the space of imaginations in support of conditions of (#2).
(4) Create a *system of calculated deceptions* over the space of acceptance-rejection dualities through fake incentives and conspiracy theories to make reduced variety net benefits and increased variety net costs acceptable through promotions of gradual processes of transformation as well as advocate for *a system of institutions* through which policies may be transmitted to amplify cognitive capacity limitations of the neuro-decision-choice agents with a system of confused past-present-future connectivity.
(5) Creating a *system of emotional dispositions* to overrun reason and rational thought for cognitive agents to accept the emotional conditions and irrational computations as explaining their neuro-decision-choice actions over the space of cost-benefit dualities.
(6) Create *antagonistic social groups* with concentrations on struggle for trivialities and cost-benefit crumbs as diversions to those things that are critical to their progressive socio-economic wellbeing.
(7) Create a *system of increasing intellectual mediocracy* to reduce greater expectations for the purpose of control of cost-benefit distribution.

(8) Create a *phantom belief system* with the support of a system of fake news, disinformation, conspiracy theories and misinformation to keep some cognitive agents in imaginative bubble supported by fake hope and cost-benefit apparitions in impossible world contained in the space of possible-world-impossible-world dualities, where the impossible-world dual dominates the possible-world dual, where some cognitive agents take residence in the space of impossible world with dualistic-polar structure of imagination and reality.

(9) Create a *system of disbelief dispositions* in truth and science and knowledge with the support of a phantom system of outcomes with a system of phantom net benefits residing in the space of possibility-impossibility dualities.

(10) Create a *system of mediocre individual personalities* with inferiority complex, self-doubt and schizophrenia from the national personality, with beliefs that somebody or an authority knows what is good for them to increase their net benefits if they accept the personal sacrifice of pain and suffering; and

(11) Create a *system of media supporters* of a core believers with the same preferences over the space of cost-benefit dualities to promote the disinformation, misinformation, propaganda, conspiracy theories, fake news and disbeliefs in truth science and knowledge.

The information manipulations are effective since the inputs of all neuro-decision-choice actions are information that must be processed into knowledge. By altering the quality of the information, the knowledge abstracted from the information may or may not be affected if the individual personality dispositions are not affected. The information manipulations amplify cognitive capacity limitations as well as magnify the space of certainty-uncertainty dualities, strengthen and increase the variety cost duals against the variety benefit duals over the space of cost-benefit dualities relative to the space of identification-transformation dualities in knowing and information-knowledge production in all areas of human actions. In this way confusion is placed on neuro-decision-choice agents over the space of variety ranking with direction to choice unfavorable varieties due to information-knowledge confusion back by institutional enforcement through the manipulation of the elements in the space of incentive-disincentive dualities where rewards are delivered as benefits to the obedient and punishments are delivered to the stubborn and disobedient as cost and where compliance through neuro-decision-choice actions are created by manipulating the varieties in the space of cost-benefit dualities.

Let us understand some essential aspects of knowing and information-knowledge accumulation over the space of ontology-epistemology dualities and how economic production approach through production-consumption and input-output theoretic system help to the understanding of principles of diversity and unity of information-knowledge accumulation in disequilibrium dynamics. The understanding must be judiciously connected to input-output and supply-demand information conditions of neuro-decision-choice actions in relation to information-knowledge conditions of the space of cost-benefit dualities, where any neuro-decision-choice action is both information-knowledge consumption and information-knowledge production in the space of variety knowing as intellectual investment flows and information-knowledge accumulation as intellectual capital-accumulation stocks. The information-knowledge consumption uses the intellectual capital-stock services and information-knowledge production generates intellectual investment flows to update the available intellectual capital-stock services over the global space of production-consumption dualities, where every variety is simultaneously an input and output relative goals and objectives of cognitive agents and as transformations over the space of process-technology dualities as abstracted from the space of actual-potential dualities through the space of imagination-reality dualities.

We have worked on the concepts of ontology and epistemology as phenomena for intellectual actions through neuro-decision-choice activities. Let us now examine the concepts and phenomena of *ontological space* and *epistemological space* and their relationality as they are connected to knowing as a production of intellectual investment flows and information-knowledge accumulation as intellectual capital stocks. Here, the intellectual capital stocks function as intellectual factories developed by knowing as intellectual investment flows to provide intellectual capital input services into the space of neuro-decision-choice actions over the space of problem-solution dualities. We shall end this chapter by discussing the concept and phenomenon of the ontological space of problem-solution dualities. The concept and phenomenon of epistemological space of problem-solution dualities and polarities will be taken up in Chap. 5.

4.3 The Ontological Space of Problem-Solution Dualities

The ontological space is a collection of all-natural variety problem-solution dualities over the space of actual-potential dualities which exist dependently or independently of the ontological elements such as cognitive agents, where the varieties in the space may be expanded by the activities of socio-natural actions. Here, there are all kinds of active symbiotic relationships that are developed between the actual varieties and the potential varieties, among the actual varieties themselves and among the potential varieties themselves in the spaces of natural input-output dualities and destruction-replacement dualities. The symbiotic relationship may be mutually beneficial, mutually defensive, mutually transformational and many others, where actual varieties are transformed to variety potential and variety potentials are transformed to actual by the elements in the space of natural process-technology dualities. There is no conceivable space of cost-benefit dualities in the ontological processes, but only the space of input-output processes for continual variety transformations. In this respect the ontological problem-solution dualities cannot be defined by cost-benefit conditions. Over the ontological space, natural transformations are systems of natural corrections and restructuring for environmental stability and variety cooping characteristics, where variety extinction and variety creation are part of the natural diversity and organicity. The conditions of the ontological varieties in terms of changes in properties and relations are part of the space of ontological destruction-replacement dualities that constitutes what is ontological in the time trinity of the past, present and the future for continual creations. The ontological varieties are the objects of ontological transformation decision-choice actions on transforming matter-energy inputs into matter-energy outputs through natural technological processes acting on the varieties, transforming variety actuals to variety potentials and vice versa in continuous processes defining new properties and relations. The awareness of the conditions of ontological variety existence is revealed by the variety characteristic-signal dispositions through the space of source-destination dualities for mutual sending-receiving processes where the signal dispositions are the messages and the characteristic dispositions are the contents of the messages to reveal the identities of the ontological varieties. These are the input-output transformation processes for the internal natural transformations, where the outcomes of the input-output processes present themselves as varieties, each of which is composed simultaneously of matter, energy and information with different matter-energy characteristic dispositions that establish the information contents for variety identities.

The transformations of matter-energy inputs into matter-energy outputs are the activities that maintain the matter-energy equilibrium with continual variety changes and rearrangements of ontological characteristic-signal dispositions that create information-knowledge disequilibrium processes under the telescopic past through the telescopic present and to the telescopic future of the sankofa-anoma information-time traditions. The variety changes and transformation may lead to either an expansion or a contraction of the space of varieties through the disappearance of some varieties, (extinction)and emergence of new varieties (growth) in the space of extinction-preservation dualities. Cognitive awareness finds expressions of variety existence in information-knowledge structure through the space of acquaintance in the epistemological space, where ontological signal dispositions are sent to destinations for decoding.

The ontological elements and their transformations are also the objects of epistemological identification through the knowing processes at static and dynamic states. The knowing process, at the static level of existence, is the epistemic activities over the *space of identification problem-solution dualities* of matter-energy varieties to acquire information-knowledge structures about *what there was* (the past telescopic) and *what there is* (the present telescopic), while the knowing process, at the dynamic level of existence, is the epistemic activities on the *space of transformation problem-solution dualities* of matter-energy varieties to also acquire information-knowledge structure about *what would be* and *what would not be* (future telescopic), given *what there is,* where *what there is* an offspring of *what there was* as well as a parent of *what there would be* under the Sankofa-anoma information-time dynamics where the past produces the present and the present will produce the future, and the past produces the future the present. It is at this level of present-future processes over the space of transformation problem-solution dualities of natural varieties, that there have been beliefs and claims that some cognitive agents could discern what kind of mathematics will correspond to physical laws, and use these mathematical representations to exactly predict the future natural variety outcomes of behavior of *what there is*. It is always forgotten in all theoretical abstractions with mathematics that mathematics as a member of the family of abstract languages (FAL) is a cognitive creation and refinements from the family of ordinary languages (FOL) by a thinking-reasoning processes with neuro-decision-choice agents constrained by all elements of cognitive capacity limitations over the space of acquaintance-description dualities. Not only that. but it is based on the science of numbers which constitutes the primary category of abstract languages while other abstract languages are derived categories. We shall produce the descriptive structure that connects the relationships among the science of numbers, diversity and the unity of knowing and information-knowledge accumulation with the unity of science in the sense that in oneness is dividedness and in organization is disorganization.

Every type of mathematics derived on the basis of the science of numbers to abstract information-knowledge items is constrained by all the conditions of cognitive capacity limitations where conditions of ill-posed problems, non-computability, undecidability and contradictions, in information-knowledge misrepresentations are present to generate phantom problems and paradigmatic paradoxes that challenge the cognitive sense of the epistemic trust in relation to all kinds of approaches such as the formulation of quantum theory, chaos, neuro-decision actions and many others to come from our defective classical paradigm of thought with the excluded middle. Some questions that arise concern what kind of information and mathematical language can carry us over the space of source-destination dualities and in what space can they be constructible. The classical paradigmatic process with the excluded middle and non-acceptance of contradiction in the knowing process excludes the bulk of what can be known through a properly constructed thought system that embraces variety conditions over the spaces of quality-quantity and subjective-objective dualities under the principles of relational continuum and unity that diversities entail.

Given the defined concepts of ontological and epistemological spaces, the central driving force connecting the core of my investigative and research activities, leading to the developments of a number of monographs that I have produced, involves an epistemic attempt to understand and make explicit the *universal connectivity principle* that relates to the universal statics and dynamics of things, object, states and processes which create the universal environment of mutual existence and non-existence, mutual interdependencies and independence of all forms, mutual destruction and non-destruction of all forms of cognitive and non-cognitive agents in terms of information-knowledge structures at the levels of ontological and epistemological spaces. The investigative and research activities toward this end require us to establish the essential connecting cords between the ontological and epistemological spaces that allow us to construct from the space of ontological varieties the corresponding space of epistemological varieties and to how these connecting cords relate to knowing as a production of intellectual investment flows and information-knowledge accumulation as a production of intellectual capital stocks as well as connecting them to human creative activities and conditions of the common socio-natural environment of existence over the space of human-nonhuman dualities. The connecting codes are the set of neuro-decision-choice actions.

Each ontological variety contributes to the common environment in accordance with the nature of its negative-positive duality. At the epistemological space, each variety is a problem-solution duality in the sense that it is composed of a *problem dual* in the characteristic disposition and a solution dual also in the characteristic disposition in terms of behavioral identity, and transformation, and hence its knowing reveals its cost and benefit to conditions of existence. It also serves simultaneously as a variety problem and a variety solution within the common environment of variety identifications and variety transformations relative to goal-objective elements of neuro-decision-choice agents as conceived from the epistemological space. The individual variety contributes to the environment of category of its belonging and overall common environment in asserting the relational continua and unity of knowing and information-knowledge accumulation. The individual and categorial contributions by varieties to the environment and the space of the variety transformations will vary in accordance with the relative importance of their internal negative-positive characteristic dispositions that will affect the transformational dynamics over the ontological space as well as over the elements in the *space of cost-benefit dualities* to generate the space of variety problem-solution dualities within the epistemological space.

At the epistemological space where cognitive agents operate, there are internal cost-benefit dispositions within any variety to establish the intra-variety cost-benefit duality. There are also external cost-benefit dispositions within the space of varieties to establish its inter-variety cost-benefit duality. The intra-variety and inter-variety cost-benefit dualities relate to intra-inter-variety conflict-resolution dualities, destruction-construction dualities, intra-inter-variety conditions in progress-retrogressive dualities and intra-inter variety change-replacement dualities in variety transformations and info-dynamics that tend to support infinite transformations within matter-energy structures, where the beginning resides in the end which finds residence in the beginning to support variety parent-offspring dynamics as pointed out in Adinkra tradition with *nyansa-po,* (the wisdom knot), where the beginning resides in the end and the end resides in the beginning, such that the end and the beginning are cognitively indistinguishable in the sense that the end is the beginning and the beginning is the end to affirm the universal dualistic-polar structures of diversity and unity with oneness and dividedness, organization and disorganization in knowing, where there are the concepts and knowledge of unity from which the concepts and knowledge of diversity are derived by means of dividedness through which epistemic organization is universally obtained.

The diversity-unity conditions of the knowing process and information-knowledge accumulation may then be seen in terms of analytics in the space of universal-particular dualities of variety identifications and transformations irrespective of whether an element is classified as science or non-science since the knowing of all identifications and transformations have one thing in common which is elimination-replacement activities over the space of problem-solution dualities, where a variety problem is destroyed and replaced by a variety solution with a seed of destruction for a new variety problem replacement for analytical works over the space of ignorance-knowledge dualities. The collective existence of matter-energy varieties creates the individual and collective environments for individual and collective variety identifications and transformations through acquaintance over the epistemological space.

The individual-collective relation belongs to the ontological space of particular-universal dualities as well as the epistemological space of the epistemic principle of individual-community analytics over the space of varieties under the principles of opposite. The static and dynamic characteristic dispositions of each variety are also involved in understanding changed actions or the transformative conditions at both the levels of ontology and epistemology, where the activities over the epistemological space are cost-benefit controlled to manage knowing, neuro-decision-choice actions and stock-flow disequilibrium dynamics of information-knowledge accumulation to manage intellectual stocks that provide capital-service inputs into the neuro-decision-choice processes to connect the epistemological varieties to the elements in the ontological space and to act over the epistemological space for games of life and existence over the space of cost-benefit dualities.

4.3.1 Ontological Information-Creation System and Information-Processing Systems

Let us keep in mind that ontological varieties are *what there was* (the past), *what there is* (the present) and *what would be* (the future) which exist as identities with perfect information-knowledge equality defined by their characteristic dispositions, where all varieties belong to the space of quality-quantity dualities called the ontological information. The epistemological varieties of *what there was*, *what there is* and *what would be* are derived from ontological signal dispositions to abstract the characteristic dispositions under cognitive capacity limitations and information-knowledge inequality also defined in terms of characteristic dispositions, the results of which are called the epistemological information. In the whole process of interactions between the ontological and epistemological varieties, the cognitive agents are seen as belonging to the space of subject-object dualities for understanding, where such understanding involves the activities of knowing and the development of information-knowledge stocks from the ontological space to act as an input-output system for enhancing the continual survivability of cognitive agents in a mode of variety destruction-creation dualities as the conditions of environment are affected by socio-natural activities.

The input-output process maintaining the ontological food chain in the space of production-consumption dualities is the system of input-output process or a system of destruction-creation dualities. It is this system of processes of variety destruction-creation dualities to affect the conditions of common the environment where the concept of leading categories arises from in terms of key variables as determining factors in the directions of destruction-creation processes in variety transformations where such leading categories have resulting impact on socio-natural transformations. It is also the relational effects in unity among varieties where inter-impact-variety studies become important for the understanding of variety sustainability, extinction and change, where human activities of all forms may have impacts on climatological and geomorphological changes and behaviors over the ontological space.

It must be kept in mind that the socio-political disagreements on the scientific information of climatological and geomorphological responses in the space of acceptability-unacceptability dualities to the effect of human activities will not change the ontological responses to epistemological actions delivered by neuro-decision-choice actions. It is, here, that the relevance of scientific information-knowledge stock reveals its input-output powers in neuro-decision-choice actions. It may be noted that misinformation and disinformation of scientific results through pollicization, politicization, fake news and other such forms may lead to conditions of irreversible disasters where the effects of misinformation and disinformation are to change the cost-benefit distribution over the social space and not over the natural space of construction-destruction, organization-disorganization, oneness-dividedness dualities without the applicational assessments of cost-benefit conditions and relativity.

The universal survivability of ontological elements and cognitive agents involves a continual system of problem-solution dynamics, especially living agents to ensure *universal self-continuity,* fight for individual and collective existence and against *variety finality* and *variety extinction,* as well as fight for the variety beginning and *permanence of universal existence* from the infinitely permanent universe with continual variety creation and destruction within the space of variety nothingness-somethingness dualities and rational management of the socio-natural environment based on information-knowledge conditions in the space of realities and not on disinformation, misinformation and fakeness in the space of imagination-reality dualities. The socio-natural varieties are both inputs and outputs of socio-natural transformations, where the variety inputs are turned into variety outputs which are then turned into variety inputs in the space of input-output dualities under matter-energy-information dynamics with neutrality of time.

Every Variety act as an input to generate output as well as an output derived from another input depending on the natural environmental conditions over the ontological space and on the goal and objective of the users acting over the *action space* within the epistemological space. Agents of transformation differ substantially over the ontological space and the epistemological space. The ontological transformation decision-choice actions are understood in input-output conditions of primary-derived variety processes without cost-benefit considerations since nature works directly with information-knowledge equality destroying variety actuals and transforming them into variety potentials while creating variety potentials and transforming them into variety actuals as replacements to maintain the matter-energy equilibrium and information-knowledge disequilibrium. The cognitive agents, on one hand, are part of socio-natural *information creation system* (ICS) through neuro-decision-choice activities over the action space of the epistemological space and, on the other hand, the cognitive agents constitute a *knowledge-information-processing system* (KIPS) at the epistemological space for their neuro-decision-choice actions over the space of the problem-solution dualities at the guidance of real cost-benefit conditions and rationality seeking answers to questions and generating questions to answers.

Both the (ICS) and (KIPS) find expressions over the *space of problem-solution dualities* (SPSD)$\left(\mathfrak{R}\right)$. The (SPSD) is a creation of cognitive agents under the continual inefficient command-control actions of individual and collective decision-choice modules which form the system of central activities over the epistemological space, where varieties are input-output dualities through the ontological and epistemological destruction-construction processes that maintain matter-energy stock-flow equilibrium and information-knowledge stock-flow disequilibrium where every variety input and variety output have cost-benefit implications to cognitive agents in defining problems and finding solutions. We shall find out that every problem is cos-benefit defined and every solution is cost-benefit defined in the space of neuro-decision-choice actions relative to goals, objective and visions of cognitive agents.

It is important to note that every cognitive activity over the space of problem-solution dualities is either information discovery at the level of acquaintance, information generation or knowledge discovery at the static and dynamic states. It is in the space of the problem-solution dualities $\left(\mathfrak{R}\right)$ that we find areas of diversities and the links of all areas of knowing, including sciences, into a relational unity. The relational unity is enhanced by neuro-decision-choice activities with information-knowledge acquisitions on matter-energy conditions over the epistemological space which provides a permanent bridge between natural transformations and social knowing over the space of the variety problem-solution dualities where problems just like solutions reveal information at the level of acquaintance and knowledge at the level of description as conceptually unifying forces of all areas of knowing.

The concept and phenomenon of unity imply the existence of the concept and phenomenon of diversity, just as the concept and phenomenon of negative imply the existence of the concept and phenomenon of positive, and similarly the concept and phenomenon of oneness imply the existence of the concept and phenomenon of dividedness. Generally, the concepts and phenomena of negative duals in dualities and negative poles in polarities imply the existence of positive duals and poles under the general principle of opposites. The analytical and epistemic disagreement is whether the opposites exist in relational separations and disunity without connectivity to give rise to the paradigm of thought with the principle of excluded middle, where opposites cannot exist in the same space or exist in relational continua and unity with unbreakable connectivity to give rise to the principle of acceptance of logical contradictions, where opposites exist in the same space as a natural order. In the theories of knowing and information-knowledge system, the concept and phenomenon of unity also imply the concept and phenomenon of categorial diversity in terms of multiplicity of relational continuum of information-knowledge existence but not in relational disunity. The concept and phenomenon of diversity are enhanced by the principle of dividedness from oneness to generate new varieties or multiple varieties from one variety, and with continual creation of corresponding characteristic dispositions as information containing knowledge to establish their identities and transformations.

The understandings of the existence of diversities, continual destructions of varieties and creation of new varieties from the existing varieties are enhanced by neuro-decision-choice activities with information-knowledge acquisitions on matter-energy conditions which containing knowledge of processes and transformation technologies under the principle of dividedness over both the ontological and the epistemological spaces, where the neuro-decision-choice actions provide a permanent cognitive bridges among the natural dividedness and ontological varieties on one hand, and social dividedness, epistemological varieties and social knowing of natural technologies and processes over the spaces of the natural variety transformations and social variety problem-solution dualities on the other hand. We must always keep in mind that in the knowing processes as the production of intellectual investment flows and information-knowledge accumulation processes as the production of intellectual capital stocks, there are two conceptual cores of the set of *ontological varieties* and the set of *epistemological varieties*. These two sets with defined characteristic dispositions are linked together by neuro-decision-choice actions to generate input-output disequilibrium dynamics. The differences between ontological variety dispositions and epistemological variety dispositions are variety *epistemic distances* which reveal the degrees of variety ignorance under knowing with cognitive capacity limitations to establish degrees of knowledge about varieties in the space of ignorance-knowledge dualities under an epistemic conditionality from the spaces of certainty-uncertainty, doubt-surety and decidability-undesirability dualities which together have given rise to the developments of possibility theory, probability theory, statistical theory, information theory and data science.

4.3.2 Ontological Information, Telescopic Present, Telescopic Past and Telescopic Future

To understand the concept and phenomenon of variety distances, we must fist understand the concepts of ontological information and the epistemological information, their differences and similarities, their relations to the ontological and epistemological spaces in cognition, the works to know and accumulate knowledge and how the knowing and knowledge accumulation relate to the varieties in the spaces of ignorance-knowledge, doubt-surety, decidability-undesirability and problem-solution dualities as extended to question-answer dualities with relational continuum and unity. In the ontological space, the matter-energy conditions of the static state of *what there is* (telescopic present), and the dynamic state *of what there was* (telescopic past) and *what would be* (telescopic future) may be viewed as the ontological conditions of static-dynamic processes for continual matter-energy stock-flow equilibrium and information-knowledge stock-flow disequilibrium processes with reference to universal and particular varieties. The ontological static and dynamic processes may be viewed as ontological activities over the ontological space of ontological problem-solution dualities in response to the changing conditions of ontological environment where ontological activities are simply to maintain certain natural laws of stability within the space of static-dynamic dualities. The development of information-knowledge structures about the static and dynamic conditions of ontological varieties requires a cognitive development of an *abstract space* to represent the space of ontological identification-transformation dualities consistent with the ontological space of info-statics and info-dynamics with connecting principle of neuro-decision-choice actions to generate knowing as intellectual investment flows and information-capital accumulation as intellectual capital stocks to maintain the neuro-decision-choice dynamics.

The abstract space developed by cognitive agents to abstract information and knowledge from the ontological space as an identity is the surrogate space of problem-solution dualities of information-knowledge conditions about ontological matter-energy varieties under dualistic-polar restrictions such that each duality or polarity is in a relational continuum and unity with itself and with other varieties and the collectives under all forms of symbiotic relational continuity and unity. In this framework, every variety solution to a variety problem generates a new variety problem requiring a search for a variety solution from within the space of problem-solution dualities. The concept of *problem-solution duality* is central to all existence in the ontological and epistemological spaces whether acknowledged or not. Cognitive agents have no way of learning and knowing except through the conditions of the system of sense organs over the space of problem-solution dualities. The dynamics of the ontological space of problem-solution dualities define the individual and collective paths of internal ontological transformations in terms of *what there was, what there is* and *what there would be* which together define the past-present-future information structures of matter-energy varieties and categorial varieties with corresponding characteristic dispositions. The dynamics of the epistemological space of problem-solution dualities present two directions of development of the information-knowledge structure about the ontological existence and the utilization of the acquired information-knowledge structure to transform the conditions of social existence through the second dimension of problem-solution processes which finds expressions in engineering, planning, technology and prescriptive sciences. It is at this point that we speak of primary and derived conditions of knowing where the primary varieties relate to the conditions of fundamentalism and the derived varieties relate to the conditions of the applied within the space of fundamentalism-applied dualities as embedded in the epistemological space.

The ontological space presents two interdependent sets of information conditions of knowing in the development of knowledge about the static and dynamic states of varieties. They are the *set of static ontological information* conditions and the set of *dynamic ontological information* conditions. The set of ontological static conditions from the acquaintance at the epistemological space reveals itself through the system of variety *identification problem-solution dualities* from the ontological signal dispositions. The set of ontological dynamic information conditions reveals itself at the epistemological space through a system of *variety transformation-identification problem-solution dualities* in relation to the totality of life and the socio-natural environment that contains the dualities with dualistic-polar struggles for dominance.

The knowing and information-knowledge processes of cognitive agents, whether they are about science or non-science, are about either variety identification, variety transformation or both for the development of information-knowledge accumulation about the universe, where the variety identifications initialize conditions for variety transformations through neuro-decision-choice activities of cognitive agents to provide conditions to understand the meaning and essence of life relative to necessity and freedom. At the epistemological space, therefore, the *meaning of life* of cognitive agents finds expressions in the neuro-decision-choice actions over the space of problem-solution dualities in relation to variety identifications.transformations and usage. The *essence of life* of cognitive agents finds expressions in the manner of executing the neuro-decision-choice actions over the space of problem-solution dualities to obtain solutions to problems in relation to varieties in the space of destruction-replacement dualities. Both the meaning of life and the essence of life find complete expressions in the neuro-decision-choice activities in the space of problem-solution dualities. The dynamic conditions of the space of problem-solution dualities constitute the foundations of variety transformations as well as the behavior of info-dynamics and stock-flow conditions of variety information-knowledge processes that provide conditions of *diversity creation, category formation* and *unity analytics* of general knowing and science within the categories of explanatory and prescriptive sciences under the universal principle of oneness containing dividedness that helps to establish diversities.

At the level of the epistemological space, the results of social variety transformations are generated by social neuro-decision-choice processes involving problem-solution dynamics in relation to variety preferences governed by real *cost-benefit dynamics* under conditions of non-satiations. The conditions of the results of the variety transformations become information inputs into the mechanism of the social neuro-decision-choice processes in a dynamic complexity of social existence and social transformations in a continual search for new-information-knowledge items and the understanding of variety behaviors to be used as further inputs into subsequent neuro-decision-choice actions for continual variety transformations towards state of an increasing preferences as subjectively defined over generations seeking the state of ultimate satisfaction, happiness and joy through actions on the elements in the space of imagination-reality dualities as a subspace of actual-potential dualities. At the level of the ontological space, however, the results of variety transformations are generated by *natural technological processes* within the forces of nature for continual disequilibrium of stock-flow information-knowledge dynamics, destroying some existing ontological varieties and creating new ontological varieties to expand the space of *what there was*, as well as creating the possibility of either increasing or reducing the space of *what there is* relative to *what there would be* in the space of actual-potential dualities, where cognitive agents, with the acquired information-knowledge stocks remain in the state of evasiveness in the journey to the ultimate state of preference without knowing its characteristic dispositions except through the space of variety imagination with continual dualistic-polar war in the space of finite-infinite dualities, where finite relates to the conditions of telescopic present and infinite relates to the conditions of telescopic past and telescopic future.

Chapter 5
The Epistemological Space of Problem–Solution Dualities, Unity of Knowing, Unity of Science and Diversity in Knowledge

Abstract This chapter is used to present the structural conditions of the epistemological space for the understanding of neuro-decision-choice activities in knowing as intellectual investment flow, cognitive actions over the space of problem–solution dualities and how such actions relate to information-knowledge accumulation as intellectual capital stocks, the service flows of which become inputs into neuro-decision-choice actions. It presents a theoretical understanding of how the space of problem–solution dualities is embedded in the space of imagination-reality dualities which is a sub-space of the space of actual-potential dualities. The concepts and phenomena of problems, solutions and the space of problem–solution dualities are defined through the conditions of cost–benefit dualities over the space of relativity. The concepts of criterion, criteria space, neuro-decision-choice space are verbally and algebraically introduced and defined, and their relationships to knowing, information-knowledge processes and their connectivity to the space of problem–solution dualities are specified and explained. The important and prominent role played by conditions of cost–benefit duality is made explicit and linked to the criteria space. The problem–solution process is argued to link the knowledge by acquaintance to knowledge by description through a general paradigm of thought irrespective whether an area is defined as science or non-science, where the classifications of areas of knowing and problem–solution dualities are seen as fuzzy-set closure with fuzzy conditionality. The chapter also presents the nature of the analytical connectivity within the ontological-epistemological duality and how the ontological space constitutes the primary identity of all variety existence in the past-present-future telescopic structure functioning with information-knowledge equality constraints, while the epistemological space constitutes the derived identity of variety existence, in the a derived past-present-future telescopic structure functioning with information-knowledge inequality constraints.

K. K. Dompere, *The Theory of Problem-Solution Dualities and Polarities*,
Studies in Systems, Decision and Control 405,
https://doi.org/10.1007/978-3-030-90279-7_5

5.1 Inroduction to the Epistemological Space

In Chap. 4, we discussed some important conditions and characteristic dispositions of the ontological space of the space of problem–solution dualities in the knowing and information knowledge production. The extensive discussion on ontology and epistemology on one hand and the ontological and epistemological space on the other hand are given in chapter two. It was pointed out that the ontological space is the universal identity containing all varieties of past, present and future defined by the conditions of the space of actual-potential dualities containing the three information structures of telescopic past, telescopic present and telescopic future producing the Sankofa-anoma tradition where the past is the parent of the present which is the parent of the future and hence the future is always the ancestor of the past through the present. Everything we need to know about variety existence and variety changes are housed in the ontological space as ontological characteristics where their transformations are due to the actions of nature which are not dependent on the will and preferences of ontological varieties but are mutually independent under symbiotic conditions of give-and-take sharing modes. The variety transformation processes in the ontological space are not driven by cost–benefit conditions in terms preferences of cognitive agents. The ontological transformations are not neuro-decision-choice actions over the space problem–solution dualities as we are familiar. In fact, there is no space of problem–solution dualities and there is no way to define one since the conditions for establishing the existence of problem–solution dualities are the variety information-knowledge disparities and the variety cost–benefit disparities which are not characteristic dispositions of the ontological space. The ontological space is a union of the space of actuals and the space of the potentials as have been discussed in chapters two and three of this monograph. The concept and the phenomenon of ontology contain the concept and the phenomenon of ontic, and hence the concept and phenomenon of ontological space contain the concept and phenomenon of ontic space where the ontic relates the actual variety characteristic dispositions while the ontology relates the conditions of variety and categorial existence in the actual and potential spaces harboring the information-knowledge of information-knowledge trinity under the Sankofa-anoma principles. The ontic space contains the variety characteristic dispositions of the space of the actual but nor the variety characteristic dispositions of the potential space.

The similarities and differences of ontology and epistemology are not well understood or well defines to establish their characteristic dispositions. As such, the concepts and phenomena of ethnic knowledge, ontologies, epistemologies reside in phantom analytical space in our attempt to understand neuro-decision-choice actions over the space problem–solution dualities and the activities of intellectual investments as flows to contribute to information-capital accumulations as intellectual capital stocks to generate intellectual capital services as inputs into neuro-decision-choice actions over the space of problem–solution dualities relative to variety identifications and transformations. In the theories of knowing and information-knowledge accumulation as human productive activities, very little is known of ontological and epistemological spaces and what role do they play in knowing and the development of philosophical consciencism. To guide against a process of falling into phantom spaces of thinking and problem–solution dualities, it is useful to note that the concept of epistemological space that has been presented in this monograph is different from the concept of epistemic space which is relate to the discussions of the concepts of possibility and possible worlds where epistemology is distinguished from epistemics in the space of knowing. Similarly, there are differences and similarities between the concepts of ontic and ontology one hand and ontic space and ontological space as established by their characteristic dispositions on the other hand.

There is only one ontological space and ontic space containing the characteristic dispositions of varieties and categorial varieties that may be connected to the epistemological space containing the space of acquaintance. There are different spaces of acquaintances creating different experiential information structures for different ethnicities and geographies leading to the concepts and phenomena of different ontic spaces, where the epistemic activities on the experiential information from ethnicity is viewed as ethnic epistemologies. This is not a good way for the understanding of the space of ontology-ontic dualities in knowing and information-knowledge production. Knowledge is one and the process to reach it from the knowledge by acquaintance to knowledge by description may vary from the reasons of cognitive and geographical capacity limitations. Given the discussions on the ontological space of ontological-ontic dualities and the corresponding problem-solutions dualities, we turn attentions to the epistemological space of problem–solution dualities and how it relates to the conditions of the epistemological space discussed in chapter two and three of this monograph.

5.2 The Epistemological Space of Problem–Solution Dualities

In the previous section, the conditions of the relationship between the space of ontological-ontic dualities and space of problem–solution dualities were discussed. Here attenpts were made to point out the similarities and differences between ontology and ontic and the corresponding spaces relation to the space of actual and potential varieties and categorial varieties. An attempt is now turned to the nature of the concepts and phenomena of epistemological space, epistemic space and the problem-solving dynamics and how they relate to knowing and information-knowledge stocks to provide a framework for the understanding of the particular unity of science and the general unity of knowing. There are important epistemic differences and similarities among the concepts and phenomena of problem-solving, problems, solutions, problem–solution duality, neuro-decision-choice actions, and information-knowledge system within the general activities of cognitive agents over the epistemological space, epistemic space and the space of epistemological-epistemic dualities in relational conditions of diversity and unity of intellectual investment flows and intellectual capital stocks. These concepts and phenomena are interdependent, unifying and yet different in terms of constructs and applications as they relate to the characteristic dispositions of technology, process, transformation, change and neuro-decision-choice actions.

All of them find meanings in the epistemological space of problem–solution dualities which is composed of the sub-space of problems and the sub-space of solutions and the space of the development of abstract and ordinary languages which are connected to the concepts and phenomena of epistemics and epistemic spaces defined over the space of imagination-reality dualities. The following questions must be answered relative to the first principle of knowing. What are the meanings of the concepts and phenomena of problems, problem sub-space, solutions, solution sub-space, problem–solution duality, space of problem–solution dualities and the space of decision-choice actions? The relationship between the space of problem–solution dualities and the space of neuro-decision-choice actions are always taken as known at best and as understood at the extreme by scientists, non-scientists, mathematicians, philosophers and specialists in all areas of knowing and information-knowledge accumulation and human endeavors. Complicating the understanding of the relationships among the spaces is an analytically epistemic failure to make explicit the concept of information in support of knowledge as some form of either justified or non-justified belief system or both. The definition of the concept of information, knowledge, fact and evidence is extensively discussed in the monographs [670, 671]. It is important to note that information is not knowledge and knowledge is not information over the epistemological space even though they are derivatively interdependent. It may be reiterated that information is defined as characteristic-signal disposition, data is defined as the recording variety occurrences, knowledge is defined as paradigmatically derived from information with fuzzy-stochastic conditionality under the conditions of cognitive capacity constraints from the space of certainty-uncertainty dualities.

5.2.1 Epistemological Information-Creation Systems and Information-Processing Systems

Over the epistemic space, the concept of information is taken at all areas of knowing as a primitive concept which is exactly exist in the ontological space for knowing, learning analysis and as input into epistemic analysis to understand variety identifications and variety transformations over the space of static-dynamic dualities. In the traditional approach, the conditions of the *first principle* in knowing and understanding are violated in the sense that a distinction is not established between information and knowledge creating a situation in knowing where information and knowledge are interchangeably used with the confusion on which of the information or knowledge is the correct input into neuro-decision-choice actions. It is the active and continual interactions among the varieties of the problem–solution dualities and the elements of neuro-decision-choice actions within the information space that give rise to the knowing, the *concept of thinking* and the *concept of reasoning* as decision-choice actions to resolve conflicts over the space of problem–solution or ignorance-knowledge dualities under the *principle of opposites* with relational continua and unity over the epistemological space, where varieties and categorial varieties are distinguished by their characteristic dispositions. The question is: what is information and how does it relate to input–output processes in cognition, knowing, information-knowledge development and event recording to keep the past connected to the present through knowing and learning and present connected to the future through transformation?

The concepts in linguistic theory in relation to knowing, knowledge and communication over the space of source–destination dualities cannot be understood without an explicit meaning of information. For example, the language of classical mathematics has limited use when some form of variety information is attached to it. Decision-choice actions are about activities to resolve goal-objective conflicts over the space of problem–solution dualities. Thinking is about the epistemic development of algorithmic tactics and strategies to effect direction of reasoning in the space of decision-choice actions in order to abstract solutions and affect the desired outcomes relative individual and collective preferences. We shall now give explicit definitions and explications to these concepts of problem–solution duality, problem, solution and the corresponding spaces in verbal and algebraic representations in terms of their uses in this monograph. The first principle involves the following questions: What is x and what is y and what are their differences and similarities? What are their information contents, what are the types of relationality among them and how do their information contents enter the neuro-decision-choice actions as either inputs or outputs? Similarly, what enters as inputs into the decision-choice modules as actions are taken over the elements of the space of problem–solution dualities and what comes out as outputs? Explicitly, are decisions based on either information or knowledge? The definitions of the essential concepts in establishing problems, solutions and problem–solution dualities will be given around the concepts and phenomena of real costs and real benefits involving the negative–positive dualities of varieties under the existing preference scheme and philosophical consciencism that has taken hold of the society at the generation time point to examine the variety cost–benefit differences in variety knowing as intellectual investment flows and information-knowledge accumulation as intellectual capital stocks.

Definition 5.1.1.1A: Real Cost, Real Benefit and Real Cost–Benefit Duality (Verbal) A real cost is a set of variety characteristics the presence of which makes a variety repulsive for a selection by a decision-choice agent or collective operating over the space of varieties. It is called a *repulsive characteristic disposition*, where every repulsive characteristic disposition is a cost-characteristic disposition. Similarly, a real benefit is a set of variety characteristics the presence of which makes a variety attractive for selection by a decision-choice agent or collective operating over the space of varieties. It is called an *attractive characteristic disposition*, where every attractive characteristic disposition is a benefit characteristic disposition. The joint occurrence of cost disposition and benefit disposition in a variety is an *internal cost–benefit duality*. The intra-categorial and inter-categorial joint appearances of cost disposition and benefit disposition is the *external cost–benefit duality*.

Definition 5.1.1.1B: Real Cost, Real Benefit and Real Cost–Benefit Duality (Algebraic) Given the space of varieties, $(\mathbb{V})$ with a generic element $(\nu \in \mathbb{V})$ and the space of characteristics $\mathbb{X}$ with generic element $(x \in \mathbb{X})$ where $(\mathbb{C} \subset \mathbb{V})$ is a category with corresponding characteristics $(\mathbb{X}_{\mathbb{C}} \subset \mathbb{X})$ then $\{\mathbb{X}_\nu^{C} | \nu \in \mathbb{V}\}$ is cost characteristic disposition with C representing cost and $\{\mathbb{X}_\nu^{B} | \nu \in \mathbb{V}\}$ is benefit characteristic disposition where B represents benefit. The pair structure $\{\mathbb{X}_\nu^{C}, \mathbb{X}_\nu^{B} | \nu \in \mathbb{V}\}$ is the *internal cost–benefit duality* for variety in $\mathbb{V}$ and the pair structure $\left\{\left(\mathbb{X}_{\nu_i}^{C}, \mathbb{X}_{\nu_i}^{B}\right), \left(\mathbb{X}_{\nu_j}^{C}, \mathbb{X}_{\nu_j}^{B}\right) | \nu_i \in \mathbb{C}_i \subset \mathbb{V} \text{ and } \nu_j \in \mathbb{C}_j \subset \mathbb{V}, i, j \in \mathbb{I}^{\mathbb{V}}\right\}$ is the *external cost–benefit* duality, where the internal cost–benefit duality reveals the identity of the variety and the external cost–benefit duality reveals the comparative positions of the varieties (ν_i, ν_j) over the space $(\mathbb{V})$. The collection of all the variety internal and external cost–benefit dualities constitutes the space of cos-benefit dualities with cost information $\left(\mathbb{Z}_\nu^{C} = \mathbb{X}_\nu^{C} \otimes \mathbb{S}_\nu^{C}\right)$ and benefit information $\left(\mathbb{Z}_\nu^{B} = \mathbb{X}_\nu^{B} \otimes \mathbb{S}_\nu^{B}\right)$.

Note 5.1.1.1: Concepts of Cost–Benefit Duality and the Space of Cost–Benefit Dualities The concepts and phenomena of cost, benefit and cost–benefit duality are not only analytical essentiality but they are the epistemic foundations of all neuro-decision-choice actions where they establish the space of incentive-disincentive dualities as connected to the ordinary reward-punishment systems for deterrence and encouragement over the action spaces in intra-societies and inter-societies. The concept and the phenomena of a variety cost–benefit duality contain cost dual and benefit dual in an internal dualistic conditions and external dualistic-polar conditions such that every variety is either a cost or a benefit if the cost or the benefit dual dominates the benefit or the cost dual, where the union of the cost dual and the benefit dual constitutes the cost–benefit duality providing the variety's identity in existence, ranking and neuro-decision-choice space to form the information-knowledge conditions for preference ordering and neuro-decision-choice action to generate new information for new knowledge.

Definition 5.1.1.2A: Preference and Preference Order (Verbal) A preference is a neuro-cognitive arrangement of available varieties over the space of quality-quantity dualities in a sequence of an increasing or a decreasing desirability or undesirability under reward-punishment modes and the variety defining characteristics as identity revealing. The schedule of such an increasing or decreasing sequence of desirability or undesirability is the *preference order* or *preference scheme* that must obey the law of relational comparability of varieties in the space of varieties with relational regularity made up of *reflexivity,* where every variety is preferred to itself, *anti-symmetry* in that the comparability relational order between two different varieties is non-reversible and the sequence is continuous through relational *transitivity* in that the variety disability sequence is order preserving in that if the variety one is more desirable than variety two and variety two is more desirable than variety three then the variety one is more desirable than variety three.

Definition 5.1.1.2B: Preference and Preference Order (Algebraic) A preference is a neuro-cognitive arrangement over the elements in the space of varieties such that if $\left(\nu_i, \nu_j, \nu_\ell \in \mathbb{V} | i, j, \ell \in \mathbb{I}^{\mathbb{V}}\right)$, $\mathbb{I}^{\mathbb{V}}$ is an index set of $\mathbb{V}$ and $(\succsim)$ represents a symbol of preference order then it must be the case that there is a ranking of triple free postulate such that for $\left(\nu_i, \nu_j, \in \mathbb{V} | i, j, \in \mathbb{I}^{\mathbb{V}}\right)$ then 1), (a) $\nu_i \succsim \nu_j$ or (b), (1) $\nu_j \succsim \nu_i$, c)$\nu_i \sim \nu_j$ indicating that the elements are comparable and ordered supported by (2) $\nu_1 \succsim \nu_1$(reflexivity), (3)$if\, \nu_i \succsim \nu_j$ and $\nu_j \succsim \nu_i \Rightarrow \nu_i = \nu_j$ (anti-symmetry), and if $\nu_i \succsim \nu_j$ and $\nu_j \succsim \nu_\ell \;\Rightarrow\; \nu_i \succsim \nu_\ell$(transitivity). In this respect, if the preference is expressed over $(\nu_1, \nu_2, \ldots \nu_i, \ldots \in \mathbb{V}_1 \subset \mathbb{V})$ then we have a *preference order* in the form $\left(\nu_1 \succsim \nu_2, \succsim \cdots \succsim \nu_i, \ldots \in \mathbb{V}_1 \subset \mathbb{V}\right)$.

Note 5.1.1.2: On Preferences and Preference Order Preferences arise under conditions of varieties distinguished by different characteristic dispositions as seen by cognitive agents in the space of decision-choice actions in relation to goals, objectives and visions that reside in the space of imagination-reality dualities under the principle of non-satiation over the spaces of quality-quantity and cost–benefit dualities in relation to the elements of increasing–decreasing cost–benefit dualities. The establishment of preference ordering is thus defined over the space of variety problem–solution dualities as established over the space of relativity, where comparative analytics can be exercised under static-dynamic conditions of behaviors of varieties, where the varieties are established hierarchy of preferences in ascending or descending order toward a destination. Such preferences are not defined under conditions of sameness and lack of varieties. They are established over the space of relativity and are difficult to be established over the space of absoluteness. Preferences are very important not only for neuro-decision-choice actions over the space of problem–solution dualities but for the construction of any theory of human progress through knowing and information-knowledge accumulation under the socio-psychological principle of non-satiations. The criteria for preference and preference ordering are established by some measure of cost–benefit conflict with a rationality under the *asantrofi-anoma* principle, where every variety contains simultaneously *repulsive characteristics*, that are unwanted due to undesirability and *attractive characteristics* that are wanted due to desirability to establish the identity of the variety over the space of neuro-decision-choice actions.

Every variety in the space of imagination-reality dualities is a *relevant alternative* for preference ordering of varieties and subject to neuro-decision-choice action in the action space towards a destination, while every element not in the space pf imagination-reality dualities is an *irrelevant alternative* for preference ordering of varieties and not subject to neuro-decision-choice action in the action toward a destination. The collection of all the relevant alternatives the *relevant action set*, while the collection of all the irrelevant alternatives are *irrelevant action set.* The relevant action set is a union of imaginary action set and reality action set which belongs to the attainable space of possibility-probability dualities. The irrelevant action set belongs to the complement of the space of possibility-probability dualities which is contained in the space of impossibility-improbability dualities in reference of goal-objective destination. This is a cost–benefit approach in defining the relevant and irrelevant alternatives and the corresponding sets over the action space as may be related to the space of goal-objective varieties as contained in the space of imagination-reality dualities.

In this cost–benefit approach, the set of the repulsive characteristics constitute the *cost dual* while the set of the attractive characteristics constitute the *benefit dual* of the internal duality, where any choice of variety is defined over the space of attractiveness-repulsiveness dualities which is mapped onto the space of incentive-disincentive dualities. The rationality of neuro-decision-choice action is that the attractive disposition must dominate the repulsive disposition. The preference ordering for neuro-decision-choice actions has a minimal element that is determined relative to the conditions of an unwanted actual variety and a maximal element where the stopping rule of the maximal element is resource availability for its actualization.

The individual, collective and generational preferences are affected by the intellectual stock-flow disequilibrium dynamics in such a way that there are intergenerational continual shifting towards new variety preferences and evaluative conditions of what varieties constitute problems and solutions over the space of destruction-destruction dynamics affecting the social conditions for knowing as the production of intellectual investment flows and information-knowledge accumulation as the intellectual capital stocks, the service of which enters as input–output varieties into the space of neuro-decision-choice actions. The changing nature of the problem–solution relationality affects the techniques and methods required to be exercised over the space of problem–solution dualities. The important thing to note is the nature of the relational interactions between the space of preferences and the space of information-knowledge systems over the space of static-dynamic dualities affect the social philosophical consciencism which has taken hold of the society and the generational timepoint continually shifting the generational preference ordering, goals, objectives and visions creating intergenerational differences in relevant and irrelevant alternatives creating new and different problems requiring new and different knowing and information-knowledge stocks to locate corresponding solutions and locate corresponding solutions. As the information-knowledge stock-flow system expands, it changes the size and quality of the social philosophical consciencism which then affects the neuro-decision-choice activities in the evaluation of variety preferences over the space of problem–solution dualities as guided by the elements of the space of incentive-disincentive dualities which is defined by cost–benefit dualities as seen over the system of time dualities of past-present dualities, present-future dualities and past-future dualities as well as a system of information dualities of past-present information, present-future information and past-future information as derivatives from the Sankofa-anoma conditions and principle.

In other words, the understanding of the human progress, decision-choice activities, knowing and information-knowledge production are shaped by the time trinity (sankofi-anoma principle of time as the fourth dimension of existence) from the past to the present, from the present to the future and from the past to the future.

Where the expansions of the information-knowledge accumulation alter the social philosophical consciencism and hence the preferences, judgments, vision and neuro-decision-choice activities over all areas of human actions including knowing and information-knowledge accumulation to aspirations under the principle of non-satiation. The degree of alternation will depend on the effective interactions between the information-knowledge system and the social philosophical consciencism that have taken root in thee social set-up and the generational interpretations with the space of source–destination dualities relative to the generations decision-choice position in the problem–solution dynamics.. The important thing to note is that that preferences, the information-knowledge system, neuro-decision-choice and social philosophical consciencism are interdependent over the space of identification-transformation dualities, where variety problems and solutions are to be identified and decision-choice actions are to set the variety solutions against variety problems relative to elements in the goal-objective space. The important analytical concerns center around how do problems and solutions arise and how do they relate to knowing and information-knowledge production in any social set-up.

Note 5.1.1.3: On Philosophical Consciencism and the Problem–Solution Process We have made some notes on the concept and phenomena of costs, benefits, cost–benefit dualities, preferences and preference order, relevance-irrelevant dualities and how their uses are affected by social philosophical consciencism of social set-ups, where the social philosophical consciencism provides a continual expansions of *echo-chamber* of organicity for information-knowledge input–output services into thinking and reasoning over the action space. It is important to be familiar and understand the characteristic disposition of the nature of social philosophical consciencism in its generality and specificity and the role that it plays in the space of neuro-decision-choice systems, knowing and information-knowledge accumulation. The generality refers to the concept and phenomenon of its characteristic disposition while the specificity refers to its development and use in different national setting with changing contents over the intergenerational paths of national histories. The organicity of the philosophical consciencism, like any variety, is composed of the positive characteristic disposition that establishes the positive dual and the negative characteristic disposition that establishes the negative dual. The positive dual and the negative dual reside in the space of cost–benefit dualities in reference to the elements in the goal-objective space as conceived within the space of imagination-reality dualities. The positive dual simultaneously contains a cost disposition and a benefit disposition as viewed over the space of neuro-decision-choice space. Similarly, the negative dual simultaneously contains a cost disposition and a benefit disposition as also viewed over the space of neuro-decision-choice space.

The organic echo chamber of the philosophical consciencism is also a collection of competitive echo sub-chambers composed of partitioned organic characteristic disposition of the philosophical consciencism into sub-system of the general philosophical consciencism and continually amplified for socio-political cost–benefit advantage in cost and benefit distributions over the space of socio-political input–output processes. Corresponding to each sub-chamber is a social sub-group with similar mindedness to acquire the same information input into their decision choice process with a complete disregard of fuzzy-stochastic conditionality in terms of intersectionality, group intentionality and the conditions of cost–benefit distributions as assessed through the contents of the social philosophical consciencism. The neuro-decision-choice activities over the space of problem–solution dualities depend essentially on the contents of the social philosophical consciencism. The evolving contents of the philosophical consciencism guide information selections, knowledge approval, problem definition, thinking development, reasoning execution, solution development, decision-choice acceptance, goals, objectives and vision over the space of imagination-reality dualities. For extensive development of the theory social philosophical consciencism and it role in socio-economic and socio-political and socio-legal transformation see [670] and its relationship to the understanding of necessity and initial conditions of change see [668, 669], where categorial conversion and social philosophical consciencism present the conditions of the space of necessity-freedom dualities where the degrees of necessity establish the external conditions for change and the degrees of freedom defines the internal conditions for change in the space of external-internal dualities.

The competing social groups will coalesce into socio-political duality to establish a socioeconomic polarity relative to the distribution questions over the space of cost–benefit dualities. It is within this framework that systemic ills and goods develop and hence must be examined relative to the telescopic history of past-present duality as one views the direction of telescopic history of the present-future dualities and the lessons from telescopic history of past-future dualities can be learned for neuro-decision choice corrections to set the direction of social change and where the systemic ills and goods are teachable feedback conditions for social improvements toward preferred states. The epistemic geometry of relational connectivity of categorial conversion, philosophical consciencism, systemic ills, goods, cost and benefit is shown in Fig. 5.1.

5.2.2 The Problem, Problem Space, Solution and the Solution Space

Let us connect the concepts of preference, cost, benefit and cost–benefit duality to the concepts of problems, solutions and problem–solution duality in terms of definitions and conditions of diversity and unity of science through the diversity and the unity of knowing and information-knowledge accumulation. The objective, here, is to establish the claim that problems and solutions are established by cost–benefit conditions over the space of human actions relative to the space of goals, objectives and vision. Let us define these concepts and phenomena to connect them to the process of knowing as a production of intellectual investment flows and information-capital accumulation as intellectual capital stocks.

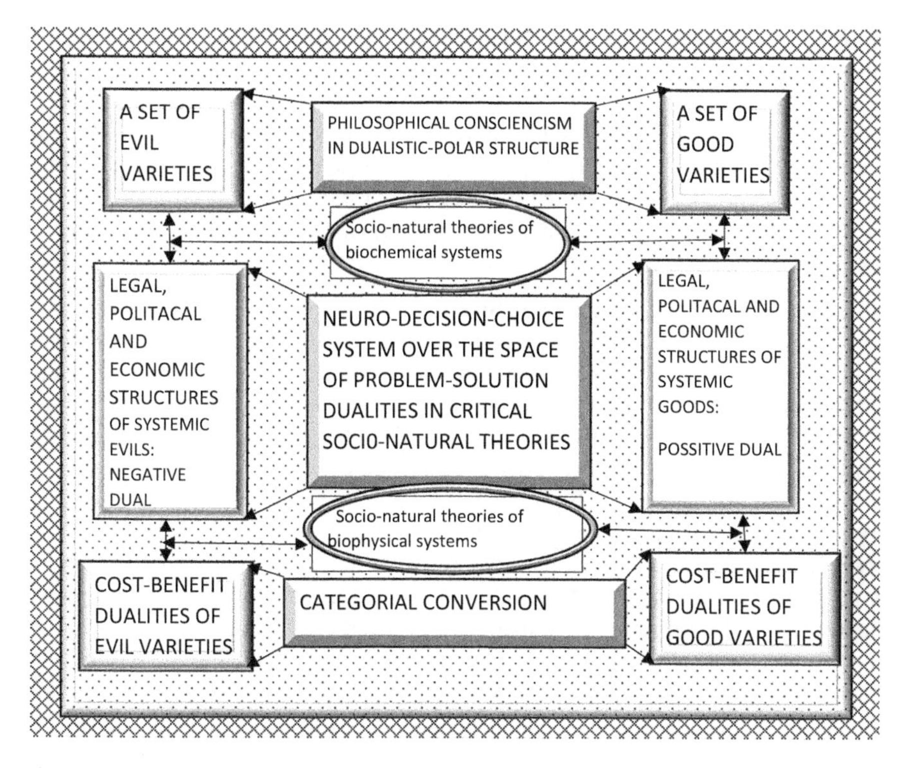

Fig. 5.1 Epistemological Geometry of relational Connectivity of categorial Sub-systems

Definition 5.1.2.1A: Problems and Problem Space (Verbal) A *problem* is an existence of a set of cost–benefit characteristics of a variety such that the relative cost–benefit (C–B) characteristic combination is assessed as undesirable variety, the existence of which is an obstruction to progress towards preferred varieties or variety which constitute(s) goals or a goal as seen by cognitive agents, and the transformation of which will open up a way for progress towards the preferred varieties or a variety, the relative real cost–benefit ($^{C}/_{B}$) characteristic dispositions of which are preferred to be obtained as a goal, objective or vision with a lower unit real cost disposition relative to a unit real benefit deposition, where the real cost disposition over the real benefit disposition of the actual variety is greater than the real cost disposition over the real benefit disposition of the potential variety to be actualized. The collection of such problems *defines the problem space* relative to the *space of goals, objectives and vision*, where every variety in the problem space is defined by dualistic principle with cost dual dominating the benefit dual.

Definition 5.1.2.1B: Problems and Problem Space (Algebraic) Given a variety $\{\nu \in \mathbb{V} = (\mathbb{V}_{\mathfrak{A}} \cup \mathbb{V}_{\mathfrak{U}})\}$ where $(\mathbb{V}_{\mathfrak{A}})$ is the space of the actuals and $(\mathbb{V}_{\mathfrak{U}})$ is the potential space of varieties with a real cost–benefit-induced preference order $(\succsim)$ in the form $(\mathbb{V}, \succsim)$ such that if $(\nu_1, \nu_2 \in \mathbb{V})$, $(\nu_1, \nu_2, \nu_3 \in \mathbb{V})$, and $(\nu_1, \nu_2, \nu_i \dots \nu_j, \dots \in \mathbb{V})$

with a corresponding space of relative real cost–benefit characteristic dispositions $\Re = \{\mathfrak{r} = {}^{\mathrm{C}}/_{\mathrm{B}} | \mathrm{C} \in \mathbb{C}, \mathrm{B} \in \mathbb{B}\}$, then $\{(\nu_i, \nu_j \in \mathbb{V}) \subset (\mathbb{V}, \succsim), \forall i, j \in \mathbb{I}_{\mathbb{V}}^{\infty}\}$, $\{(\mathbb{V}, \succsim) \subseteq ((\mathbb{V} \otimes \Re), \succsim)\}$ and $\{(\nu_i \mathfrak{r}_i), (\nu_j \mathfrak{r}_j) \in ((\Re \otimes \mathbb{V}), \succsim)\}$ then a problem exists if $(\mathfrak{r}_i > \mathfrak{r}_j)$ such that the actual variety $(\nu_i \in \mathbb{V}_{\mathfrak{A}})$ stands in the way of reaching the goal of a potential variety $(\nu_j \in \mathbb{V}_{\mathfrak{U}})$ where $(\nu_j \mathfrak{r}_j)_{\mathfrak{p}} \succsim (\nu_i \mathfrak{r}_i)_{\mathfrak{A}}$. The existence of variety $(\nu_i \in \mathbb{V}_{\mathfrak{A}})$ is said to be a problem and the collection of such varieties $(\nu_i \in \mathbb{V}_{\mathfrak{A}}, \forall i, \in \mathbb{I}_{\mathbb{V}}^{\infty})$ with corresponding $(\mathfrak{r}_i \in \Re, \forall i, \in \mathbb{I}_{\mathbb{V}}^{\infty})$ constitutes the general *problem space* $(\mathfrak{P})$ with a generic element $(\mathfrak{p} \in \mathfrak{P})$.

Note 5.1.2.1: Problem and Problem Space The definitions of the concepts of problems and problem space are very general, and, in fact, more general than what one might think in terms of ordinary linguistic usage and conditions of neuro-decision-choice activities. They use the concept of real relative cost–benefit characteristic dispositions rather than relative negative–positive characteristic dispositions in the sense that the negative (positive) characteristics may be either cost or benefit under problem–solution situations. Both the explanation and understanding of this usage are extremely important to the understanding of the existence of problems and solutions in neuro-decision-choice space where a solution variety leads a new solution-problem variety. It is not the relative dominance of either the positive or negative disposition in the variety duality that is an obstruction to either a goal, objective, vision and hence a problem. It is the relative dominance of either the cost or benefit disposition that creates conditions of a variety problem where cost is an undesirable disposition and benefit is a desirable disposition. It is useful at this point to note that the negative characteristic disposition may appear as either a real cost-characteristic disposition or as a real benefit-characteristic disposition relative to a goal-objective element from the goal-objective space. This also holds for the positive-characteristic disposition, where the variety positive-characteristic disposition may appear as real cost or benefit characteristic disposition relative to an element of a goal, objective or a vision. Problems exist when the current cost–benefit characteristic dispositions are ranked as less preferred to potential cost–benefit dispositions of potential varieties.

The definition of a problem is in terms of the dominance of cost-characteristic disposition over the benefit-characteristic disposition of the actual variety relative to cost–benefit dispositions of potential varieties. The definition of a problem space is, thus, a collection of all varieties where the cost-characteristic dispositions dominate the benefit-characteristic dispositions relative to goals, objectives or visions in the neuro-decision-choice space. Any problem is thus defined as an obstruction to reach a goal, an objective or a vision, and any neuro-decision-choice action is defined as a process to remove a problem to create a vacuum for its replacement, where a removal of a problem is part of variety transformation. In this respect, a question arises as to what is either a solution, a goal, an objective or a vision. Let us keep in mind that a problem is always an actual variety belonging to the space of reality and a solution is always a potential variety belonging to the space of imagination in relation to either a goal, an objective or a vision within an actual-potential polarity containing the space of imagination-reality duality with relational continuum, unity and with residing cost–benefit dualities under a *cost–benefit preference order*.

The analytical concepts of actual-potential duality, imagination-reality duality and the corresponding spaces are powerfully important to the understanding of neuro-decision-choice processes relative to historic past, present and the future and their importance to knowing as intellectual investment flows and information-knowledge accumulation as intellectual capital stocks and their effects on the dynamics of social philosophical consciencism, where all neuro-decision-choice actions are resolutions of conflicts between imagination and reality in the space of imagination-reality dualities as contained in the space of actual-potential dualities relative to elements in the space of problem–solution dualities in relational continuum and unity. It is the conditions of the space of imagination-reality dualities under cognitive capacity limitations that give rise to analytical structures of possibility-impossibility duality, probability-improbability duality, necessity-freedom duality, epistemic spaces, epistemic fields and possible-impossible-world duality in the search for the elements in the space of falsity-truth dualities. The cost–benefit ratio in defining the conceptual differences between variety problem and variety solution also acts as a criterion in ordering variety problems in decreasing preferences and similarly in ordering variety solutions in increasing preferences. In this respect, a question arises as to what the meaning of the concept and phenomenon of a solution are to which we turn our attention.

Definition 5.1.2.2A: Solution and Solution Space (Verbal) A *solution* is the existence of a set of cost–benefit characteristics of a imagination variety belonging to the space of potential varieties such that the relative cost–benefit characteristic combination $({}^{C}/_{B})$ is assessed as a desirable variety relative to an actual variety problem in the space of reality contained in the space of the actuals under dualistic-polar conditions, the choice and implementation of which will constitute a replacement or an enabler to remove a variety problem which stands in the way of progress towards a goal, an objective or a vision as seen by cognitive agents. A variety solution has a lower unit real cost relative to a unit real benefit as compared to that of its corresponding variety problem, where the unit real cost over real benefit of the variety problem is greater than the unit real cost over the real benefit of the potential variety as a solution. The collection of such variety solutions *defines the solution space* relative to the problem space given the *space of goals, objectives and visions.*

Definition 5.1.2.2B: Solution and Solution Space (Algebraic) Given a variety $\{\nu \in \mathbb{V} = (\mathbb{V}_{\mathfrak{A}} \cup \mathbb{V}_{\mathfrak{U}})\}$ where $(\mathbb{V}_{\mathfrak{A}})$ is the space of the actuals and $(\mathbb{V}_{\mathfrak{U}})$ is the potential space of varieties with a real cost–benefit-induced preference order $(\succsim)$ in the form $(\mathbb{V}, \succsim)$ such that if $(\nu_1, \nu_2 \in \mathbb{V})$, $(\nu_1, \nu_2, \nu_3 \in \mathbb{V})$, and $(\nu_1, \nu_2, \nu_i \ldots \nu_j, \ldots \in \mathbb{V})$ with a corresponding space of relative real cost–benefit characteristic dispositions, $\mathfrak{R} = \{\mathfrak{r} = {}^{C}/_{B} | C \in \mathbb{C}, B \in \mathbb{B}\}$, such that $\left\{(\nu_i, \nu_j \in \mathbb{V}) \subset (\mathbb{V}, \succsim), \forall i, j \in \mathbb{I}_{\mathbb{V}}^{\infty}\right\}$, $\left\{(\mathbb{V}, \succsim) \subseteq ((\mathbb{V} \otimes \mathfrak{R}), \succsim)\right\}$ and $\left\{(\nu_i \mathfrak{r}_i), (\nu_j \mathfrak{r}_j) \in ((\mathfrak{R} \otimes \mathbb{V}), \succsim)\right\}$ then a variety solution exists if $(\mathfrak{r}_i < \mathfrak{r}_j)$, such that the potential variety $\{\nu_j \in \mathbb{V}_{\mathfrak{U}}\}$ helps to remove a problem variety $\nu_i \in \mathbb{V}_{\mathfrak{A}}$ where $(\nu_j \mathfrak{r}_j)_{\mathfrak{p}} \succsim (\nu_i \mathfrak{r}_i)_{\mathfrak{A}}$. The existence of variety $\nu_j \in \mathbb{V}_{\mathfrak{U}}$ as a replacement is said to be a solution and the collection of such varieties $\{\nu_j \in \mathbb{V}_{\mathfrak{U}}, \forall j \in \mathbb{I}_{\mathbb{V}}^{\infty}\}$ with corresponding $(\mathfrak{r}_j \in \mathfrak{R}, \forall j, \in \mathbb{I}_{\mathbb{V}}^{\infty})$ constitutes the general *solution space* $\mathfrak{S}$ with a generic element $(\mathfrak{s} \in \mathfrak{S})$ relative to a problem $(\mathfrak{p} \in \mathfrak{P})$.

Note 5.1.2.2: Solutions and Solution Space The definitions of the concepts of variety solution and solution space, just like the variety problem and the problem space, are very general. They are applicable over all areas of human endeavor. They use relative real cost–benefit characteristic dispositions rather than relative negative–positive characteristic dispositions to establish their identities. The explanation and understanding of this approach based on the principle of opposites are extremely important to the understanding of the existence of both problems and solutions in the spaces of variety identification and neuro-decision-choice actions. Every variety simultaneously contains positive and negative characteristics in a relative proportion that defines the existential identity of the variety in knowing and information-knowledge accumulation in the space of input–output space. Similarly every variety contains simultaneously benefit and cost characteristics in a relative proportion that defines the decision-choice identity of the variety in ranking, preference and selection. It is important to note the similarity and differences between existential identity and the neuro-decision-choice identity in the space of variety static-dynamic dualities. Both the problem and solution existing with dualistic-polar conditions are varieties distinguished by their relative characteristic dispositions provide the framework for preference ranking order and neuro-decision-choice actions. All problems and corresponding solutions are subjectively defined in terms of real cost–benefit dispositions relative to goals, objectives and visions where the problems exist in the space of realities as contained in the space of the actuals while the solutions exist in the space of imaginations as a subspace of the space of the potentials which are executed as a goal, objective, mission or vision.

Generally, it is not the relative dominance of either the positive or negative disposition in the variety duality that is an obstruction to either a goal or objective and hence a solution, but the relative dominance of either cost or benefit disposition as assessed by cognitive agents. The driving forces of the destruction-replacement processes are the results of the interactively distributional dynamics of real cost–benefit dispositions and neuro-decision-choice actions over the space of problem–solution dualities with information-knowledge inputs from the services of the intellectual capital stocks. The variety destruction-replacement processes in the areas of knowing are information-knowledge processes required for the identification-transformation processes over the space of variety ignorance-knowledge (*ohnim-sua-ohu*) dualities, where corresponding to every variety ignorance is a variety knowledge and corresponding to every variety knowledge is a variety ignorance under relational continuum and unity, where the unity of knowing is through the spaces of acquaintance, methodological construction-reduction dualities, certainty-uncertainty dualities, doubt-surety dualities and decidability-undecidability dualities irrespective of the category of knowing whether classified as science or non-science in cognitive activities over the space of imagination-reality dualities.

5.3 The Goals, Objectives and the Goal-Objective Space for Decision-Choice Actions

The problems, problem space, solution and the solution space must all be related to a goal, an objective, and the goal-objective space in the space of human actions containing visions whether explicitly or implicitly stated. It is useful to note that the negative characteristic disposition may appear as either cost-characteristic disposition or as benefit-characteristic disposition or both relative to a goal-objective element from the goal-objective space. This asymmetric relation induced by preference ordering also holds for the positive-characteristic disposition. Under the principle of opposites, the definition of a variety solution is in terms of the dominance of the real benefit-characteristic disposition over the real cost-characteristic disposition of a potential variety relative to the cost–benefit conditions of the variety problem. The definition of a solution space is, thus, a collection of all potential varieties where the benefit-characteristic dispositions dominate the cost-characteristic dispositions relative to the goal-objective varieties in the space of imagination-reality dualities as a subspace of the space of actual-potential dualities. Any variety solution is thus defined as an *enabler* to reach a goal, an objective or a vision while any variety problem is thus defined as an *inhibitor* to reach a goal, an objective or a vision.

In other worlds, any problem–solution duality is seen as an inhibitor-enabler duality over the action space towards a goal, objective or a vision. A question, therefore, arises as to what is either a goal, an objective or a vision around which the concepts of problems and solutions are defined. Let us keep in mind that a variety problem is a variety reality and a variety solution is a variety imagination where any goal-objective element has a residing cost–benefit duality within an actual-potential polarity, and where every variety is defined by its negative–positive duality, where every neuro-decision-choice action is defined by a cost–benefit preference scheme towards preferred varieties under the principle of non-satiation constrained by resource availability. These conditions will be used to define the concepts of goals and objectives and extended to ideas, missions and visions as well as their relationships to information and knowledge in these discussions. It must be pointed out that the concepts of cost–benefit duality and polarity are the epistemic foundations and justification of cost–benefit analytics as the general theory of neuro-decision-choice actions. Let us turn our attention to the concepts and phenomena of goals, objectives and visions.

Definition 5.2.1A: Goals, Objectives and the Goal-Objective Space (Verbal) A goal or an objective is an imaginative potential variety which a cognitive agent sees as defining a preferable state with a unit of real benefit(cost) disposition relative to a unit of real cost(benefit) disposition, as assessed in the actual or present state, to have greater (lower) inferred future real benefit (cost) disposition relative to real unit cost (benefit) disposition, and where the agent *strives to achieve* it by transforming the *current variety* state or varieties qualitatively, quantitatively or both to the imaginatively *potential variety* or state through the destruction-replacement process. The collection of such preferred potential varieties in terms of things, states or processes with preferred cost–benefit (benefit-cost) dispositions is the *goal-objective space*.

Definition 5.2.1.B: Goals, Objectives and the Goal-Objective Space (Algebraic) Let $\left((\mathbb{V} \otimes \Re), \succsim\right)$ be a cost–benefit preordered variety space in the universe $\Omega = (\mathfrak{A} \cup \mathfrak{U})$ as a union of the space of the actual varieties $(\mathfrak{A} = \mathbb{V}_{\mathfrak{A}})$ and potential varieties $(\mathfrak{U} = \mathbb{V}_{\mathfrak{U}})$. A goal or an objective is an existence of a *variety in the space of imagination* $(\mathfrak{G} = \mathbb{V}_{\mathfrak{G}})$ as a subspace of the *potential space* where $(\mathfrak{G} = \mathbb{V}_{\mathfrak{G}} \subset \mathfrak{U} = \mathbb{V}_{\mathfrak{U}})$ with relative real cost–benefit value $\left\{\left(\nu_j \mathfrak{r}_j\right) \in \left((\mathbb{V} \otimes \Re), \succsim\right)\right\}$ and $\left(\nu_j \in \mathfrak{U} = \mathbb{V}_{\mathfrak{U}} | j \in \mathbb{I}_{\mathfrak{U}}\right)$ with lower relative real cost–benefit disposition $(\mathfrak{r}_j \in \Re)$ among the attainable varieties relative to a real variety in the space of real varieties $(\mathfrak{H} = \mathbb{V}_{\mathfrak{H}})$ as a dualistic-polar subspace of the space of actual variety $(\nu_i \in \mathfrak{A} = \mathbb{V}_{\mathfrak{A}} | i \in \mathbb{I}_{\mathfrak{A}})$ with $(\mathfrak{r}_i \in \Re \ni \mathfrak{r}_i > \mathfrak{r}_j)$ and $\left(\nu_j \succsim \nu_i\right)$ where $(\mathfrak{H} = \mathbb{V}_{\mathfrak{H}} \subset \mathfrak{A} = \mathbb{V}_{\mathfrak{A}})$ The collection of all $\left(\nu_j \succsim \nu_i, \forall i, j \in \mathbb{I}_{\mathbb{V}}^{\infty}\right)$ with $(\mathfrak{r}_i \in \Re \ni \mathfrak{r}_i > \mathfrak{r}_j)$ constitutes the *goal-objective space*$(\mathbb{G})$ with generic element $(\mathfrak{g} \in \mathbb{G})$ defined in the space of imagination-reality dualities $((\mathfrak{H} \otimes \mathfrak{G}) = (\mathbb{V}_{\mathfrak{H}} \otimes \mathbb{V}_{\mathfrak{G}}))$.

Note 5.2.1: Goals, Objectives and the Goal-Objective Space The concepts of *strive to achieve* and variety transformation imply actions, processes and expenditures of energy to create variety inputs and variety outputs. The variety inputs are information items and the variety outputs are information-knowledge item about other variety dispositions over the general spaces of input–output and production-consumption dualities. They imply also problem-solving activities in the *input–output space* in terms of destruction-replacement processes, where every problem–solution process has input(s) and output(s). The variety is a general concept that relates to information-knowledge activity as a problem in static states and dynamic states in relation to continual social transformations of organizational and institutional arrangements with innovations in continual problem-generating processes and solution-generating processes where every problem is a solution and every solution is a problem relative to the varieties in the space of goals, objectives and visions and the system of preferences. In other words, the goal-objective space is intimately connected to the space of problem–solution duality which finds meaning over the goal-objective space. There are no problems if there are no alternative varieties and subjective preferences under the principle of non-satiation which is constrained by resource availability of intellectual and non-intellectual types. A goal is either a protected or changed condition of existing varieties in the space of realities to varieties which are preferable where protected conditions relate to info-statics in variety identifications and changed conditions relate to info-dynamics of variety transformations. These analytical structures connect all areas of knowing into the unity of science and non-science, intra-science unity and intra-non-science unity as constituting the unity of universal knowledge and the processes of knowing and accumulation as intellectual stock-flow processes over the space of imagination-reality dualities. The characteristic dispositions of the space of imagination is $\mathbb{X}_{\mathfrak{G}}$, with the corresponding signal disposition $(\mathbb{S}_{\mathfrak{G}})$, the space of reality is $\mathbb{X}_{\mathfrak{H}}$ with a corresponding signal disposition ($\mathbb{S}_{\mathfrak{H}}$) and the space of imagination-reality dualities is $\left(\mathbb{X}_{(\mathfrak{H}\otimes\mathfrak{G})} = \mathbb{X}_{(\mathbb{V}_{\mathfrak{H}}\otimes\mathbb{V}_{\mathfrak{G}})} = \mathbb{X}_{\mathfrak{L}}\right)$ where $\mathfrak{L} = \mathfrak{H} \otimes \mathfrak{G} = \mathbb{V}_{\mathfrak{L}} = \mathbb{V}_{\mathfrak{H}\otimes\mathfrak{G}}$ with the corresponding signal disposition $\left(\mathbb{S}_{(\mathfrak{H}\otimes\mathfrak{G})} = \mathbb{S}_{(\mathbb{V}_{\mathfrak{H}}\otimes\mathbb{V}_{\mathfrak{G}})} = \mathbb{S}_{\mathfrak{L}}\right)$.

In fact, the concepts and phenomena of epistemology, epistemological space, epistemics and epistemic space unlike the concepts and phenomena of ontology, ontological space, ontic and ontic space, have no existence without the existence of cognitive agents with activities over the space of goals, objectives and vision. The processes of achieving this goals, objectives and visions create variety problems, solutions and problem–solution dualities which have no logical existence without cognitive preferences that establish goal-objective conditions for neuro-decision-choice activities over the action space to define the space of variety destruction-replacement dualities for continual transformations over the space of imagination-reality dualities as a subspace of actual-potential dualities. The variety preferences in turn have no meaning and existence without the existence of cognitive agents who constitute a social set-up with conflicts, cooperation and information-knowledge needs as input–output processes over social-natural dualities. The special property about the preference is the conditions of non-satiation in the sense that the preferences over the space of individual-collective dualities have no upper bound creating the *universal principle of non-satiation* where greater and greater benefits are sought and lower and lower costs are sought by cognitive agents acting over the space of problem–solution dualities in search for more varieties through the spaces of production-consumption and input–output dualities.

There is no need for the information-knowledge system and its development if there are no neuro-decision-choice actions which in turn have no existence without the existence of problems and solutions and the space of problem–solution dualities. Every decision-choice activity in cognitive existence is an action over the space of problem–solution dualities to achieve a path to a state of preference with anticipated dominances of benefit duals over the cost duals on the aggregate in the space of imagination-reality dualities. Problem-solving is simply a variety destruction-replacement process over the space of variety imagination-reality dualities as a subspace of the space of actual-potential dualities. The dualistic-polar conditions are such that the space of imaginations is a subspace of the space of the potentials relative to the space of individual-collective dualities. This way of seeing the problem–solution process must be understood. It provides an understanding of continual transformations to maintain the matter-energy equilibrium as well as information-knowledge disequilibrium, where the information-knowledge disequilibrium is intimately connected to the elements in the space of goals, objectives and visions. It also provides an understanding of conditions of continual existence of neuro-decision-choice actions, problems, solutions, conflicts, progress and retrogress and many other conditions. The individual, collective and social imaginations vary with degrees of curiosity under the social philosophical consciencism under the generational past-present-future telescopic enveloping.

How are the concepts and phenomena of goals, objectives and the goal-objective space related to the concepts and phenomena of problems, problem space, solutions, solution space, the space of problem–solution dualities, decisions, choices and decision-choice space? The interdependent nature of these concepts, in the study of spaces of the human retrogress-progress, peace-war, justice-injustice, freedom-necessity and many others dualistic-polar conditions are yet to be fully understood in terms of elements in the problem–solution polarities, where every pole has a residing duality to generate energy for transformation dynamics. How must problems and solutions be conceived and interpreted in relation to the development and use of information-knowledge system as input–output phenomena over the action space relative to neuro-decision-choice activities? Similarly, how must the concepts of *preference order* and *cost–benefit duality* be interpreted in a broad abstract way to understand the diversity principle within the unity principle of knowing that is being presented here. It is important to keep in mind that the researcher's choice of a problem–solution duality is also governed by *preference orderings* under rationality defined over the space of real cost–benefit dualities for neuro-decision-choice activities over the action space under the principle of the asantrofi-anoma tradition as guided by the social philosophical consciencism that has taken hold of the society at a generation point.

These conditions of preferences and a choice of problem–solution duality cover all areas of knowledge to unite all categories of knowing including science, non-science and engineering, non-engineering, and religion and have nothing to do with methods and techniques of actions over the space of problem–solution dualities. The problem–solution dualities are not selected by available techniques and methods since such an approach will destroy the continual development of methods and techniques of problem-solving, the growth of knowledge and the conditions of the information-knowledge problem in the reduction of ignorance, doubt and uncertainty, and in continual improvements of knowledge, surety and certainty of neuro-decision-choice activities. The idea of tailoring problem to technique and methods is substantially discouraged. Such an approach leads to ill-posed problems and phantom problems and paradoxes.

Categories or departments of knowing are not established by common tools and methods of neuro-decision-choice actions over the problem–solution dualities. Departments of knowing, research, learning and teaching are established by conditions of commonness of problem–solution dualities, where, for example, the department of physical sciences has a set of problem–solution dualities with common characteristic dispositions that are different from those in the set of problem–solution dualities of biological systems, leading to the establishment of the department of biological sciences and many others. The managerial and controlling departments in national affairs or organizational affairs are not established by common methods and techniques of neuro-decision-choice actions but by the conditions of common problem–solution dualities in economics, politics and law. The unity of science through the unity of knowing is established by the general principles of information-knowledge production with the use of epistemic instruments of the spaces of acquaintance, methodological constructionism-reductionism processes over the space of decidability-undecidability dualities as connected to the space of acceptance-rejection dualities. These general principles constitute the methodology or paradigm of thought in the information-knowledge stock-flow disequilibrium dynamics for self-correcting and self-exiting over the spaces of neuro-decision-choice activities and action, where knowing is an intellectual investment flow as additions to social intellectual capital stocks and information-knowledge accumulation is an intellectual capital stock as additions to the contents of the social philosophical consciencism as well as to provide intellectual capital services as inputs into the space of neuro-decision-choice actions.

The *unity and diversity principles of knowing* are not methods, concepts, language or other conditions of information-knowledge constructs that have been presented in the literature on the unity of science. The *unity and diversity principles of science* through the *unity and diversity principles of knowing* are the cognitive activities over the *space of problem–solution dualities* ($\mathfrak{R}$) in relation to goals, objectives, the goal-objective space and visions, where the space of cognitive actions is defined by the spaces of identification and transformation problem–solution dualities in static and dynamic states. The neuro-decision-choice activities encompass the processes of integrating the problem–solution dualities with the elements in the goal-objective space relative to the space of cost–benefit dualities. Every neuro-decision-choice action is about a transformation of an actual variety (*what there is*) in the space of realities ($\mathfrak{H} = \mathbb{V}_{\mathfrak{H}}$) which is assessed to have high real cost per real benefit to a potential variety (*what there is not*) in the *space of imagination* ($\mathfrak{G} = \mathbb{V}_{\mathfrak{G}}$) which is assessed to have lower real cost per real benefit in present-future relative terms. In other words, every neuro-decision-choice action is about the non-destruction or destruction of an actual variety ($\mathfrak{H} = \mathbb{V}_{\mathfrak{H}}$) and the actualization of a potential variety in the *space of imagination* ($\mathfrak{G} = \mathbb{V}_{\mathfrak{G}}$) that resides in the future with more preferred value as defined by its characteristic disposition and to potentialize an actual variety that resides in the present with less preferred value defined by its characteristic disposition. The neuro-decision-choice activities in the space of imagination-reality processes contained in the space of actual-potential process are information-knowledge generating. The actualization of potential varieties and the potentializing of actual varieties are all problem–solution polarities that are central to all cognitive activities to manufacture categorial convertors to change relational terms of imagination-reality varieties of the actual-potential varieties. In this respect, every problem-solving is a change in relation or a change in property to the dualistic-polar existence as well as information-knowledge generating and hence connected to knowing and information-knowledge accumulation.

The goals, objectives and the goal-objective space ($\mathbb{G}$) relate to the problems and the elements of the problem space as well as the solutions and elements in the solution space as specified in the information-knowledge structure at any static state which presents info-static conditions in terms of variety identifications of *what there was* and *what there is*. The variety identification of *what there is* presents itself as a problem in the space of knowing, and this *variety identification problem* exists in every area of knowing irrespective of the nature of the variety. The variety identification of *what there is* or *what there was* does not imply *what there is not* and *what there was not*. In fact, the variety identification of any element finds its meaning and conditions in the problem space relative to an element in the goal-objective space as seen from the experiential information structure under dualistic-polar conditions. Variety identifications and the identifications of categorial variety $(\mathbb{C}_\nu|\nu \in \mathbb{V})$ are information-knowledge problems in terms of individual and categorial identities. From the viewpoint of variety identification, the problem space and the goal-objective space present two sets of initial information-knowledge conditions for evaluations by cognitive agents in terms of the desirability of their state of being.

The understanding of the structure and implementation of concepts of evaluation, desirability and preference ordering of neuro-decision-choice actions on any static or dynamic variety requires the existence of an overall *criterion space* which must relate information-knowledge conditions of real cost–benefit, input–output or production-consumption structures of the elements to the varieties in the problem space relative to the varieties of the goal-objective space. It must be kept in mind that the information-knowledge conditions of any criterion in the criterion space are reducible in the last analysis to real cost–benefit information to craft the cost–benefit criterion to rank varieties in different categories and varieties in the same category as well as ranking categories if there are needs for preference ordering and neuro-decision-choice actions that must be related to the contents of the evolving philosophical consciencism. The concepts of ranking, order and comparison find meanings and definitions in the space of relativity which presents conditions of neuro-decision-choice actions and makes possible the practice of the principle of non-satiation constrained by intellectual and non-intellectual resources in space of cognitive actions.

The whole ideas of computability and non-computability, mathematical representation, logical completeness of representation and paradoxes relate to ill-posed problem–solution dualities under information-knowledge stock-flow disequilibrium dynamics within a paradigm of thought, where the inputs of neuro-decision-choice actions are information-knowledge capital-stock service flows and the outputs are information-knowledge intellectual investment flows to update the existing intellectual capital stocks through knowing as intellectual investment flows. In this respect, the characteristic dispositions of both failure and successes are information-knowledge outcomes. There are no limits to decisions and choice, organization and disorganization, oneness and dividedness, unity and diversity, ignorance and knowledge in that all these cognitive activities contained in neuro-decision activities reside in an infinite space under the principle of non-satiation as guided by conditions of philosophical consciencism in the use of dualistic-polar conditions of variety costs and benefits. The limits that may be associated with these neuro-decision-choice activities reside in the cognitive capacity limitations that are essential for the system of input–output corrections and the working mechanism of the principle of non-satiation in the space of feedback dynamics, where there is the same structure of the neuro-decision system with varying information-knowledge input–output processes over the space of problem–solution dualities and the telescopic past-present-future path of generational enveloping. The unity of neuro-decision-choice actions, irrespective of the area of human endeavor, is a process with disposition to balancing the activities between duals of dualities and poles of polarities over the space of cost–benefit dualities. The apparent differences in the neuro-decision-choice actions in different areas (departments) of knowing is due to different problem–solution dualities in the space of knowing and information-knowledge accumulation, where differences of social philosophical consciencism explain the differences of generational preferences of the same society and inter-societal differences in international preferences over the variety space in terms of knowing and production-consumption dualities.

5.4 Criterion Space, Decision-Choice Space and the Action Space

All neuro-decision-choice actions take place in the space of imagination-reality dualities involving the existence of variety diversities that must meet some meaningful comparative analytics in terms their relative desirability-undesirability duality. The comparative analytics involve not simply the negative and positive duals as established by their characteristic dispositions but in terms of their benefits and costs in claims of possession of various varieties. It is not the characteristic dispositions of diversities that are useful in the neuro-decision-choice assessments in terms of order of preferences but the degrees to which each characteristic disposition contributes to the welfare of the neuro-decision-choice agents in the space of individual-collective dualities. The comparative analytics of the diversities require ranking, while the ranking requires a criterion or criteria. The criterion for ranking must be related to the activities of the neuro-decision-choice space and then to the variety space and the action space under an epistemic guidance that an action without thought from information-knowledge service inputs is blind and thought from information-knowledge service inputs without action is empty within the space of imagination-reality dualities where thought resides in the space of imaginations and action transforms thought into a variety that will reside in the space realities. We shall discuss the criterion and the criterion space, followed by decision and the decision space, then the actions and the action space.

5.4.1 The Criterion and the Criterion Space

Over the space of goals, objectives and visions, there are many varieties expressed in diversities that are presented to neuro-decision-choice agents with variety constraints from the space of intellectual-non-intellectual resource dualities. The neuro-decision-choice actions over these varieties require some form of comparative analytics with some form of judgemental instruments called criteria to assist preferences over the space of individual-collective dualities in relation to the spaces of past-present, present-future and past-future dualities of varieties with their corresponding dispositions. The preferences, in general, relate to intellectual and non-intellectual varieties in the variety spaces in the past-present, present-future and past-future dualities where the intellectual varieties involves knowing and information-knowledge resource accumulation and use, while the non-intellectual varieties involve all other varieties for use and accumulation in the space of production-consumption dualities. In non-intellectual production, the variety accumulation is called inventories while in the knowing investment flows, the variety accumulation is intellectual stocks which are inexhaustible in their use and admit of continual give-and -take sharing mode in intergeneration and inter-societies. The intellectual stocks, when they are produced, are permanent variety inventories that occupy the position of both *domestic* and *international pubic goods* with non-exclusivity in use and not private goods with usage exclusivity. The concept and the phenomenon of criterion arise when neuro-decision-choice actions are to be exercised over the space of varieties that exist in diversities, where variety diversities may be pulled into categories with common characteristic dispositions to form categorial diversities for categorial comparative analytics.

Definition 5.3.1.1A: Criterion (Verbal) A criterion is a binary order-relational operator with a characteristic disposition over the space of quantity-quality dualities which allow for comparative analytics of variety characteristic dispositions in space of diversities over the space of varieties in terms of the order of standing within the spaces of importance-unimportance, desirability-undesirability, preference-un-preference, usefulness-non-usefulness dualities relative to the action spaces of decidability-undesirability, doubt-surety, uncertainty-certainty, problem–solution dualities and other dualistic-polar existence such that the criterion is an order operator of preference ordering with a characteristic disposition that meets the conditions of reflexivity, symmetry, anti-symmetry and transitivity over the space of varieties which may be subjected to neuro-decision-choice action. The collection of all the criterion characteristic dispositions that meet these conditions constitutes the *criterion space.*

Definition 5.3.1.1B: Criterion (Algebraic) Given the definitions of the concepts of goals, objectives, and goal-objective space, a criterion is an operator ($\mathfrak{F}$) expressed over the elements of the space of imagination-reality dualities of the form $\mathfrak{L} = \mathfrak{H} \otimes \mathfrak{G} = \mathbb{V}_{\mathfrak{L}} = \mathbb{V}_{\mathfrak{H}\otimes\mathfrak{G}}$ where $\mathfrak{H} = \mathbb{V}_{\mathfrak{H}}$ is the space of realities, $\mathfrak{G} = \mathbb{V}_{\mathfrak{G}}$ is the space of imaginations under dualistic-polar conditions and $\mathbb{I}^{\mathfrak{L}}$ is the index set of the elements in the space of imagination-reality dualities such that if $\left\{(\nu_i, \nu_j, \nu_l) \in \mathfrak{L} = \mathfrak{H} \otimes \mathfrak{G} = \mathbb{V}_{\mathfrak{L}} = \mathbb{V}_{\mathfrak{H}\otimes\mathfrak{G}} | i, j, l \in \mathbb{I}^{\mathfrak{L}}\right\}$ then there is (1) the triple fee postulate where $\mathfrak{F}(\nu_i, \nu_j) \Rightarrow$ (a)$\nu_i \succsim \nu_j$, (b)$\nu_j \succsim \nu_i$ or (c)$\nu_i \approx \nu_j$ for comparability conditions, (2) $\mathfrak{F}(\nu_i, \nu_i) \Rightarrow (\nu_i \approx \nu_i)$ for reflexivity condition (3) $\mathfrak{F}(\nu_i, \nu_j) = \mathfrak{F}(\nu_j, \nu_i) \Rightarrow \nu_i = \nu_j$ as a condition of anti-symmetry (4) if $\mathfrak{F}(\nu_i, \nu_j)$ and $\mathfrak{F}(\nu_j, \nu_l) \Rightarrow \mathfrak{F}(\nu_i, \nu_l)$ in that $\nu_i \succsim \nu_j$ and $\nu_j \succsim \nu_l \Rightarrow \nu_i \succsim \nu_l$ as a condition of transitivity with $(\succsim)$ implying preferred to or indifferent and $(\approx)$ implying indifferent to. The collection of all criteria constitutes the criteria space ($\mathfrak{C}$) with a generic element $(\mathfrak{F} \in \mathfrak{C})$, where every element in the criteria space is reducible to dualistic-polar space of cost–benefit criterion, $(\mathfrak{R})$ with a generic element, $(\mathfrak{r} \in \mathfrak{R})$.

Note 5.3.1: Criterion and Criterion Space The criterion and criterion space intimately relate to the preference and the space of preferences in order establishing as information-knowledge input into the neuro-decision-choice actions. We must now connect the criteria space to the problem–solution dispositions and then to the space of decision-choice activities and then to the cognitive action space where actions are thought dependent and thoughts are action motivated. The *criterion space*, is now restructured for general comparative analytics as $\mathfrak{C} = \{\mathfrak{R} \cup (\mathfrak{P} \otimes \mathfrak{S})\}$ that relates problem–solution dualities to the general decision-choice space $(\mathfrak{D})$ which is the union of *relative cost–benefit space* $(\mathfrak{R})$ with a generic element $(\mathfrak{r} \in \mathfrak{R})$ and Cartesian product of the *problem space* $(\mathfrak{P})$ with a generic element $(\mathfrak{p} \in \mathfrak{P})$ and solution space $(\mathfrak{S})$ with a generic element $(\mathfrak{s} \in \mathfrak{S})$ that is available to cognitive agents in terms of subjective assessments of the distribution of degrees of desirability of varieties over all states through the subjective ranking of inter-categorial and intra-categorial varieties, where $(\mathfrak{P} \otimes \mathfrak{S}) = \mathbb{V}_{\mathfrak{P}\otimes\mathfrak{S}}$ defines the space of problem–solution dualities. The general decision-choice process is the vehicle for defining a problem and problem space. It is also the vehicle for defining a *solution* and the *solution space under a* real cost–benefit process as the operator on the problem–solution dualities meeting the conditions of the principle of non-satiation. Definitions and explications of decision-choice actions and the decision-choice space will be analytically useful in this epistemic approach to understand the relational structure of human neuro-decision-choice actions under information-knowledge processes that bring about transformations in the social space changing ignorance to knowledge or knowledge to ignorance in the space of intellectual investment-capital processes over the space of stock-flow dualities as well as spin the space of problem–solution dualities and as vehicles for the production of information-knowledge systems in disequilibrium dynamics relative to matter-energy equilibrium dynamics.

Let us keep in mind that all problem varieties and all solution varieties are defined in the space of varieties $(\mathbb{V} = \mathbb{V}_{\mathfrak{A}} \cup \mathbb{V}_{\mathfrak{U}})$ such that problems belong to the space of realities contained in the space of actuals $(\mathfrak{P} \subset \mathbb{V}_{\mathfrak{H}} \subset \mathbb{V}_{\mathfrak{A}}) \Rightarrow (\mathfrak{H} \subset \mathfrak{A})$ and solutions belong to the potential space $(\mathfrak{S} \subset \mathbb{V}_{\mathfrak{G}} \subset \mathbb{V}_{\mathfrak{U}}) \Rightarrow (\mathfrak{G} \subset \mathfrak{U})$ as containing the space of imaginations with dualistic-polar conditions such that the space of actual-potential dualities $\Omega = (\mathfrak{A} \cup \mathfrak{U})$ contains the *space of imagination-reality dualities* $(\mathfrak{H} \cup \mathfrak{G}) \subset (\mathfrak{A} \cup \mathfrak{U}) = \Omega$ as the a subspace. It has been pointed out that nature works directly in the space of actual-potential dualities transforming actuals to potentials and potentials to actuals with information-knowledge equality in relational continuum and unity to generate variety transformations over the ontic space $(\mathfrak{O})$ containing variety characteristic dispositions such that $(x \in \mathbb{X} \subseteq \mathfrak{O})$. Cognitive agent with neuro-decision-choice actions work in the space of imagination-reality dualities that is contained in the space of actual-potential dualities in that $(\mathfrak{H} \cup \mathfrak{G}) \subset (\mathfrak{A} \cup \mathfrak{U}) = \Omega$ under cognitive capacity limitations transforming realities $(\mathfrak{h} \in \mathfrak{H} \subset \mathfrak{A})$ to potential $(\mathfrak{u} \subset \mathfrak{U})$ and imaginations $(\mathfrak{g} \in \mathfrak{G} \subset \mathfrak{U})$ to realities $(\mathfrak{h} \in \mathfrak{H} \subset \mathfrak{A})$ with information-knowledge inequality under cognitive capacity limitations over the epistemic spaces $\mathbb{E} = (\mathbb{X}_{\mathfrak{E}} \subset \mathbb{X}_{\Omega})$ containing epistemological characteristic dispositions.

Definition 5.3.1.2A: Decision-Choice and Decision-Choice Space (Verbal) A decision-choice activity is a mental action over the problem–solution space to reconcile the conflicts over the real cost and benefit information structures as revealed by the inherent negative–positive characteristic dispositions in alternative actual and potential varieties to solve the ranking-selection problems in the cost–benefit and criterion spaces to bring about transformations and new information outputs. It is also a cognitive action in the problem–solution space which encompasses all aspects of human existence over the actual and potential spaces in the epistemological space where undesirable actual varieties are destroyed and replaced with desirable varieties under the principle of non-satiation. The collection of all the decision-choice actions on the varieties of the space of negative–positive dualities constitutes the general decision-choice space which is a union of the *general problem space* and *the general solution space* given the *criterion space*. The problems, solutions, preferences, and decision-choice activities are all defined in the space of varieties where each variety is simultaneously a cost and a benefit and where the relative cost–benefit information becomes an information input in the space of variety ranking and decision-choice activities.

Definition 5.3.1.2B: Decision-Choice and Decision-Choice Space (Algebraic) Let the general *decision-choice space* be represented by $(\mathfrak{D})$ with a generic element $(\mathfrak{d} \in \mathfrak{D})$, the general problem space be $(\mathfrak{P})$ with a generic element $(\mathfrak{p} \in \mathfrak{P})$ and the general solution space $(\mathfrak{S})$ with a generic element $(\mathfrak{s} \in \mathfrak{S})$, then the general *decision-choice space* is defined as the union of the general problem space and the general solution space such that $(\mathfrak{D} = \mathfrak{P} \cup \mathfrak{S})$ and if $(\mathfrak{p} \in \mathfrak{P})$ then the solution belongs to the intersection of the problem space and solution space **is** in the form $(\mathfrak{s} \in \mathfrak{P} \cap \mathfrak{S})$. Similarly, if $(\mathfrak{s} \in \mathfrak{S})$, then the problem belongs to the intersection of the problem space and the solution space such that $(\mathfrak{p} \in \mathfrak{P} \cap \mathfrak{S})$. In this respect, the *decision-choice space* is analytically $\mathfrak{D} = \{\mathfrak{d} = (\mathfrak{p}, \mathfrak{s} | \mathfrak{r}) | \mathfrak{p} \in \mathfrak{P}, \mathfrak{s} \in \mathfrak{S}, \mathfrak{r} \in \mathfrak{R}\}$ and the space of problem–solution dualities $(\mathcal{R})$ is analytically the union of *problem space* $(\mathfrak{P})$ and the *solution space* $(\mathfrak{S})$ which is $(\mathcal{R} = (\mathfrak{P} \cup \mathfrak{S}) \subseteq \mathfrak{E})$ with a fuzzy conditionality of $(\mathfrak{P} \cap \mathfrak{S} \neq \emptyset)$ which is a non-zero intersectionality and $(\mathfrak{R})$ is the *relative cost–benefit space* such that under dualistic-polar conditions $(\mathcal{R} = (\mathfrak{P} \cup \mathfrak{S}) = (\mathfrak{P} \otimes \mathfrak{S}) \subseteq (\mathfrak{H} \otimes \mathfrak{G}) \subset \mathfrak{E})$, where a variety problem is a variety solution relative to the elements in the space of goals, objectives and visions as seen in the space of imagination-reality dualities.

5.5 The Problem–Solution Duality, the Space of Problem–Solution Dualities and the Criteria

The concepts of problems, problem space, solutions and solution space have been defined in relation to cost–benefit conditions of relative dominance within the varieties in the space of imagination-reality dualities as contained in the space of actual-potential dualities. An important attention must be paid to the similarities among the concepts and phenomena of imagination-reality duality and actual-potential dualities as they relate to the concepts of imagination and potential on one hand and reality and actual on the other hand. Imagination and reality are associated with epistemological space leading to the concept of *epistemic spaces,* while actual and potential are associated with ontological space leading to the concept of *ontic spaces* in the understanding of the differences and similarities of ontological varieties and epistemological varieties and the role of neuro-decision-choice activities for ontological-epistemological connectivity in intellectual production through knowing and information-knowledge stock-flow disequilibrium dynamics.

5.5.1 Definitions of the Concepts and Phenomena of Problem and Solution Duals

The definitions of the concepts of problems, solutions and the corresponding spaces are not enough to develop an increasing understanding of cognitive behavior over the space of knowing and information-knowledge accumulation. Further understanding of diversity and the unity of science requires us to relate variety problems and variety solutions to the space of the principle of opposites with relational continuum and unity. The relationship must connect the problem and solution to the concepts of duality and polarity in the space of relativity for the understanding of the infinite decision-choice process in conditions of statics, dynamics, generation and intergeneration for continual development of knowing process, information-knowledge disequilibrium dynamics as well as problem–solution disequilibrium dynamics, where a solution to a problem generates a new problem where it may be more difficult to find its corresponding variety solution over the space of human action. The variety problem and variety solution must be related to decision-choice actions over the space of varieties and then linked to the developments of the theory of knowing and the theory of knowledge, as have been discussed in previous chapters, and then to examine the diversity dispositions and unity dispositions contained in science through knowing. Every variety problem has a corresponding variety solution which then generate a new variety solution induced by the elements over the goal-objective-vision space under the principle of non-satiation.

Definition 5.4.1.1A: Problem–Solution Duality and the Space of Problem–Solution Dualities (Verbal) A problem–solution duality is a conceptual representation of a variety with characteristic disposition that simultaneously contains problem characteristics as a cost disposition and solution characteristics as a benefit disposition relative to a goal or a set of goals, where the set of the problem characteristic acts as an obstruction to progress towards a preferred variety, while the set of solution characteristics acts as a facilitator towards a preferred variety or a set of varieties, The set of problem characteristics is the problem dual while the set of solution characteristic is the solution dual, where the unity of the *problem dual* and *solution dual* of varieties constitutes the problem–solution duality with deferential combination of problem characteristics and solution characteristics that establish the decision-choice identity of the variety. Any variety is said to be a problem variety towards a goal if the problem dual dominates the solution dual. It is said to be a variety solution towards a goal if the solution dual dominates the problem dual. The collection of these problem–solution dualities constitutes the *space of problem–solution dualities* over the space of varieties.

Definition 5.4.1.1B: Problem–Solution Duality and the Space of Problem–Solution Dualities (Algebraic) A space $(\mathfrak{R})$ is said to be a *space of problem–solution dualities* if $(\exists \mathfrak{P} \subseteq \mathfrak{E})$ and $\{\mathfrak{S} \subseteq \mathfrak{E} \ni ((\mathfrak{P} \cup \mathfrak{S}) \subseteq (\mathfrak{H} \otimes \mathfrak{G}) \subset \mathfrak{E})\}$ with $(\mathfrak{R} = (\mathfrak{P} \cup \mathfrak{S}) \subseteq \mathfrak{E})$ and $(\mathfrak{P} \subseteq \mathfrak{S} \subseteq \mathfrak{P})$, where every element $\lambda = (\mathfrak{p} \cup \mathfrak{s})$ with $\{(\mathfrak{p} \cap \mathfrak{s}) \neq \emptyset\}$ and $(\lambda \in \mathfrak{R})$ defines a *problem–solution duality* in that λ is simultaneously a problem $(\mathfrak{p} \in \mathfrak{P})$ and a solution $(\mathfrak{s} \in \mathfrak{S})$ and hence $\lambda = \mathfrak{p} = \mathfrak{s}$ with $\lambda = (\mathfrak{p} \cup \mathfrak{s}) = (\mathfrak{p} \cap \mathfrak{s}) = \mathfrak{p} = \mathfrak{s}$ relative to a goal-objective element $(\mathfrak{g} \in \mathbb{G})$ where $(\mathfrak{P} \cup \mathfrak{S} = \mathfrak{P})$ and $(\mathfrak{P} \cap \mathfrak{S} = \mathfrak{S})$ with $((\mathfrak{P} \cup \mathfrak{S}) = (\mathfrak{P} \cap \mathfrak{S}))$ and λ, $\mathfrak{p}$ and $\mathfrak{s}$ are defined by characteristic dispositions for identities and distinctions.

5.5.2 The Basic Characteristics of the Space of the Problem–Solution Dualities

The concepts of problem–solution duality and the space of problem–solution dualities are defined and established by characteristic dispositions that present distinguishable identities and integrated system of decision-choice actions, cost–benefit conditions, goal-objective conditions, destruction-replacement conditions and new-old conditions. The conditions of the distinguishable identities of the space of problem–solution dualities contain the following:

1. All the activities of cognitive agents are a collection of decision-choice actions about varieties $(\nu \in \mathbb{V})$ in the past, present and future, where these cognitive actions involve problem–solution dualities on the bases of real cost–benefit distributions under subjective preferences of the individual and the social collectivity guided by the principle of non-satiation and constrained by resource availability under the matter-energy equilibrium conditions.

2. Every variety production belongs to a transformation system of varieties and every variety transformation belongs to the space of problem–solution dualities in the decision-choice space to generate new information to update the information-knowledge stock-flow disequilibrium dynamics within the space of imagination-reality dualities.
3. Every variety is a duality composed of negative and positive duals which are expressed individually as cost–benefit dualities with cost dual and benefit dual that show themselves as problem–solution dualities depending on the preference ordering of cognitive agents over the space of varieties relative to the elements in the space of goals, objectives and visions.
4. Every variety is simultaneously a problem and a solution in the space of problem–solution dualities relative to the variety space, where the choice of variety imagination, in the form of a variety potential, as a solution to the destruction of an actual variety reality in the form of variety actual, as a problem, is either individual-decision-choice specific or collective-decision-choice specific under the real cost–benefit rationality of the asantrofi-anoma tradition of the decision-choice agents with given preferences and sociopsychological principle of non-satiation.
5. The general space of varieties ($\mathbb{V}$) is the union of the space of actual varieties ($\mathbb{V}_{\mathfrak{A}}$) and the space of potential varieties ($\mathbb{V}_{\mathfrak{U}}$) in that ($\mathbb{V} = (\mathbb{V}_{\mathfrak{U}} \cup \mathbb{V}_{\mathfrak{A}})$) at any given time. The space of imagination-reality dualities ($\mathbb{V}_{\mathfrak{L}}$) is a union of the space of imaginations ($\mathbb{V}_{\mathfrak{G}}$) and the space realities ($\mathbb{V}_{\mathfrak{H}}$), where $\mathfrak{P} \subset \mathbb{V}_{\mathfrak{H}}$ and $\mathfrak{S} \subseteq \mathbb{V}_{\mathfrak{G}}$ in that every solution is a variety imagination and every variety imagination is a variety solution relative to either a goal, an objective or a vision. However, every problem is a variety reality but not every variety reality is a problem. Similarly, every goal-objective element or vision is a variety imagination but not every variety imagination is a goal-objective element or a vision, so that $\mathbb{G} \subset \mathbb{V}_{\mathfrak{G}}$ and $\mathbb{G} \cap \mathbb{V}_{\mathfrak{H}} = \emptyset$ in the space of human problem–solution dualities (R). Let us also keep in mind that under dualistic-polar conditions, the union of spaces is the same as the Cartesian products of spaces such that ($\mathbb{V} = (\mathbb{V}_{\mathfrak{U}} \cup \mathbb{V}_{\mathfrak{A}}) = (\mathbb{V}_{\mathfrak{U}} \otimes \mathbb{V}_{\mathfrak{A}})$) with a fuzzy-stochastic conditionality ($(\mathbb{V}_{\mathfrak{U}} \cap \mathbb{V}_{\mathfrak{A}}) \neq \emptyset$) and similarly $(\mathbb{V}_{\mathfrak{H}} \cup \mathbb{V}_{\mathfrak{G}}) = (\mathbb{V}_{\mathfrak{H}} \otimes \mathbb{V}_{\mathfrak{G}})$ with $(\mathbb{V}_{\mathfrak{H}} \cap \mathbb{V}_{\mathfrak{G}}) \neq \emptyset$.
6. Every decision-choice action is a problem–solution duality in relation to things under destruction-replacement processes to bring about transformations and to generate new varieties and new information. The space of problem–solution dualities forms the primary category of conditions for the establishment of the principle of diversity and the principle of the unity of knowing with extension to science and non-science under problem–solution analytics and the principle of opposites relative to the principle of non-satiation to maintain the information-knowledge disequilibrium dynamics and acknowledgement of matter-energy disequilibrium.

5.5.3 *Reflections on the Basic Characteristics of Problem–Solution Duality*

The epistemic structure of the problem–solution analytics as the primary categorial foundation of the unity of knowing, and by extension the unity of science, is such that the elements of the space of the problem–solution dualities are interchangeable as problems and as solutions under real cost–benefit imputations relative to any goal-objective ($\mathfrak{g} \in \mathbb{G} \subset \mathbb{V}_{\mathfrak{G}} \subset \mathbb{V}_{\mathfrak{U}}$). It is important to note that all goal-objective elements are variety imaginations abstracted from the space of potential varieties but not all variety imaginations from the space of space of potential varieties are goal-objective elements which are also defined over the space of attainable-unattainable dualities under resource constraint. Whether a variety is a benefit or a cost, viewed in terms of a solution or a problem depends on the goal-objective element of the decision agent and the preference assessment of a cost–benefit relativity within the cost–benefit duality of the variety.

The problem–solution process for any decision agent is a search process not over the problem space or the solution space but over the space of problem–solution dualities embedded in the variety space ($\mathbb{V}_{\Omega} = (\mathbb{V}_{\mathfrak{U}} \cup \mathbb{V}_{\mathfrak{A}}) = (\mathbb{V}_{\mathfrak{U}} \otimes \mathbb{V}_{\mathfrak{A}})$). In this respect, there is no absolute problem space nor absolute solution space but the space of problem–solution dualities since every variety is a problem–solution duality in the space of relativity. When a variety problem is identified, it is transformed into a pole of a problem–solution polarity, where the pole has a residing problem–solution duality with the problem dual dominating the solution dual.

The process in knowing now is to search for a solution pole which has a residing problem–solution duality with a solution dual dominating the problem dual. The claim of the existence of the space of problem–solution dualities with relational continuum denies the independent existence of problem space and solution space. The problem-solving process is a search process to find a variety solution pole with a residing problem–solution duality, where the solution dual dominates the problem dual. It is this problem–solution interdependency that maintains the continual dynamics of the infinite problem–solution enveloping and *parent–offspring process* for continual human development towards greater and greater welfare and a continuous search for a development path toward a perfect state of welfare under the politico-economic principle of non-satiation in the space of variety cost–benefit dualities. This continual search for variety solutions involves continual information-decision-choice interactive processes over the space of problem–solution dualities leading to the creation of the space of variety destruction-construction dualities of the same matter, and a continual generation of information-knowledge systems in stock-flow disequilibrium states where the parent–offspring processes are completely consistent with the dynamics of natural varieties.

The meaning of human problem-solving is about variety destructions and replacements while the essence of human problem-solving is about the nature and manner of variety destructions and replacements, where new variety solutions generate new variety problems which then demand a search for new solutions in the dynamics of neuro-decision-choice actions over the space of problem–solution dualities in continual disequilibrium processes. In the epistemics of diversity-unity analytics, it is useful to keep in mind the analytical toolbox of the principle of opposites with duality, polarity, relational continuum and unity, where the problem–solution dualities and the problem–solution polarities are the uniting factors of all areas of information-knowledge productions involving science and non-science without an exception from the *knowledge of acquaintance* to the *knowledge of description.* These dualistic-polar conditions of the principle of opposites with relational continuum and unity reflect an idea in the African conceptual system that each individual or each generation comes to solve the problems essential to its being, but not to solve all problems. The solutions to the generational problems may expand the space of problem–solution dualities just as they will expand the space of information-knowledge systems.

The corresponding information structure of the principle of opposites with relational continua and unity is the *fuzzy information* with a corresponding toolbox, where the methodology for information processing is the *fuzzy paradigm of thought* within which a corresponding toolbox of methods and techniques is developed for the neuro-decision-choice activities over the space of problem–solution dualities. The alternative information structure is the *classical information structure* that corresponds to the principle of opposites with relational separation, an excluded middle and disunity. The methodology for classical information processing is the *classical paradigm of thought* within which a corresponding toolbox of methods and techniques is developed for the neuro-decision-choice activities over the space of problem–solution dualities. In the problem–solution analytics, there are two cost–benefit (*asantrofi-anoma* principle of Adinkra symbolism) conditions that must be analyzed. There are the conditions of *internal cost–benefit duality* of every variety and the conditions of *external cost–benefit duality* of each variety relative to other varieties. All neuro-decision-choice actions are guided by these cost–benefit conditions in relation to the elements in the space of problem–solution dualities with input–output processes that generate continual *information-knowledge* stock-flow *disequilibrium* dynamics. These conditions of cost–benefit duality or the *asantrofi-anoma* rationality form the epistemic foundations for the theory of cost–benefit analysis as comparative analytics for all neuro-decision-choice actions in the space of relativity where the internal cost–benefit dispositions show the neuro-decision-choice values of the individual varieties and the external cost–benefit dispositions show the comparative values of the varieties in ranking. The resolution of the cost–benefit conflicts is the *asantrofi-anoma problem* where every variety is simultaneously attractive and repulsive, while the intelligence of resolving the asantrofi-anoma neuro-decision-choice problem is the *asantrofi-anoma* rationality or cost–benefit rationality.

Here, arises the following interesting dichotomy in the dualistic-polar processes over the space of the problem–solution dualities in intellectual stock-flow dynamics: Does a problem cause a neuro-decision-choice action or does a neuro-decision-choice action cause a problem? In other words, does an unwanted state, where the real variety cost dominates the real variety benefit in the preference space, create conditions of a problem or the existence of the conditions of an undesirable state is neuro-decision-choice created since the problem is an artificially manufactured variety in the sense that the conditions of the state inhibit the progress on a path to a goal, objective or vision? About this question, the following statement by Afanasyev is useful:

> Decisions are caused by problem situations, that is, by situations in which the existence of a particular state of the system produces a need for another state, because the continuation of the current state interferes with its normal functioning, improvement and development.
>
> The scope and contents of such problems, their intensity, concrete forms and methods for solving them will differ from social systems that have reached different levels of social organization. The higher the level obtained by that system along the ladder of social organization and the more it is complex, the greater will be the volume and diversity of problem situation and the greater will be the importance and responsibility attaching to decisions [320, p. 168].

In this quote, Afanasyev thinks that decisions take place because there are problems that create impediments to normal functioning, improvement and development. The difficulty, here, is what are the conditions that give rise to a problem and identifying what is a problem. The neuro-decision-choice action is to remove the impediment, where the removal of the impediment will constitute a variety solution replacement. The difficulty, here, is defining what is impediment and explaining what and why should it be removed. These statements may apply to a small set of problems. Furthermore, how will a state produce a need for another state? Every impediment removal is a variety solution in relation to either a goal or an objective. It is also a variety problem in relation to the variety replacement. The complexity of neuro-decision-choice actions finds expressions in the relationships between the space of variety negative–positive dualities and the space of variety cost–benefit dualities. This complexity is also the foundation of neuro-decision-choice actions over the spaces of doubt-surety, decidability-undecidability and wrong-right dualities evaluated around conditions of cost–benefit conflicts. It is this complexity that gives rise to the *asantrofi-anoma problem* of Adinkra tradition of the African conceptual system containing the principle of opposite from the Africa's antiquity. This situation is also reflected on by Gaius Valerius Catullus (c. 84–c. 54 BC), where every decision-choice element belongs to the space of attractiveness-repulsiveness dualities in the sense that it is simultaneously attractive and repulsive [688].

A question, therefore, arises as to what the relationship between the set of elements in the space of neuro-decision-choice actions and the set of elements in the space of problem–solution dualities is. In a sense, does a problem generate a neuro-decision-choice action? Does a neuro-decision-choice action generate a problem or are problems and neuro-decision-choice actions dualistically interdependent, mutually self-creating and self-destructing? The answers to these questions are complex, the roots of which *neuro-decision-choice* actions are based, involving the conditions and activities in the space of unconscious-conscious dualities. It will become clear that the relational conditions between the space of neuro-decision-choice actions and the space of problem–solution dualities constitute the foundations to human social existence relative to dualistic-polar structures such as progress-retrogress, war-peace, destruction-construction, evil-good, ignorance-knowledge, imagination-reality dualities and the conditions of relational interdependencies of give-and-take sharing variety structures that divide them into diversities, as well as unite them into the meaning and essence of individual, collective and intergenerational lives in information-knowledge systems in search for better solution varieties in the sense of greater satisfaction as seen through the relativity space of cost–benefit dualities under the conditions of non-satiation.

5.5.4 Solutions, Solution Space and the Space of Decision-Choice Actions

All the defined concepts and discussions help to understand diversity-unity conditions of all areas of knowing in the variety space and cognitive existence involving science and non-science, justified beliefs, unjustified beliefs and other forms of knowing as intellectual information-knowledge investment flows to update intellectual capital stocks. All problems and problem space are hindrances on the path to elements in the goal-objective space. The removal and replacement of variety problems are neuro-decision-choice determined, where the replacements are variety imaginations from the space of potentials which constitute the variety solutions as generated by neuro-decision-choice actions that must be defined within the space of destruction-replacement dualities as contained in the space of imagination-reality dualities. Given the concepts of problems and problem space, we now turn our attention to the concepts and phenomena of solutions, solution space and their relationship to the space of neuro-decision-choice activities over the action space. From the definition of epistemological space in Chap. 2, the decision space may be specified as:

$$\begin{aligned} \mathfrak{D} &= (\mathbb{V} \otimes \Pi \otimes \mathbb{Z}_{\mathfrak{E}} \otimes \mathfrak{D} \otimes \mathfrak{K}) \\ &= \left\{ \mathfrak{d} = (\nu, \eta, z_e, \mathfrak{d}, \mathfrak{k}) | \nu \in \mathbb{V}, \eta \in \Pi \otimes \mathbb{Z}_{\mathfrak{E}}, \mathfrak{d} \in \mathfrak{D}, \mathfrak{k} \in \mathfrak{K} \right\} \end{aligned} \tag{5.3.1}$$

The point, here, is to reconcile the apparent differences among these analytical definitions within the discussions. It is noted that the *space of decision-choice actions* is the union of *problem space* and *solution space* ($\mathfrak{D} = \mathfrak{P} \cup \mathfrak{S}$) defined in terms of varieties with supporting sets of characteristic-signal dispositions within the problem–solution analytics, where the decision-choice space may be defined as a Cartesian product of the problem–solution spaces, $\mathfrak{D} = \{\mathfrak{d} = (\mathfrak{p}, \mathfrak{s}) | \mathfrak{p} \in \mathfrak{P}, \mathfrak{s} \in \mathfrak{S} \& (\mathfrak{P} \otimes \mathfrak{S}) \subset (\mathfrak{H} \otimes \mathfrak{G}) \subset \mathfrak{E}\}$, where every problem is a *variety reality* from the space of the actuals and every neuro-decision-choice action is a *variety imagination* from the space of potentials. The similarity and equality of these definitions of decision-choice action and decision-choice space will become clear as we proceed in the epistemic process. Let us proceed with the alternative definitions of variety solution and solution space.

All decision-choice actions are related to the removal of obstructions to goals, and hence are explainable in the problem space which relates to all the elements in the spaces of realities and potentials. A neuro-decision-choice action is about a problem-solving event, while a problem-solving event involves either variety identification or variety transformation or both. A variety identification is about either *what there was* (past-history) or *what there is* (the present history) in the field of info-statics for variety knowing at static states [472, 669, 671]. It is, therefore, a discovery of information-knowledge items through neuro-decision-choice actions over the space of variety ignorance-knowledge dualities in dynamic states. The variety transformation is about *what would be* (future history) in the field of the info-dynamics for variety knowing in terms of a change, either in relation, in property or both of present-future varieties, which relate to information production in the field of info-dynamics in terms of dynamics of history which represents stock-flow disequilibrium dynamics [669, 670, 672]. Transformation is, therefore, related to the discovery of variety information-knowledge changes as neuro-decision-choice actions over the space of variety ignorance-knowledge dualities in dynamic states. Let us provide definitions to the concepts of *variety solution* and *solution space* in terms of neuro-decision-choice actions over the space of problem–solution dualities and how they relate to the theory of knowing and the theory of knowledge.

Definition 5.4.2.1A: Solutions and Solution Space (Verbal) A *solution* is a neuro-decision-choice construct from $\mathfrak{D}$ to transform an existing set of cost–benefit characteristics of a variety, the relative cost–benefit characteristic combination of which is assessed as a problematic obstruction on the path to a more *preferred set* of cost–benefit characteristics of a variety imagination with a preferred relative cost–benefit characteristic combination as a goal or an objective. The collection of all potential solutions to the elements of the problem space relative to the elements of the goal-objective space is the *solution space*, where the benefit characteristic dispositions dominate the cost characteristic dispositions relative to the elements in the *space of goals and objectives* in terms of individual or collective existence.

Definition 5.4.2.1B: Solutions and Solution Space (Algebraic) Given a variety $(\nu \in \mathbb{V})$ and a preference order $(\succsim)$ in the form $(\mathbb{V}, \succsim)$ such that if $(\nu_1, \nu_2 \in \mathbb{V})$, $(\nu_1, \nu_2, \nu_3 \in \mathbb{V})$, and $\left(\nu_1, \nu_2, \nu_i \cdots \nu_j, \cdots \in \mathbb{V}\right)$ with a corresponding space of relative real cost–benefit characteristics $\mathfrak{R} = \{\mathfrak{r} = {}^{\mathrm{C}}/_{\mathrm{B}} | \mathrm{C} \in \mathbb{C}, \mathrm{B} \in \mathbb{B}\}$, if $\left\{\left(\nu_i, \nu_j \in \mathbb{V}\right) \subset \left(\mathbb{V}, \succsim\right), \forall i, j \in \mathbb{I}_{\mathbb{V}}^{\infty}\right\}$, $\left(\mathbb{V}, \succsim\right) \subseteq \left((\mathbb{V} \otimes \mathfrak{R}), \succsim\right)$ and $(\nu_i \mathfrak{r}_{\mathrm{i}})$, $\left(\nu_j \mathfrak{r}_{\mathrm{j}}\right)_{\mathrm{j}} \in \left((\mathfrak{R} \otimes \mathbb{V}), \succsim\right)$ then a solution is a neuro-decision variety with $(\mathfrak{r}_{\mathrm{j}} < \mathfrak{r}_{\mathrm{i}})$ where $(\nu_i \in \mathbb{V})$ is a decision-destroyed, with information retention $(z_i \in \mathbb{Z})$, and $\left(\nu_j \in \mathbb{V}\right)$ is a decision-created, and with information $\left(z_j \in \mathbb{Z}\right)$ as an realized in its replacement in the way of reaching the goal of ν_j. The existence of variety ν_j with information $z_j \in \mathbb{Z}$ is said to be a solution in $(\nu_i \in \mathbb{V})$ with information $(z_i \in \mathbb{Z})$. The collection of all such variety $\left(\nu_j, \forall j \in \mathbb{I}_{\mathbb{V}}^{\infty}\right)$ with corresponding $(\mathfrak{r}_{\mathrm{j}} \in \mathfrak{R})$ and $\left(z_j \in \mathbb{Z}\right)$ constitutes the general solution space $\mathfrak{S}$ with a generic element $(\mathfrak{s} \in \mathfrak{S})$.

5.5.5 *Reflections on Decision-Choice Dependent Definitions*

This chapter will be ended with a reflection on the decision-choice-dependent approach to the definitions. The existence of a solution and a solution space implies the existence of a problem and a problem space which implies the existence of goals, objectives and goal-objective space to the elements of which cognitive agents seek to reach individually or collectively by the means of variety transformations through *neuro-decision-choice actions* with past-present-future information-knowledge inputs of general existence, destruction of problem varieties and the creation of future varieties as solutions under the conditions of the space of certainty-uncertainty dualities as generated by cognitive limitations of decision-choice agents, where the existence of neuro-decision-choice actions implies the existence of neuro-decision-choice space under the principle of non-satiation. It is important to notice from the definitions the existence of problem–solution symmetry rather than asymmetry. Over the variety space, and within the epistemological space, a problem as expressed in terms of a variety is a solution to a variety or some varieties as seen from the principle of opposites with polarities and residing dualities. In other words, a problem is a solution, and a solution is a problem in the spaces of info-statics, info-dynamics and information-knowledge certainty structure as defined by fuzzy-stochastic entropy. There is a problem–solution symmetry which is complemented by cost–benefit symmetry in transformation neuro-decision-choice processes where real costs are real benefits and real benefits are real costs in ex-ante and ex-post analytics. The problem–solution and cost–benefit symmetries are such that whether a state of being is a problem a real cost a solution or a real benefit they are always seen in relation to a goal-objective element given an individual and collective preference system, where a variety solution may be a variety problem and a variety problem may be a variety solution relative to goal-objective varieties under preference and preference order within the space of relative real cost–benefit dualities that reside in the space of opportunity in variety conditions over the space of static-dynamic dualities relative to the space of imagination-reality dualities as a subspace of actual-potential dualities.

The problem–solution symmetry just like real cost–benefit symmetry will provide us with important conditions to unite the concepts of the problem space and the solution space into the concepts of problem–solution duality and the space of problem–solution dualities under neuro-decision-choice dynamics over the space of cost–benefit dualities under preference ordering. It will enhance the argument that the *unity principle of science* through the unity of knowing is the existence of the problem–solution duality $(\lambda \in \mathfrak{R})$ and the space of problem–solution dualities $(\mathfrak{R})$ under the asantrofi-anoma tradition within the space of static-dynamic dualities. It will also provide an explanatory understanding for the existence of the principle of diversity in the process of knowing and the rise of differential toolboxes in categories of diversity. The explanatory conditions of the existence of differential toolboxes over the space of problem–solution dualities suggest that the unity principle does not find meaning and support in the use of methods or techniques of knowing including acquaintance, description and the methodological constructionism-reductionism duality since different categories of varieties will require either different or same methods, techniques and approaches to deal with either the corresponding identification or transformation problem–solution dualities over the space of problem–solution dualities.

The unity of methods and techniques of science through knowing is induced by the specificities of varieties to be identified or transformed in knowing, while the existence of the space of problem–solution dualities is induced from the universal *concept of non-satiation of preferences* as discussed in the economic theory of choice which maintains continual variety transformations in the variety space for the continual state of stock-flow disequilibrium info-dynamics in the space of information-knowledge accumulation. The principle of non-satiation points to the condition that the optimum welfare is not reachable and is in support of the idea of infinite problem–solution dynamics and information-knowledge development towards the perfect-welfare state that resides in the potential space which contains the space of absolute. If the perfect-welfare state is reached by either the individual, collective or a generation, then there will be no problems to be solved by the individual, the collective or the generation, and there will be neither social progress nor life, and there will be no need for knowing leading to *information flow equilibrium* and *knowledge stock equilibrium*. Under this state of ultimate perfection, the problems of knowing and the unity of science are *phantom problems.*

The concept of preference ranking $\left(\succsim\right)$ over the space of imagination and varieties and categorial varieties within the epistemological space has no lower and upper bounds in that $(\nu_i \mathfrak{r}_\mathrm{i}), \left(\nu_j \mathfrak{r}_\mathrm{j}\right)_\mathrm{j} \in \left((\Re \otimes \mathbb{V}) \subseteq \left(\nu_{-\infty}, \nu_{+\infty}, \succsim\right)\right)$ and if $\nu_j \succsim \nu_i, \Rightarrow \exists \nu_{j+1} \succsim \nu_j$, and $\exists \nu_{\mathrm{i}-1} \ni \nu_i \succsim \nu_{\ominus i-1} \ldots$. In this respect, every variety in the universal set of the form $(\nu \in \mathbb{V} \subseteq \Omega \subseteq \mathbb{V})$ is both a problem $(\mathfrak{p} \in \mathfrak{P})$ and a solution $(\mathfrak{s} \in \mathfrak{S})$ by preference ordering $\left(\succsim\right)$ in the general neuro-decision-choice space $\mathfrak{D}$, where the elements are ranked with both internal and external cost–benefit dispositions relative to an existing preference scheme. The unboundedness of the preference ranking meets the general neuro-decision-choice postulate of the *principle of non-satiation* in backward-forward transformations of individual, collective and generational progress. It is important to note that the ranking for the neuro-decision-choice actions over the space of varieties to identify problem varieties and search for solution varieties is also a ranking for goal-objective elements that induce the unity of science through the unity of knowing.

The neuro-decision-choice actions, the problem–solution dualities and the information-knowledge systems are interdependent, mutually creating and inseparably connected seen in terms of dualistic-polar conditions. The cognitive actions that operate over the neuro-decision-choice space are also the same cognitive actions that operate over the space of problem–solution dualities, the space of knowing and the space of information-knowledge development in relational continuum and unity under differences of cognitive environments. Characteristic dispositions and information-knowledge structures may vary over different varieties, but all varieties are united by cognitive actions of knowing through different methods and techniques under an accepted general principle over the spaces of problem–solution, certainty-uncertainty and acceptance-rejection dualities.

The definitions of the concepts of problems, solutions, goals, objectives and decision-choice actions are all in relation to cost–benefit dispositions over the space of relativity, where the cost–benefit dispositions define the foundations of variety destruction-replacement or transformation-substitution processes relative to the goal-objective element of a vision in the space of imagination-reality dualities contained in the space of actual-potential dualities. In fact, it must be understood that the *theory of knowing* composed of neuro-decision-choice actions over the acquaintance space, the knowledge by acquaintance and knowledge by description, is implicitly and explicitly driven by cost–benefit implications and rationality as cognitive agents strive for better states of existence. In other words, the theory of knowing is part of the theory of the *asantrofi-anoma* problem and rationality where every variety exist as simultaneously a cost disposition and a benefit disposition in the space of rejection-acceptance dualities. The development of methods and techniques, and the selection of the appropriate ones are themselves neuro-decision-choice driven under the cost–benefit rationality, where the methods and techniques are themselves subject to information-knowledge identifications, the information contents are the characteristic dispositions and the messages are the signal dispositions within the language of their development in the space of relativity [218, 219].

The whole space of problem–solution dualities is under cost–benefit conditions and corresponding rationality which find expressions in information as a matter-energy property, information-knowledge as input–output phenomena knowledge as input into decision-choice actions and decision-choice actions as generators of information and knowledge. The point of general emphasis is that the space of problem–solution dualities, as the space of destruction-replacement dualities, under neuro-decision-choice actions, partitions the area of knowing into categorial diversities, and then relationally unites all areas of knowing and the resulting theory of knowledge into oneness under neuro-decision-choice actions. In this respect, the theory of knowledge, as the theory of *intellectual information-capital accumulation* through the theory of knowing as *intellectual investment*, is a *theory of unity* in the sense that all variety knowing must proceed from universal unity to create dividedness into diversities, under a fuzzy-set closure and then return to the unity through its relationally fuzzy dividedness under the principle of dualistic-polar conditions, where in unity resides diversity and where in diversity we find unity, where unity is undefinable without diversity which is conditions for unity [832].

Chapter 6
The Theory of Categorial Conversion and the Problem-Solution Dualities in Knowing and Knowledge

Abstract The entry point to this chapter is a discussion on the meanings of the concepts and phenomena of category and categorial conversion and their uses in the understanding of knowing, knowledge and the principles of diversity and unity that they may entail. The concepts are introduced to deal with transformations and changes within the space of quantitative–qualitative dualities as they relate to info-statics and info-dynamics of varieties in all areas of knowing as intellectual investment flows and information-knowledge accumulation as intellectual capital stocks. Alongside the theory of categorial conversion is an introduction of the theory of Philosophical Consciencism and the discussions on its relations to neuro-decision-choice processes, education, preference formations, varieties of human progress, shifting of generational preferences and increasing individual and social abilities in the space of variety creation-destruction dualities as seen in terms of individual and collective curiosities over the space of imagination-reality dualities and continual variety transformations. The theories of categorial conversion and philosophical consciencism are mutually interdependent, inter-supportive and inseparable, just like the theories of info-statics and info-dynamics are mutually interdependent, inter-supportive and analytically inseparable. Similarly, the theories of categorial conversion and info-statics are mutually creating, while the theories of Philosophical Consciencism and info-dynamics are also mutually interdependent, creating and inseparable. The epistemic frameworks of both the theories of categorial conversion and info-statics are are shown to be linked to the epistemic frameworks of *possibility* and *necessity*, while the epistemic frameworks of the theories of philosophical consciencism and info-dynamics are shown to be linked to the frameworks of *probability* and *freedom* over the spaces of neuro-decision-choice activities and the action space as they relate to the conditions established over the spaces of identification-transformation and imagination-reality dualities within which the concepts and the phenomena of phantom, ill-posed, unresolved, unintended and other problems are defined and discussed in relation to paradigms of thought and cognitive capacity limitations in knowing. The knowing and information-knowledge processes are shown to be cognitive journeys between variety ignorance to variety knowledge in the space of

K. K. Dompere, *The Theory of Problem-Solution Dualities and Polarities*,
Studies in Systems, Decision and Control 405,
https://doi.org/10.1007/978-3-030-90279-7_6

ignorance-knowledge dualities as a sub-space of the space of actual-potential dualities, where the journeys are constrained by informogy, categorial conversion is an epistemic ation-knowledge inequality conditions with uncertainty and risks.

6.1 Introduction to the Concept of Categorial Conversion

The epistemic objective, in this chapter, is to link the concepts and phenomena of categorial conversion and Philosophical Consciencism to the conditions and foundations of critical thinking, knowing and information-knowledge development and how the ideas contained in these frameworks help to establish diversity and the unity of knowing and by extension, diversity and unity of science, as well as to create foundations to link information-knowledge accumulation to economic production as an element in the input–output space. The theories of categorial conversion and info-statics provide the conditions of necessity within the possibility space contained the space of imagination-reality dualities to establish conditions of necessity for change in terms of variety identities, while the theories of Philosophical Consciencism and info-dynamics provide the conditions of freedom within the probability space also contained in the space of imagination-reality dualities to establish conditions for freedom under neuro-decision-choice actions over the space of actual-potential dualities, where variety transformations are induced by the principle of dividedness into diversities, and preferences over varieties are induced by real cost–benefit conditions. The discussions of the theorem of dividedness is provide in the manuscript on entropy [832, 837, 841].

6.2 Problem–Solution Duality and the Space of Problem–Solution Dualities

The discussions in Chap. 5 suggest that the symmetry of the problem–solution structure is such that, every variety in the universal set of the form ($\nu \in \mathbb{V} \subseteq \Omega \subseteq \mathbb{V}$) is both a problem ($\mathfrak{p} \in \mathfrak{P}$) and a solution ($\mathfrak{s} \in \mathfrak{S}$), when the varieties enter the space of imagination-reality diversities, given the space of goals and objectives, by preference ordering $(\succsim)$ within the general or specific decision-choice space $\mathfrak{D}$. From the epistemics and mathematics of cost–benefit foundations of preference orderings, every real cost disposition is a real benefit disposition and every real benefit disposition is also a real cost disposition in relation to the varieties in the goal-objective space. Similarly, from the epistemics and mathematics of foundations of the principle of opposites, composed of duality and polarity to establish the general dualistic-polar conditions, every variety is simultaneously a problem as well as a solution in the space of decision-choice activities relative to some elements in the goal-objective space.

Generally, the decision-choice foundations are such that every potential variety imagination is real as well as a potential, but not necessarily, a goal and a real opportunity in relation to other imagination and potential varieties. It is within this conceptual understanding that the economic theory of *opportunity cost* and cost–benefit analytics find universal meaning, understanding and application in the transformation space of varieties as the varieties relate to information-knowledge systems over the space of certainty-uncertainty dualities and neuro-decision-choice activities over all areas of knowing and problem–solution dualities, doubt-surety dualities and acceptance-rejection dualities under national Philosophical Consciencism and its influence on individual and social preference scheme [218, 219]. It is within this epistemic framework that cost–benefit analysis may act as the general neuro-decision-choice theory defining the foundational techniques and methods for the construct of all types of decision theory. Every decision theory can be reduced to real cost–benefit analytics under mini-max conditions where every event has a minimum and a maximum over the space of acceptance-rejection dualities [273, 275, 278].

Every variety reality is an *opportunity cost* in transformations, where the opportunity cost is a real benefit forgone in the imagination-reality-variety transformations for realizing the imagination as an actual and potentializing the variety actuality as a variety actual through the *construction-destruction* or *destruction-construction process*. The opportunity cost must be seen as net benefit forgone in that it is not only the variety benefit forgone but also the cost of the variety forgone, where every variety contains real cost characteristics which are unattractive characteristics, and realbenefit characteristics which are attractive characteristics in an unequal but relative combination. By getting rid of the attractiveness, one also gets rid of the repulsiveness of the variety in the *asantrofi-anoma* tradition, where one cannot take the benefit and leave the cost because the benefit and cost are contained in the same variety. Any variety is composed of a set of characteristics which is partitioned into a sub-set of negative characteristics to constitute the negative dual, and a sub-set of positive characteristics to constitute the positive dual as it is discussed in [671, 672]. The sub-set of the negative characteristics reside in cost–benefit duality where the negative characteristics are also divided into cost and benefit characteristics. Similarly, the sub-set of positive characteristics resides in cost–benefit duality relative to a goal-objective element, where the positive characteristics are also divided into cost and benefit characteristics, the structure of which will depend on a given goal-objective disposition. These characteristics are specified in a fuzzy information space over the epistemological space as seen over the space of imagination-reality dualities. Every variety exists individually as *internal cost–benefit duality* and collectively as *external cost–benefit dualities* to each other in the decision-choice actions over the space of the problem–solution dualities where the neuro-decision-choice actions continually change the problem–solution terms of relations in the space of static-dynamic dualities over the space of imagination-reality dualities contained in the space of actual-potential dualities.

6.2.1 Variety Analytics, Categorial Analytics, Fuzzy Decomposition, Fuzzy-Statistical Decomposition, Epistemic Unity and Internal and External Dualities in Knowing

The whole process of knowing is about neuro-decision-choice analysis of internal and external negative–positive dualities of varieties in terms of negative–positive characteristic dispositions in information structures over the space of actual-potential dualities as input into neuro-decision-choice activities over the space of imagination-reality dualities contained in the space of actual-potential dualities which constitutes the primary identity while the space of imagination-reality dualities constitutes the derived identity in the neuro-decision-choice processes over all areas of human actions. The negative–positive dualities are projected as real cost–benefit dualities of individual varieties, while the negative–positive dispositions are projected as cost–benefit dispositions in terms of information structures over the spaces of imagination-reality and actual-potential dualities. The cost–benefit dualities are projected as problem–solution dualities of varieties while the cost–benefit dispositions are projected as problem–solution dispositions in terms of information structures over the goal-objective space in relation to the elements in the spaces of varieties under a preference ordering of varieties over the goal-objective space. The contents of information are defined in terms of characteristic dispositions while the variety messages of the information are defined as signal dispositions on the basis of which knowledge is abstracted and tested over the spaces of uncertainty-certainty, doubt-surety, decidability-undecidability and acceptance-rejection dualities.

The varieties are seen by cognitive agents as things, objects, technologies, processes, states and conditions for which their relative real cost–benefit dispositions may be assessed for neuro-decision-choice actions in the space of destructive-creative dualities. In this respect, every preference ordering in relation to neuro-decision-choice actions is simply reducible to real cost–benefit conditions viewed in terms of the economic theory of the real production-consumption process or the input–output process. Every set of conditions of a variety of knowing finds meaning in the space of problem–solution dualities, where every variety problem is an information-knowledge element and every variety solution is also an information-knowledge element that provides the needed information-knowledge system about goal-objective elements for neuro-decision-choice actions. Every knowing area is an economic production with an input–output relation, where such production is a neuro-decision-choice action on problem–solution duality whether the area of knowing is seen as either science, non-science, social science or non-social science where classifications and category formations are analytical strategies in creating simplicities in the space of knowing through neuro-decision-choice actions over the general space of problem–solution dualities where the classifications are under fuzzy dualistic-polar closed-open sets in the space of inclusion–exclusion dualities.

In general, the varieties exist with each identity defined by negative characteristic sub-set in combination with a positive characteristic sub-set. At the level of neuro-decision-choice actions, the negative (positive) characteristic sub-set exists as either cost (benefit), benefit (cost) characteristic sub-sets or both to the neuro-decision-choice agent over the elements of the goal-objective space as seen in terms of varieties depending on either the individual preference relation or collective preference relation at the level of either classical or fuzzy aggregation. The cost characteristics translate into problem characteristics while the benefit characteristics translate into solution characteristics in relation to varieties or things as determined by a goal-objective element of decision-choice agents. These characteristics become information-knowledge conditions as inputs into decision-choice actions to establish variety identities and behavior (identification information-knowledge) as well as foundations for purposeful variety transformation over the space of problem–solution dualities.

The conditions of knowing are such that every variety exists in a dualistic-polar situation in a negative–positive, cost–benefit and problem–solution structures, where negatives reside in positives and positives reside in negati8ve, cost is benefit and benefit is cost, and problem is a solution and a solution is a problem under certain conditions of variety transformation or neuro-decision-choice action in the goal-objective space. All these dualistic-polar representations must obey the principle of opposites with relational continuum and unity of existence of every variety. No identity of a variety exists solely as positive without the negative, as negative without the positive, as benefit without cost and vice versa, and as a solution without a problem and vice versa. There are always two dualities in the neuro-decision-choice process of varieties. There is the *internal duality* in relation to variety identity for knowing and information-knowledge development. There is also the *external duality* in comparative relations with other dualities over the preference space for decision-choice actions. The analytical methods and techniques of cost–benefit analytics over the space of real-nominal dualities find no justification in non-dualistic conditions of varieties in the neuro-decision-choice space or the transformation space of socio-natural varieties.

It is important to see the indispensable role of cost–benefit conditions in the understanding of information-knowledge production in terms of the reduction process of uncertainty and doubt over the decision-choice space and the space of problem–solution dualities as uniting all areas of knowing irrespective of whether one classifies the areas into science and non-science or economics and non-economics. The decision-choice activities over the action space with the nature of the simultaneous existence of variety cost–benefit and negative–positive dualities are imbedded in the *asantrofi-anoma principle* that tends to generate *paradoxes of decision-choice actions,* where every variety has a dualistic-polar disposition of attractive characteristics and repulsive characteristics under continual qualitative motions for transformations of properties and variety relations producing conditions of unintended consequences and surprises of outcomes of neuro-decision-choices to produce information-knowledge of actualized varieties. Here, the following question arises: what is a paradox and what is an unintended outcome?

Definition 6.1.1.1: The Paradox of Decision-Choice Action The paradox of decision-choice action is a characteristic disposition, where an individual or the collective decision-choice activity for transforming an existing un-preferred variety due to the internal cost dominance over the internal benefit into the variety imagination for an increasing variety benefit over the cost disposition in the action space and into the state of increasing better-off over the preference space, leads to a realization of a variety imagination residing in the space of potentials with an increasing internal variety-cost disposition over the internal variety-benefit disposition in the action space causing either the individual or the collective to be worse off over the preference space such that the neuro-decision-choice action, instead of moving the system from its current undesirable state to a better desirable state actually moves it to a worse state of being.

Definition 6.1.1.2: Unintended Consequence and Collateral Damage of Decision-Choice Action An unintended consequence of decision-choice action is either an outcome or a joint outcome of an intended characteristic disposition, where an individual or the collective decision-choice activity over the space of problem–solution dualities, for transforming a variety problem to a variety solution, leads to a variety solution that is un-preferred, due to its internal cost dominance over its internal benefit, into a potential variety for increasing variety benefit disposition over cost disposition in the action space, and into a state of increasing *better off* over the preference space, leading to an actualization of a potentially wanted or unwanted variety with joint varieties where such joint varieties contain an increasing internal variety cost disposition over the internal benefit disposition in the action space, causing either the individual or the collective to be *worse off* over the preference space, such that the decision-choice action moves the system to either a preferred state with extra outcome characteristics which may produce undesirable cost or desirable benefit conditions that will affect directly or indirectly the welfare of the decision maker or other individuals as well as the elements of the surrounding environment, instead of moving the system from its current undesirable state to a better desirable state, it actually moves it to either a worse state of being or a more desirable state with joint outcomes of cost–benefit duality that may be increasing-welfare enabling or decreasing-welfare enabling.

6.2.2 Reflections on the Paradox and Unintended Outcome of Decision-Choice Action

It is important to note that the decision-choice paradox finds explanation under the *asantrofi-anoma principle* (cost–benefit principle) and decision-choice information-knowledge entropy from the effects of cognitive capacity limitations regarding real cost–benefit balances over the spaces of certainty-uncertainty dualities, doubt-surety dualities and decidability-undesirability dualities as well as over the space of either the problem–solution dualities or question–answer dualities that relates to destruction-replacement processes. The unintended consequence find expression under the *asantrofi-anoma* principle over the space of problem–solution dualities constrained by cognitive capacity limitation over the space of cost–benefit dualities specified in their characteristic dispositions in their destination interpretation of signal dispositions. The paradoxes and unintended consequences in the decision-choice space under destruction-replacement processes are cognitive failures due to information deficiencies on correct balances on varieties in the space of cost–benefit dualities.

The paradoxes and unintended consequences are the results of the use of classical paradigm on fuzzy information to evade vagueness and inexactness. They are paradigmatic in nature that may not be resolved in the same paradigm with classical stochastic conditionality. They must be resolved in another paradigm with nonclassical conditionality. They may be resolved by reducing the information deficiencies in the space of cost–benefit dualities. The decision-choice paradox and unintended consequences also unite all areas of knowing where variety solutions become variety problems and variety problems become variety solutions in destruction-replacement processes over the space of input–output dualities. The unintended consequences are the negative and positive externalities of either cost disposition or negative disposition that were not anticipated, but, affect the cost–benefit balances of decision-choice outcomes to change relational terms of inter-welfare positions of the members in the society.

Under the principle of opposites, composed of sub-principles of polarity and duality, all universal varieties are separated from each other by information for identity-identification and yet are linked together into families of categories and into universal unity by information, where diversities are established in unity by information, and unity is established in a set of diversities by information to maintain the *universal give-and-take sharing process* for mutual existence, mutual destruction and mutual creation in relational continua and unity under the transformation-substitution principles in real engineering-economic productions in the information space of cost–benefit relativity and not in the space of cost–benefit absoluteness. There is no absolute cost and there is no absolute benefit. Every variety is simultaneously a benefit and a cost in multiplicity of mutual relation over the space of cost–benefit dualities, where without real cost there will be no real benefit and without real benefit there will be no real cost relative to matter-energy stock-flow equilibrium conditions. Similarly, without inputs there will be no outputs and without outputs there will be no inputs over the space of production-consumption dualities and inventory stock-flow dynamics over the space of excess-shortage dualities.

A unity of any form has no meaning without diversities which acquire meaning in unity through which totality and oneness are expressed to allow operations of addition, subtraction, division and multiplication, union, intersection, symmetric difference, exclusion and inclusion and other operations to be expressed and performed on varieties in qualitative and qualitative forms over the space of problem–solution dualities which, in the last analysis, finds an expression over the space of identification-transformation activities on problems of knowing to produce information-knowledge structures that unite different areas of *basic science*, *applied science* and engineering sciences in the enterprise of knowing and information-knowledge production as expressed over the space of imagination-reality dualities as a subspace of actual-potential dualities. The apparent rise of decision-choice paradoxes may also be explained by the imaginational potentiality of cost–benefit outcomes of the dynamics of the problem–solution dualities and the paradigm of thought as an input–output processor over different areas of knowing. It may be kept in mind that in all areas of knowing a solution to a current problem leads to an unanticipated new problem or problems, the full cost–benefit dispositions of which may have escaped the best epistemic analysis and computation. The existence of neuro-decision-choice paradoxes may help the understanding of unintended consequences, collateral damages and socio-economic decay as seen in terms of information-knowledge dynamics of elements in the imagination-reality dualities.

The information-knowledge structures of varieties are presented in an interactive mode in the space of problem–solution dualities within the space of imagination-reality duality as a subspace of the space of actual-potential dualities, where every knowing, at the space of static-dynamic dualities, is a cognitively problem-solving action in the space of certainty-uncertainty duality with a relational continuum and unity. At the static state, knowledge involves solutions to variety identification problems while at the dynamic state, knowledge involves solutions to variety transformation problems. Over the space of static-dynamic dualities, knowledge involves simultaneously the variety identification-transformation problems with fuzzy-stochastic conditionality and fuzzy-set closure. A neuro-decision-choice activity to find a variety solution is a *thinking process* for cost–benefit balances to set a variety solution against a variety problem through a resolution of conflict conditions of internal and external cost–benefit dualities. The problem–solution analytics requires us to keep in mind the relativity conditions of internal duality and external duality which impose structural boundaries for all decision analytics that establish diversities through the principle of dividedness to create diversities as well as unite all areas of sciences through the unity of knowing. The internal cost–benefit duality reveals the identity of the variety while the external cost–benefit duality reveals collective comparative conditions of preference ordering for decision-choice actions over the space of varieties and diversities.

In other words, knowledge is defined in relation to variety problem–solution identifications and a search of solutions to problems in the space of problem–solution dualities ($\mathfrak{R}$) under information conditions of an element or some elements in the goal-objective space, constrained by information processes, where the distance between an epistemological characteristic disposition and ontological characteristic disposition of a variety is the variety *epistemic distance* which finds expression in certainty-uncertainty dualities, doubt-surety dualities, and decidability-undesirability dualities as they relate to paradoxes and unintended consequences (for an extended exposition and analytical definition of epistemic distance see [467, 671, 672]). The test of the epistemic distance over the space of acceptance-rejection dualities is met with fuzzy-stochastic entropic conditions as control conditions over the space of quality-quantity dualities.

The epistemic distance of any variety, translated into conditions of certainty-uncertainty duality, provides us with the essence of a fuzzy-stochastic conditionality for the accuracy of solutions to variety information-knowledge structures of identification and transformation problems in both science and non-science and degrees of surety and confidence attached to information-knowledge variety identifications and transformations in the enterprise of knowing within the neuro-decision-choice activities over the space of imagination-reality dualities. There are critical analytical questions that arise over the space of knowing in relation to the principles of diversity and unity. They may be stated as:

(1) How is diversity defined and structured for variety distinctions in the unity of knowing and use?
(2) How is unity defined and structured to show the existence of the internal partition that allows the unity to house the contents of the diversity of the elemental varieties?
(3) What are the dividing factors of variety distinctions and their connecting factors into an epistemic unity in the knowing process? In other words, what separates the internal structural unity into functional varieties and what binds the functional varieties into an active unity for functional existence, variety input–output, and creation-destruction dynamics in the space of parent–offspring processes?

The answers to these questions are relevant to understand the unity of science through the unity and diversity of knowing and the practices that may be required of the diversities in the relational give-and-take processes for establishing departments of research, teaching and learning and translation of information-knowledge into continual cognitive variety creations over the space of imagination-reality dualities. These questions and the corresponding answers are translated and mapped onto the *space of problem–solution dualities* ($\mathfrak{R}$), where each variety exists as an internal dualistic-polar mode for identity and in an *external dualistic-polar mode* as simultaneously a problem and a solution relative to the elements in the goal-objective space in relation to cognitive preferences. The *external dualistic-polar mode* is related to the *internal dualistic-polar mode* where any variety has negative–positive and cost–benefit dualistic-polar existence under a universal give-and-take sharing in a mutual existence and destruction in the space of input–output dualities. The negative–positive duality is naturally innate in defining the diversities and identities while the real cost–benefit duality is social in an essence- in terms of real opportunity cost of creative destruction as assessed by neuro-decision-choice agents, where existing varieties are destroyed by neuro-decision-choice actions to give rise to new varieties in transformation-substitution dynamics as seen from the generating efficient set of the production possibility frontier.

Every solution to a problem is a neuro-decision-choice action over the space of realities in a destruction of the existing variety while, every solution is also a neuro-decision-choice action over the potential space in a creation of a new variety as a replacement in terms of information-knowledge conditions and trade-off dynamics. It is important to note that over the epistemological space, information and knowledge are generated by decision-choice actions over the space of acquaintance-description dualities, as one passes through the space of problem–solution dualities under the categorial unity in knowing. The categorial unity of existence is the primary category of the unity of knowing that offers us the conditions of either fuzzy partition with fuzzy numbers, statistical partition with probabilistic numbers or fuzzy-stochastic partitions with hybrid numbers, and the awareness and non-awareness of its contents which constitute the existence of the space of variety *ignorance- knowledge dualities* (which is represented by an Adinkra conceptual tradition of *nea-ohnim-sua-ohu* duality) in a relational continuum and unity under the principle of opposites that contains the principle of unity and the principle of diversity in terms of elements in the space of unity-diversity dualities.

The space of problem–solution dualities is the universal connector of science and non-science, of categories of science and categories of non-science in all areas of knowing. Life is undefinable without decision-choice action, and decision-choice action is undefinable without the space of problem–solution dualities that shapes information-knowledge structures, human progress and its directions of qualitative and quantitative motions. There are no problems and solutions without life and there are also no problems and solutions without information-knowledge structures which reveal varieties and their identities, and there are no information-knowledge structures without matter-energy varieties. It is the disconnecting force of the space of the problem–solution dualities and the interdependency of problem–solution processes that establish the peculiarity of the unity of science and the unity of non-science and their categorial interdependencies through the connecting force of the general unity of knowing as seen in neuro-decision-choice activities. When the universality of this connecting force through the family of decision-choice actions is understood relative to the fact that problem–solution activities are the essential essence of the mind, where the mind converts every potential or actual element of existence into a problem-solving and information-knowledge development irrespective of the type of variety and categorial varieties as distinguished by their characteristic dispositions, then one can understand the principles of diversity, interdependencies and unity of universal existence and relationality within the space of construction-destruction processes over the space of input–output dualities.

The space of problem–solution dualities is infinite in the space of variety static-dynamic dualities, where every element of socio-natural existence over the four dimensions of existence composed of matter, energy, information and time is a variety with an *internal duality* for the practice of science and non-science through the enterprise of identity knowing as well as *an external duality* for comparative analytics, preference ordering, decision-choice action, destruction-replacement processes and the practice of prescriptive science composed of planning and engineering of all forms. The diversity and unity of science, through the unity of knowing, are made up of nothing more than the space of problem–solution dualities and are not in the space of methods and techniques of knowing within a methodology. In other words, the principle of diversity and the unity of science through the unity of knowing are the principle of dynamics of the space of problem–solution dualities and are not in the space of methods and techniques in the general study, composed of research, teaching and learning about varieties and categorial varieties or the space of the material essence of varieties under knowing. The categories of problem–solution dualities establish required and needed methods and techniques of decision-choice activities over the space of problem–solution dualities. The diversities of the areas of knowing are not established by methods and techniques of knowing but rather by differences of the problem–solution dualities. The methods and technique of knowing are a derived category of a toolbox of diversity and unity from problem–solution dualities as the primary category of diversity and unity. The methods and techniques may be used to establish secondary principles of diversity and unity of knowing involving science and non-science where the structures of categories of problem–solution dualities are such that they share a common toolbox in in the neuro-decision-choice actions over the space of inefficient-efficiency dualities. In the same framework, a tertiary principle of diversity and unity may be established with other criteria and further divisions as will be explained under the theorem of dividedness. The important point, here, is that all areas or categories of knowing are established by common problem–solution dualities and are united by universal problem–solution dualities and the need for specific and general information from the space of acquaintance to the space of description through a paradigm of thought.

6.2.3 Internal Duality and External Duality in the Space of Problem–Solution Dualities

The concepts of internal and external dualities were introduced in the previous sections. In the process, it was pointed out that every variety exists simultaneously as a sub-set of negative characteristics, as a negative dual, as a sub-set of positive characteristics, and as a positive dual the relative combination and the unity of which establish the identity of the variety for identification. The subset of the negative characteristics functions as either a sub-set of cost characteristics, the cost dual or benefit characteristics, benefit dual which are relative to an element in the goal-objective space. Similarly, the sub-set of the positive characteristics functions as either a sub-set of cost characteristics, the cost dual or a sub-set of benefit characteristics, and the benefit dual which are relative to an element in the goal-objective space. The negative–positive relativity of the variety characteristics and the relational nature to cost–benefit relativity provide the epistemic justification for the use of techniques and methods of cost–benefit analysis as a general neuro-decision-choice framework in the sense that every variety broadly defined exists in a simultaneity of real cost and real benefit for individual and collective conditions in the general decision-choice space where cost–benefit criteria affect the development of the theories of knowing and knowledge and neuro-decision-choice actions. Let us keep in mind that reduction in uncertainty, increasing certainty, reduction of distance per minute and many others are benefits and the opposites are cost in the space of neuro-decision-choice analytics. Let us define the concept of internal and external dualities:

Definition 6.1.3.1A: Internal Duality (Verbal) The negative–positive characteristic dispositions of varieties and their translations into cost–benefit dispositions in relational continua and unity constitute a system of internal dualities under the principle of opposites, where the negative disposition functions as either an internal cost dual or an internal benefit dual and the positive disposition also functions as either an internal benefit dual or an internal cost dual in reverse under decision-choice analytics, where the internal negative–positive characteristic disposition reveals the identity of the variety and the cost–benefit disposition reveals the decision-choice value of the variety.

Definition 6.1.3.1B: Internal Duality (Algebraic) Given a set of characteristics $(\mathbb{X}_\nu \subset \mathbb{X})$ of a variety $(\nu \in \mathbb{V})$ with $\left(\mathbb{X}_\nu^{\mathrm{N}}\right)$ as the negative dual and $\left(\mathbb{X}_\nu^{\mathrm{P}}\right)$ as the positive dual where $\left(\mathbb{X}_\nu = \mathbb{X}_\nu^{\mathrm{N}} \cup \mathbb{X}_\nu^{\mathrm{P}}\right)$ with $\left(\mathbb{X}_\nu^{\mathrm{N}}=\left(\mathrm{B}_\nu^{\mathrm{N}} \cup \mathrm{C}_\nu^{\mathrm{N}}\right)\right)$ and $\left(\mathbb{X}_\nu^{\mathrm{N}}=\left(\mathrm{B}_\nu^{\mathrm{P}} \cup \mathrm{C}_\nu^{\mathrm{P}}\right)\right)$, where $\left(\mathrm{B}_\nu^{i}\right)$ is the internal real benefit dual and $\left(\mathrm{C}_\nu^{i}\right)$ is the internal real cost dual, defined in characteristic dispositions (i = N,P), then $\left(\mathbb{X}_\nu^{\mathrm{N}} \cup \mathbb{X}_\nu^{\mathrm{P}}\right)$ with $\left(\mathrm{B}_\nu^{i} \cup \mathrm{C}_\nu^{i}\right)$ constitute the *internal duality* with a system of internal cost–benefit dualities, and the collection of all internal dualities constitutes the space of internal dualities, $\mathbf{D}_{\mathbf{i}}$ with a generic element $\left(\mathbf{d}_{\mathbf{i}}^{\nu} \in \mathbf{D}_{\mathbf{i}}\right)$ where $(\mathbf{i})$ represents internal and $\mathbf{d}_{\mathbf{i}}^{\nu} = \left(\mathbb{X}_\nu^{\mathrm{N}} \cup \mathbb{X}_\nu^{\mathrm{P}}\right)$ is the variety's internal duality.

Note 6.1.3.1: Internal Duality The internal duality presents each variety as either a combination of *real benefit dual* and a *real cost dual* for its decision-choice identity relative to other varieties viewed in terms of goals or objectives in the goal-objective space, where an actual variety is viewed as real opportunity cost or real benefit and is to be transformed to a potential variety which is also viewed as real potential benefit with real potential cost, where in decision-choice analysis one ascertains the cost and benefit information structures of each relevant varieties in the space of problem–solution dualities. A variety $(\nu \in \mathbb{V})$ is said to be a real cost (benefit) if the real cost (benefit) characteristic sub-set dominates the real benefit (cost) characteristic sub-set of its internal duality $(\mathbf{d}_{\mathbf{i}} \in \mathbf{D}_{\mathbf{i}})$ relative to a goal. In terms of decision-choice action on variety $\nu \in \mathbb{V}$, the real benefit and cost dispositions are of mutual constraints, where the selection is either a maximum benefit relative to cost or a minimum cost relative to benefit. In fact, this is the epistemic foundation for the development of the theory of optimization where the goal is either a minimization of real cost subject to a given real benefit or maximization of real benefit subject to a given real cost or optimization of real cost–benefit balances under variable cost–benefit conditions. The internal duality is the cost–benefit analytics of the variety relative to the dual conditions.

Definition 6.1.3.2A: External Duality (Verbal) The external duality refers to analytical conditions where an variety reality, with internal negative–positive characteristic dispositions and a system of cost–benefit characteristic dispositions in relational continuum and unity as paired with any variety imagination with potential negative–positive characteristic dispositions and a corresponding system of potential cost–benefit dispositions over the space of problem–solution dualities under the principle of opposites, functions as either a negative or positive dual with net cost or net benefit disposition and the paired variety imagination acts as either a positive or negative dual with net benefit or net cost disposition in reversed conditions under decision-choice analysis over the space of relativity.

Definition 6.1.3.2B: External Duality (Algebraic) Given a set of characteristics $\left(\mathbb{X}_{\nu_{\mathfrak{h}}}^{\mathfrak{H}} \subset \mathbb{X}\right)$ of a variety reality $\left(\nu_{\mathfrak{h}} \in \mathbb{V}\right)$ with $\left(\mathbb{X}_{\nu_{\mathfrak{h}}}^{N\mathfrak{H}}\right)$ as the negative real dual and $\left(\mathbb{X}_{\nu_{\mathfrak{h}}}^{P\mathfrak{H}}\right)$ as the positive real dual, in the space realities $(\mathfrak{H}|\mathfrak{h} \in \mathfrak{H})$ where $\left(\mathbb{X}\ominus_{\nu_{\mathfrak{h}}}^{\mathfrak{H}} = \mathbb{X}_{\nu_{\mathfrak{h}}}^{N\mathfrak{H}} \cup \mathbb{X}_{\nu_{\mathfrak{h}}}^{P\mathfrak{H}}\right)$ with $\left(\mathbb{X}_{\nu_{\mathfrak{h}}}^{N\mathfrak{H}} = \mathrm{B}_{\nu_{\mathfrak{h}}}^{\mathfrak{H}} \text{ or } \mathrm{C}_{\nu_{\mathfrak{h}}}^{\mathfrak{H}}\right)$ and $\left(\mathbb{X}_{\nu_{\mathfrak{h}}}^{P\mathfrak{H}} = \mathrm{C}_{\nu_{\mathfrak{h}}}^{\mathfrak{H}} \text{ or } \mathrm{B}_{\nu_{\mathfrak{h}}}^{\mathfrak{H}}\right)$, where $\left(\mathrm{B}_{\nu_{\mathfrak{h}}}^{\mathfrak{H}}\right)$ is the internal real benefit dual and $\left(\mathrm{C}_{\nu_{\mathfrak{h}}}^{\mathfrak{H}}\right)$ is the internal real cost dual defined in terms of characteristic dispositions, then $\left(\mathbb{X}\ominus_{\nu_{\mathfrak{h}}}^{\mathfrak{H}} = \left(\mathbb{X}_{\nu_{\mathfrak{h}}}^{N\mathfrak{H}} \cup \mathbb{X}_{\nu_{\mathfrak{h}}}^{P\mathfrak{H}}\right)\right)$ is an external duality, where the corresponding variety $(\nu_{\mathfrak{H}} \in \mathbb{V})$ constitutes the *real pole* of an *imagination-reality polarity* with $\left(\mathrm{B}_{\nu_{\mathfrak{h}}}^{\mathfrak{H}} \cup \mathrm{C}_{\nu_{\mathfrak{h}}}^{\mathfrak{H}}\right)$ constituting the corresponding internal cost–benefit duality of the actual pole $\left(\mathbf{p}_{\nu_{\mathfrak{h}}} \in \mathbf{P}\right)$ with a fuzzy constraint of the form $\left(\mathbb{X}_{\nu}^{N\mathfrak{H}} \cap \mathbb{X}_{\nu}^{P\mathfrak{H}} \neq \emptyset\right)$. Similarly, there is a set of characteristics $\left(\mathbb{X}\ominus_{\nu_{\mathfrak{g}}}^{\mathfrak{G}} \subset \mathbb{X}\right)$ of a variety imagination $\left(\nu_{\mathfrak{g}} \in \mathbb{V}\right)$ with $\left(\mathbb{X}_{\nu_{\mathfrak{g}}}^{N\mathfrak{G}}\right)$ as the negative dual and $\left(\mathbb{X}_{\nu_{\mathfrak{g}}}^{P\mathfrak{G}}\right)$ as the imagination positive dual in the imagination space $(\mathfrak{G}|\mathfrak{g} \in \mathfrak{G})$ where $\left(\mathbb{X}\ominus_{\nu_{\mathfrak{g}}}^{\mathfrak{G}} = \mathbb{X}_{\nu_{\mathfrak{g}}}^{N\mathfrak{G}} \cup \mathbb{X}_{\nu_{\mathfrak{g}}}^{P\mathfrak{G}}\right)$ with $\left(\mathbb{X}_{\nu_{\mathfrak{g}}}^{N\mathfrak{G}} = \mathrm{B}_{\nu_{\mathfrak{g}}}^{\mathfrak{G}} \text{ or } \mathrm{C}_{\nu_{\mathfrak{g}}}^{\mathfrak{G}}\right)$ and $\left(\mathbb{X}_{\nu_{\mathfrak{h}}}^{P\mathfrak{G}} = \mathrm{C}_{\nu_{\mathfrak{g}}}^{\mathfrak{G}} \text{ or } \mathrm{B}_{\nu_{\mathfrak{g}}}^{\mathfrak{G}}\right)$, where $\left(\mathrm{B}_{\nu_{\mathfrak{g}}}^{\mathfrak{G}}\right)$ is the internal real benefit dual and $\left(\mathrm{C}_{\nu_{\mathfrak{g}}}^{\mathfrak{G}}\right)$ is the internal real cost dual defined in characteristic dispositions then $\left(\mathbb{X}_{\nu}^{\mathfrak{G}} = \mathbb{X}_{\nu}^{N\mathfrak{G}} \cup \mathbb{X}_{\nu}^{P\mathfrak{G}}\right)$ is an *external duality,* where the corresponding variety $(\nu_{\mathfrak{g}})$ constitutes the potential pole $\left(\mathbf{p}_{\nu_{\mathfrak{h}}} \in \mathbf{P}\right)$ of an imagination-reality polarity $(\mathbf{P})$ with $\left(\mathrm{B}_{\nu_{\mathfrak{h}}}^{\mathfrak{G}} \cup \mathrm{C}_{\nu_{\mathfrak{g}}}^{\mathfrak{G}}\right)$ constituting the corresponding internal cost–benefit duality with a fuzzy constraint of the form $\left(\mathbb{X}_{\nu}^{N\mathfrak{G}} \cap \mathbb{X}_{\nu}^{P\mathfrak{G}} \neq \emptyset\right)$, where the *real external duality* constitutes the *reality pole* and the imagination external duality constitutes the *imagination pole* in the system of imagination-reality polarities $(\mathbf{P})$. The transformation process is always a dualistic conflict in a game situation between the actual pole and the potential pole in the space of the problem–solution dualities to bring about transformation of the variety problem–solution duality under the conditions of relative cost–benefit dominance and information non-elimination. The symbols $(N\mathfrak{G})$ and $(P\mathfrak{G})$ mean negative and positive imagination duals respectively while $(N\mathfrak{H})$ and $(P\mathfrak{H})$ mean negative and positive reality duals respectively ($\mathfrak{G}$ = space of imaginations) and ($\mathfrak{H}$ =space of realities).

In this analytical process, it is useful to distinguish between the spaces of the actual $\mathfrak{A}$, potential $\mathfrak{U}$ and actual-potential dualities $(\mathfrak{A} \cup \mathfrak{U}) = (\mathfrak{A} \otimes \mathfrak{U})$ on one hand and the spaces of reality $\mathfrak{H}$, imagination $\mathfrak{G}$ and imagination-reality dualities $(\mathfrak{G} \cup \mathfrak{H}) = (\mathfrak{G} \otimes \mathfrak{H})$ under dualistic-polar conditions such that $(\mathfrak{G} \cup \mathfrak{H}) = (\mathfrak{G} \otimes \mathfrak{H}) \subset (\mathfrak{A} \cup \mathfrak{U}) = (\mathfrak{A} \otimes \mathfrak{U}) = \Omega = (\mathfrak{E} \cup \mathfrak{U})$ for continuum, unity and relativity. Analytically, cognitive agents operate over imagination-reality dualities through the neuro-decision-choice actions. The space of imagination reality dualities is contained in the space of actual-potential dualities which helps to define the space of ontological-epistemological dualities connected by neuro-decision-choice actions. The space of reality is a subspace of the actuals, while the space of imagination is a sub-space of the space of potentials to allow the creation of the space of goals, objectives and visions under the general preference principle of non-satiation. Further analytical development will be given as the development of diversity and unity of knowing and science is carried on.

6.2.4 Internal Duality, External Duality and Ignorance-Knowledge Polarity

In the theory of knowing, we have the ignorance-knowledge duality and actual-potential polarity, where ignorance is the actual and knowledge is the potential with neuro-decision-choice actions connecting them to set a potential variety solution with the internal benefit dual dominating the internal cost dual against the actual variety problem, where the internal cost dual dominates the internal benefit dual to protect the actual variety with internal benefit dual dominating the internal cost dual against the potential varieties with internal cost duals dominating the internal benefit duals in the space of problem–solution dualities under the command and control of neuro-decision-choice modules with real cost–benefit rationality under the *asantrofi-anoma principle*, where variety solutions and problems are mutually constraining in algorithmic processes irrespective of the area of knowing as seen in mathematical or non-mathematical representations in either a fuzzy-information space or a non-fuzzy information space. In the theory of knowledge, we have the uncertainty-certainty duality and decidability-undesirability polarity, where uncertainty is the actual and certainty is the potential with neuro-decision-choice actions to set the potential certainty variety solution with an internal benefit dual (seen in terms of degrees of surety) dominating the internal cost dual (seen in terms of degree of doubt) against the actual uncertainty variety problem where the internal cost dual dominates the internal benefit dual in the space of problem–solution dualities under the command and control of neuro-decision-choice modules with real cost–benefit rationality always under the *asantrofi-anoma principle*, where variety solutions and problems, as well as real variety benefits and real variety costs are mutually constraining in algorithmically thinking processes such that one cannot choose benefit without choosing the corresponding cost of a variety.

At the space of external duality, the actual variety is always in a dualistic-polar relation with several potential varieties which are also in dualistic-polar relation with other elements in the potential space, where the potential net cost–benefit structures vary over different varieties as replacement varieties. This is the actual-potential dualistic-polar problem, the solution of which is found through neuro-decision-choice actions under conditions of a real cost–benefit rationality with a fuzzy-stochastic conditionality, where all qualifying potential replacements are assessed in their net cost–benefit structures relative to the net cost–benefit structure of the actual acting as an opportunity cost within the epistemological space. The actual-potential dualistic-polar problem is the uniting force of all the areas of knowing irrespective of whether they are classified as either science or non-science under the *principle of acquaintance* and the *principle of methodological constructionism-reductionism duality* given either the classical paradigm of thought or the fuzzy paradigm of thought with their logics and mathematics to generate knowledge by description constrained by the elements of the spaces of qualitative-quantitative, certainty-uncertainty decidability-undesirability and doubt-surety dualities. Let us always keep in mind that the theory of knowing is the theory of intellectual investment, while the theory of knowledge is the theory of intellectual information-knowledge capital accumulation and that the information-knowledge capital accumulation becomes an input into the knowing processes to generate new information-knowledge items as output to maintain the universal conditions projected by the space of input–output dualities under the principle of parent–offspring disequilibrium dynamics.

The individual and collective decision-choice processes for transformation under the principle of opposites of cognitive activities over the space of problem–solution dualities are such that the potential variety is always the goal to be optimized, while the actual variety forms the constraint conditions on the selection of the potential variety to be actualized. Similarly, the individual and collective decision-choice processes for the preservation of the existing variety under the principle of opposites for cognitive activities over the space of problem–solution dualities are such that the actual variety is the goal to be optimized while the potential varieties form a system of constraint conditions to decide on the destruction of the actual variety for the replacement of potential variety to be actualized under the conditions of cost–benefit rationality. The methods and techniques constituting a toolbox of representation of the problem–solution duality, optimization analytics and decision-choice actions are always defined and specified within a paradigm of thought or methodology of information-knowledge production. In this respect, big-small data analysis, mathematics, statistics, entropy, programming, optimization, systems of observations and tools of research and investigations belong to a toolbox within a paradigm of thought.

The toolbox of the *principle of acquaintance* to construct the experiential information structure will vary over epistemic categories of problem–solution dualities depending on the nature of the characteristic dispositions for their analytic constructs. The contents of any toolbox will be in relation to the nature of the variety ignorance-knowledge duality and the dynamics of the category of problem–solution dualities to which it belongs. Should the similarities in the toolboxes be used to establish the principles of diversity and the unity of science through the unity of knowing? The toolboxes of methodological constructionism-reductionism duality, for the categorial convertors through neuro-decision-choice actions, as generated by the use of either the classical paradigm of thought or the fuzzy paradigm of thought to deal with the dynamics of ignorance-knowledge dualities of categorial varieties to obtain the knowledge by description, will vary over the nature of experiential information structures of varieties and the mode of acquaintance over the epistemological space. Let us keep in mind that the two spaces of cognitive activities are the ontological space and epistemological space where the ontological space constitutes the primary categorial identity and the epistemological space constitutes the derived categorial identity. The two spaces are connected by human actions and neuro-decision-choice actions that mutually affect the socio-natural situation and variety dynamics. The ontological space provides ontological varieties as inputs into the creation of the epistemological space for the activities of knowing and transformations of epistemological varieties while the epistemological varieties cannot exist without the ontological varieties, the ontological varieties can exist without the existence of the epistemological varieties. Every epistemological variety has its identity derived from an ontological variety. However, there is an infinite number of ontological varieties that have no comparable epistemological varieties. The space of ontological variety is permanently in an equilibrium state while the space of epistemological varieties is in permanently disequilibrium states.

6.3 Categorial Conversion and Problem–Solution Dualities

One thing that is clear, is that over the epistemological space relative to the ontological space, the experiential information structures will have inter-categorial variances induced by the nature of the distribution of qualitative dispositions over the actual-potential spaces. Each experimental information structure will have intra-categorial variance induced by the nature of the distributions of the quantitative disposition over the actual-potential spaces. The development of knowing is a process of resolving conflicts within variety problem–solution relationality, where the discovery of a variety problem to a variety solution is an information-knowledge item and where knowledge is obtained by neuro-decision-choice activity over the space of variety problem–solution dualities. The dynamics of problem–solution dualities is induced by *categorial conversions* from a variety problem to a variety solution and from a variety solution to a variety problem in a never-ending process of parent–offspring dynamics between the category of ignorance and the category of knowledge.

The system of explanations where variety problems are converted to variety solutions and variety solutions generate variety problems is the theory of categorial conversion as applied to knowing and information-knowledge accumulation and other categorial varieties in disequilibrium processes. In this respect, we have problem–solution categorial conversions defined in backward-forward telescopic dualities, where the discoveries of a variety problems and their corresponding variety solutions are flows of information that go to increase the information-knowledge stocks of cognitive agents. In other words, one cannot speak of the theory of knowing and the theory of knowledge without dealing with the variety problem–solution processes composed of the information-knowledge conversion systems. The discovery of a variety problem is a discovery of an information-knowledge variety, and the discovery of variety solution is also a discovery of information-knowledge variety in the space of destruction-replacement processes, where the variety problem leaves its information-knowledge structure behind after destruction and the variety solution brings in its information-knowledge structure to update the information-knowledge stocks expanding the zone of cognition.

6.3.1 Categorial Conversion in the Logic of Problem–Solution Transformations Within the Theories of Knowing, Science, Non-science and Mathematics

The study of transformations of varieties is a study of inter-categorial movements between variety problems and variety solutions or between variety ignorance and variety knowledge in dualistic-polar dynamics, where such movements are governed by qualitative and quantity transfer functions. In other words, we must study how variety problems in the category of variety problems lose their categorial characteristic dispositions, acquire new categorial characteristic dispositions, migrate from their category of variety problems which is the parent category and into a category of variety solutions as their new parent category which is a derived category. The variety internal dynamic is such that some of the solutions may be transformed to new problems migrating from a category of solutions into a category of problems requiring new variety conversions, where mathematics as a member of abstract languages is neutral. The question then becomes: what is the meaning of *categorial conversion* and its supporting characteristic disposition? This requires a conceptual definition.

Definition 6.2.1.1: Categorial Conversion Categorial conversion is a process where derived categories emerge from a primary category by processes (constructions), and where the derived categories have direct or indirect continua with the primary category through reversal processes (reductions). At the level of ontology, categorial conversion is a natural process called *ontological categorial conversion* with a *categorial convertor* ($\mathbf{C}$) that applies to ontological transformations of varieties. At the level of epistemology, categorial conversion is an epistemic process called *epistemological categorial conversion* with a *categorial convertor* ($\mathbf{C}$) that applies to neuro-decision-choice transformations of varieties. The distinction between the *primary and derived categories* is established by *categorial differences* as revealed by their qualitative characteristic dispositions over the space of quality-quantity dualities.

Note 6.2.1.1: Categorial Conversion As defined, the ontological categorial conversion is the identity of transformation as well as the primary category for epistemological categorial conversions which constitute the derivatives in the knowledge production processes. For any given phenomenon, the epistemological categorial conversion leads to a knowledge or partial knowledge discovery if it is equal to the ontological categorial conversion under an epistemic conditionality. The work in the epistemological space is to discover the varieties of events, states and processes taking place in the ontological space where these events, states and processes are independent of the cognitive and non-cognitive elements contained in the ontological space. In the decision-choice space and over the space of problem–solution dualities, the static and dynamic behaviors of the categorial variety identification problem-solution dualities are supported by social philosophical Consciencism that has taken place in the cultural confines of the social set-up as well as created implicit-explicit bias in neuro-decision-choice actions. The philosophical Consciencism affects the freedom and sufficient conditions to indicate the external conditions for transformations, while the categorial conversion affects the necessity and initial conditions to indicate the external conditions for transformations over the space of possibility-probability dualities which is contained in the space of imagination-reality dualities as a sub-space of the space of actual-potential dualities in relation to the space of ontological-epistemological dualities.

The theory of categorial conversion that projects conditions of necessity and initialize variety transformation is developed in [669, 672] and the theory of philosophical Consciencism that projects conditions of freedom of transformation in support of the theory of categorial conversion is developed in [670]. These two volumes help to initiate the dynamics of problem–solution dualities through the categorial conversion processes that form the foundation to develop a realistic theory of socio-economic development as decision-making actions over the space of problem–solution dualities transforming one socio-economic state to another over the space of quality-quantity dualities. Epistemologically, the dynamics of the space of problem–solution dualities involve acquisition of information-knowledge conditions about varieties, technology, engineering and at least an element from the family of abstract languages (STEM]. The effectiveness of any program of (STEM) will depend on our understanding of categorial convertors and convertor moments and their applications, where the degrees of successful applications will depend on the philosophical Consciencism that has taken hold of the society to guide the individual and collective preference orderings and neuro-decision-choice actions over the space of problem–solution dualities and polarities. Let us now turn our attention to the definition and explication of categorial convertor as a transformation function as well as the development of its relationship to spaces of problem–solution and ignorance-knowledge dualities.

6.3.2 The Categorial Convertors and the Problem–Solution Dualities

In this development, $(\mathbb{C}_{\mathfrak{s}})$ is a *problem convertor* that converts a variety problem to a variety solution while $(\mathbb{C}_{\mathfrak{p}})$ is a *solution convertor* that converts a variety solution to a variety problem. Every acquisition of knowledge is obtained by a *categorial conversion* of a problem–solution duality in the space of variety ignorance-knowledge dualities, where the force of the convertor sets the variety solution against the variety problem in the form of $\{\mathbf{C}_{\mathfrak{s}}(\mathfrak{p} \to \mathfrak{s}) \Rightarrow (\mathfrak{s} \to \mathfrak{p}) | \mathfrak{p} \in \mathfrak{P} \subset \mathfrak{H}, \mathfrak{s} \in \mathfrak{S} \subset \mathfrak{G}, (\mathfrak{P} \cup \mathfrak{S}) \subset (\mathfrak{H} \cup \mathfrak{G})\}$, where the set of problem categorial convertors $\mathfrak{T}_{\mathfrak{s}}$ at any time point may be specified as in Eq. (6.2.2.1).

$$\mathfrak{T}_{\mathfrak{s}} = \{\mathbf{C}_{\mathfrak{s}\nu} = \mathbf{C}_{\mathfrak{s}}(\mathfrak{p} \to \mathfrak{s}) \Rightarrow (\mathfrak{s} \to \mathfrak{p}) | \mathfrak{p} \in \mathfrak{P} \subset \mathfrak{H}, \mathfrak{s} \in \mathfrak{S} \subset \mathfrak{G}, (\mathfrak{P} \cup \mathfrak{S}) \subset (\mathfrak{H} \cup \mathfrak{G}), \nu \in \mathbb{V}\} \quad (6.2.2.1)$$

In this discussion, $\mathfrak{P}$ is the problem space with generic element $(\mathfrak{p} \in \mathfrak{P})$, $\mathfrak{S}$ is the solution space with generic element $(\mathfrak{s} \in \mathfrak{S})$ and $\mathbb{V}$ is the space of varieties with generic element $(\nu \in \mathbb{V})$. The variety convertors change the variety problem–solution relationality in the knowing process. Corresponding to the set of variety convertors $\mathfrak{T}_{\mathfrak{s}}$ is the set of *categorial problem conversion moments* $\mathfrak{T}_{\mathfrak{s}}^{\mathrm{M}}$ of the form:

$$\mathfrak{T}_{\mathfrak{s}}^{\mathrm{M}} = \{\mathbf{C}_{\mathfrak{s}\nu}^{\mathrm{M}} = \mathbf{C}_{\mathfrak{s}}(\mathfrak{p} \to \mathfrak{s}) \Rightarrow (\mathfrak{s} \to \mathfrak{p}) | \mathfrak{p} \in \mathfrak{P} \subset \mathfrak{H}, \mathfrak{s} \in \mathfrak{S} \subset \mathfrak{G}, (\mathfrak{P} \cup \mathfrak{S}) \subset (\mathfrak{H} \cup \mathfrak{G}), \nu \in \mathbb{V}\} \quad (6.2.2.2)$$

The size of the set of the convertors at any decision timepoint is the same as the size of the variety space as seen from the imagination-reality dualities. Similarly, the solution convertor is a *problem creating process* that converts the existing variety solution into a variety problem $(\mathbf{C}_{\mathfrak{p}})$ that may be written as $\{\mathbf{C}_{\mathfrak{p}}(\mathfrak{s} \to \mathfrak{p}) \Rightarrow (\mathfrak{s} \to \mathfrak{p}) | \mathfrak{p} \in \mathfrak{P}, \mathfrak{s} \in \mathfrak{S}\}$, where the set of solution categorial convertors $(\mathfrak{T}_{\mathfrak{p}})$ may also be written as:

$$\mathfrak{T}_{\mathfrak{p}} = \{\mathbf{C}_{\mathfrak{p}\nu} = \mathbf{C}_{\mathfrak{p}}(\mathfrak{s} \to \mathfrak{p}) \Rightarrow (\mathfrak{s} \to \mathfrak{p}) | \mathfrak{p} \in \mathfrak{P}, \mathfrak{s} \in \mathfrak{S}, (\mathfrak{P} \cup \mathfrak{S}) \subset (\mathfrak{H} \cup \mathfrak{G}), \nu \in \mathbb{V}\} \quad (6.2.2.3)$$

The solution categorial convertors change the variety problem–solution relationality in knowing from solutions to problems and corresponding to it is the set of *solution categorial conversion moments* $(\mathfrak{T}_{\mathfrak{p}}^{\mathrm{M}})$ of the form as in Eq. (6.2.2.4):

$$\mathfrak{T}_{\mathfrak{p}}^{\mathrm{M}} = \{\mathbf{C}_{\mathfrak{p}\nu} = \mathbf{C}_{\mathfrak{p}}(\mathfrak{s} \to \mathfrak{p}) \Rightarrow (\mathfrak{s} \to \mathfrak{p}) | \mathfrak{p} \in \mathfrak{P}, \mathfrak{s} \in \mathfrak{S}, (\mathfrak{P} \cup \mathfrak{S}) \subset (\mathfrak{H} \cup \mathfrak{G}), \nu \in \mathbb{V}\} \quad (6.2.2.4)$$

The rate at which a variety solution is obtained to a variety problem or a variety solution is transformed into a variety problem will depend on the *size force* and value of the moment of the categorial convertor moment. The size force and value of any categorial convertor moment in terms of efficiency of conversion in any social set-up will depend on the *Philosophical Consciencism* that has taken hold of any social collectivity to give meaning and essence to the available relevant social institution, social policy, social preferences and social decision-choice actions, and the way the members can participate in the social neuro-decision-choice actions through the established institutions of the social organism. The Philosophical Consciencism defines the cluster of the social characteristics such as totality of its culture, permissible boundaries of individual and social actions and preference ordering as defined over the goal-objective space (for an extensive development of the theory of Philosophical Consciencism see [670] and its connection to information-knowledge process [672]). Every categorial convertor is a dualistic-polar structure with dualistic conversion moments residing in the corresponding poles of relevant polarity that may be written as:

$$\mathbb{C} = \left\{ \begin{array}{l} \left\{ (\mathbb{C}_{\mathfrak{s}}, \mathbb{C}_{\mathfrak{p}}) | (\mathfrak{p} \rightleftharpoons \mathfrak{s}) \& (\mathfrak{s} \rightleftharpoons \mathfrak{p}),\, \mathfrak{p} \in \mathfrak{P},\, \mathfrak{s} \in \mathfrak{S} \right\} \text{ the convertor} \\ \left\{ \mathbb{C}^{M}_{\mathfrak{s} \rightleftarrows \mathfrak{p}} = \left\{ \left(\mathbb{C}_{\mathfrak{s}^{M}}, \mathbb{C}^{M}_{\mathfrak{p}} \right) | (\mathfrak{p} \rightleftharpoons \mathfrak{s}) \& (\mathfrak{s} \rightleftharpoons \mathfrak{p}),\, \mathfrak{p} \in \mathfrak{P},\, \mathfrak{s} \in \mathfrak{S} \right\} \right\} \text{ conversion moment} \end{array} \right\} \quad (6.2.2.5)$$

The symbol ($\rightleftharpoons$) represents problem–solution generation process where a variety problem generates a variety solution which then generates a new variety problem in a primary-derived dynamics consistent with parent–offspring processes in a relational continuum and unity of information-knowledge disequilibrium system in an infinite states.

For any given social set-up, there corresponds a set of problem categorial convertors and a set of solution categorial convertors and corresponding sets of categorial conversion moments of the form:

$$\left\{ \begin{array}{l} \mathfrak{C}_{\mathfrak{s}} = \left\{ \mathbb{C}_{\mathfrak{s}\nu} = \mathbb{C}_{\mathfrak{s}} (\mathfrak{p} \to \mathfrak{s}) \Rightarrow (\mathfrak{s} \to \mathfrak{p}) | \mathfrak{p} \in \mathfrak{P},\, \mathfrak{s} \in \mathfrak{S},\, \nu \in \mathbb{V}, \text{ a set of problem categorial convertors} \right\} \\ \mathfrak{C}_{\mathfrak{p}} = \left\{ \mathbb{C}_{\mathfrak{p}\nu} = \mathbb{C}_{\mathfrak{p}} (\mathfrak{s} \to \mathfrak{p}) \Rightarrow (\mathfrak{s} \to \mathfrak{p}) | \mathfrak{p} \in \mathfrak{P},\, \mathfrak{s} \in \mathfrak{S},\, \nu \in \mathbb{V}, \text{ a set of solution categorial convertors} \right\} \end{array} \right\} \quad (6.2.2.6)$$

$$\left\{ \begin{array}{l} \mathfrak{C}^{M}_{\mathfrak{s}} = \left\{ \mathbb{C}^{M}_{\mathfrak{s}\nu} = \mathbb{C}_{\mathfrak{s}} (\mathfrak{p} \to \mathfrak{s}) \Rightarrow (\mathfrak{s} \to \mathfrak{p}) | \mathfrak{p} \in \mathfrak{P}\, \mathfrak{s} \in \mathfrak{S}\, \nu \in \mathbb{V} \right\} \text{a set of problem categorial convertion moments} \\ \mathfrak{C}^{M}_{\mathfrak{p}} = \left\{ \mathbb{C}_{\mathfrak{p}\nu} = \mathbb{C}_{\mathfrak{p}} (\mathfrak{s} \to \mathfrak{p}) \Rightarrow (\mathfrak{s} \to \mathfrak{p}) | \mathfrak{p} \in \mathfrak{P}\, \mathfrak{s} \in \mathfrak{S}\, \nu \in \mathbb{V} \right\} \text{a set of solution categorial convertion moments} \end{array} \right\} \quad (6.2.2.7)$$

The dynamic processes are such that the space of the problem–solution dualities is always contained in the space of imagination-reality dualities$(\mathfrak{P} \cup \mathfrak{S}) \subset (\mathfrak{H} \cup \mathfrak{G})$. The categorial conversion process is a problem–solution process where the solution-variety convertor is a problem solver, and the problem-variety convertor is a problem creator, all of which are always under the principle of opposites and conditions of cost–benefit dualities. The categorial conversions are the works of dualistic-polar conflicts to induce problem–solution transformations as dualistic-polar games. The categorial convertors and categorial conversion moments are the works of social conditions based on Philosophical Consciencism to induce the rate, velocity, acceleration and direction of problem–solution transformations, as well as to allow an epistemic tracking of the stages of the enveloping of decision-choice actions over the space of problem–solution dualities.

The categorial convertor ($\mathbb{C}$) with a conversion moment $(\mathbb{C}^{M})$ is constructed at the level of acquaintance with epistemic tools of cognitive observation at the level of experiential information with analytic tools of a cognitive paradigm of thought, and at the level of engineering with construction tools of the cognitive transformation technology. All these stages involve decision-choice activities based on Philosophical Consciencism of societies over the space of variety problem–solution dualities with the evolving information-knowledge system in an infinite horizon. The never-ending dynamics of the space of the variety problem–solution dualities is such that the structure of the human problem-solving process is always in the form where a variety solution to a variety problem generates a new variety problem requiring a new variety solution in problem–solution disequilibrium dynamics in support of information-knowledge disequilibrium dynamics for a continually changing socio-natural environment. The existence of problem–solution disequilibrium dynamics and information-knowledge disequilibrium dynamics implies the existence of cost–benefit disequilibrium dynamics under the general principle of non-satiation which helps to explain the meaning and essences of life and social dynamics.

The old variety problem presents a non-destructible information-knowledge structure, while the new variety solution presents information flow and a new non-destructible information-knowledge structure to update information-knowledge stocks. In this epistemic structure, all variety identities are defined in terms of characteristic dispositions in the information stock-flow disequilibrium dynamics, where the non-destructible information accumulation provides justifications for the study of the past to reveal *what there was*, the study of the present to reveal *what there is*, and the study of the future to reveal *what would be* in relation to categorial conversion processes, and social *philosophical Consciencism* that provides asocio-cultural framework of simplicity of thinking for all decision-choice activities over the space of problem–solution dualities including an information-knowledge search as socio-production activities which must obey all input–output conditions in the space of variety identification-transformation dualities. The input–output conditions include dualistic-polar conflict-resolution conditions of production-consumption dualities, negative–positive dualities and cost–benefit dualities in relation to socio-cultural preferences of cognitive existence over the variety space of social goals and objectives. Let us keep in mind the analytical definitions of variety problem and variety solution. A problem exists if the internal cost of the current variety outweighs its internal benefit. Similarly, a solution exists if the internal cost of the potential variety imagination outweighs its internal cost relative to the existence variety reality. The whole of social dynamics regarding conflicts, destruction, construction, and changes over the space of negative–positive dualities under cost–benefit and ignorance-knowledge processes are under the structure of Eqs. (6.2.2.1–6.2.2.7).

6.4 The Problem–solution Sequential Structure Over the Space of Problem–solution Dualities Under Categorial Conversions

Let us turn our attention to examine the nature of the sequential structure of the problem–solution dynamics under the parent–offspring dynamic game in the space of destruction-replacement dualities and categorial conversion and how they help to understand the diversity and unity of knowing and hence the unity of science. The parent–offspring process may also be viewed as the predecessor-successor process.

Let us keep in mind that every definition involves in creating nominal or real identity by establishing the characteristic disposition that distinguishes it from others. In other words, definitions are to assist in the communications over the space of source–destination dualities. We shall now link the defined problem–solution dualities and their sequential structure to categorial conversion in diversity and the unity of knowing and science.

6.4.1 The Problem–Solution Sequential Structure

The problem–solution categorial convertors generate information-knowledge accumulation in a stock-flow disequilibrium process. The initializing structure may be represented as in Eq. (6.2.2.1). This problem–solution structure forms the epistemic unity of knowing in science and non-science where the classification of areas of knowing into categories under condition of dividedness is also an element in the space of problem–solution dualities in the form:

$$\left\{\mathbf{C}_{\mathfrak{s}}(\mathfrak{p}\rightarrow\mathfrak{s})\Rightarrow(\mathfrak{s}\rightarrow\mathfrak{p})\mid\mathfrak{p}\in\mathfrak{P},\ \mathfrak{s}\in\mathfrak{S}\right\}\Rightarrow\exists\mathbf{C}_{\mathfrak{p}}\ni\left\{\mathbf{C}_{\mathfrak{p}}(\mathfrak{s}\rightarrow\mathfrak{p})\Rightarrow(\mathfrak{p}\rightarrow\mathfrak{s})\mid\mathfrak{p}\in\mathfrak{P},\ \mathfrak{s}\in\mathfrak{S}\right\} \tag{6.3.1.1}$$

The sequential structure of the never-ending dynamics of the space of problem–solution dualities, just like the predecessor-successor in the theory of numbers unites all areas of science and non-science in knowing without exception. The problem–solution relational structure over the space of problem–solution dualities is such that the *problem* is always the *actual* and the *solution* is always the *potential*. The never-ending dynamics of the space of the problem–solution dualities may be represented in a forward-telescopic progress with an infinite index set $\left(\mathbb{I}_{\mathfrak{R}}^{\infty}\right)$ of the space of problem–solution duality $(\mathfrak{R})$ as:

$$\left.\begin{array}{l}
\left\{\mathbf{C}_{\mathfrak{s}_0}(\mathfrak{p}_0\rightarrow\mathfrak{s}_0)\Rightarrow(\mathfrak{s}_0\rightarrow\mathfrak{p}_0)\mid\mathfrak{p}_0\in\mathfrak{P},\mathfrak{s}_0\in\mathfrak{S}\right\}\Rightarrow\exists\mathbf{C}_{\mathfrak{p}_1}\ni\left\{\mathbf{C}_{\mathfrak{p}_1}(\mathfrak{s}_0\rightarrow\mathfrak{p}_1)\Rightarrow(\mathfrak{p}_1\rightarrow\mathfrak{s}_1)\mid\mathfrak{p}_1\in\mathfrak{P},\mathfrak{s}_1\in\mathfrak{S}\right\}\\
\left\{\mathbf{C}_{\mathfrak{s}_1}(\mathfrak{p}_1\rightarrow\mathfrak{s}_1)\Rightarrow(\mathfrak{s}_1\rightarrow\mathfrak{p}_1)\mid\mathfrak{p}_1\in\mathfrak{P},\mathfrak{s}_1\in\mathfrak{S}\right\}\Rightarrow\exists\mathbf{C}_{\mathfrak{p}_2}\ni\left\{\mathbf{C}_{\mathfrak{p}_2}(\mathfrak{s}_1\rightarrow\mathfrak{p}_2)\Rightarrow(\mathfrak{p}_2\rightarrow\mathfrak{s}_2)\mid\mathfrak{p}_2\in\mathfrak{P},\mathfrak{s}_2\in\mathfrak{S}\right\}\\
\left\{\mathbf{C}_{\mathfrak{s}_2}(\mathfrak{p}_2\rightarrow\mathfrak{s}_2)\Rightarrow(\mathfrak{s}_2\rightarrow\mathfrak{p}_2)\mid\mathfrak{p}_2\in\mathfrak{P},\mathfrak{s}_2\in\mathfrak{S}\right\}\Rightarrow\exists\mathbf{C}_{\mathfrak{p}_3}\ni\left\{\mathbf{C}_{\mathfrak{p}_3}(\mathfrak{s}_2\rightarrow\mathfrak{p}_3)\Rightarrow(\mathfrak{p}_3\rightarrow\mathfrak{s}_3)\mid\mathfrak{p}_2\in\mathfrak{P},\mathfrak{s}_3\in\mathfrak{S}\right\}\\
\vdots\\
\vdots\\
\left\{\mathbf{C}_{\mathfrak{s}_{i-1}}(\mathfrak{p}_{i-1}\rightarrow\mathfrak{s}_{i-1})\Rightarrow(\mathfrak{s}_{i-1}\rightarrow\mathfrak{p}_{i-1})\mid\mathfrak{p}_{i-1}\in\mathfrak{P},\mathfrak{s}_{i-1}\in\mathfrak{S}\right\}\Rightarrow\exists\mathbf{C}_{\mathfrak{p}_i}\ni\left\{\mathbf{C}_{\mathfrak{p}_i}(\mathfrak{s}_{i-1}\rightarrow\mathfrak{p}_i)\Rightarrow(\mathfrak{p}_i\rightarrow\mathfrak{s}_i)\mid\mathfrak{p}_i\in\mathfrak{P},\mathfrak{s}_i\in\mathfrak{S}\right\}\\
\vdots\\
\vdots\\
\left\{\mathbf{C}_{\mathfrak{s}_i}(\mathfrak{p}_i\rightarrow\mathfrak{s}_i)\Rightarrow(\mathfrak{s}_i\rightarrow\mathfrak{p}_i)\mid\mathfrak{p}_i\in\mathfrak{P},\mathfrak{s}_i\in\mathfrak{S}\right\}\Rightarrow\exists\mathbf{C}_{\mathfrak{p}_{i+j}}\ni\left\{\mathbf{C}_{\mathfrak{p}_{i+j}}(\mathfrak{s}_i\rightarrow\mathfrak{p}_{i+j})\Rightarrow(\mathfrak{p}_{i+j}\rightarrow\mathfrak{s}_{i+j})\mid\mathfrak{p}_{i+j}\in\mathfrak{P},\mathfrak{s}_{i+j}\in\mathfrak{S}\right\}\\
\vdots\\
\vdots\\
\qquad\qquad....\forall i,j\in\mathbb{I}_{\mathfrak{R}}^{\infty},...(\mathfrak{P}\cup\mathfrak{S})\subset(\mathfrak{H}\cup\mathfrak{G})\subset(\mathfrak{U}\cup\mathfrak{A})
\end{array}\right\} \tag{6.3.1.2}$$

Equation (6.3.1.2) represents an infinitely sequential dynamics of any element in the space of problem–solution dualities in time processes, where each problem–solution duality exists in a dualistic-polar mode of negative–positive structure, under a natural process, which is then translated into real cost–benefit dualistic-polar structure as cognitively assessed in the transformation decision-choice process. This sequential time-induced process is also the variety transformation dynamics such that $\left(\forall \nu_{it}, \nu_{jt} \in \mathbb{V}\right)$ there is a preference ordering for an individual or collective where $\left(\nu_{it} \succsim \nu_{jt}\right)$ or $\left(\nu_{jt} \succsim \nu_{it}\right)$, and where $\nu_{it} \in \mathfrak{H} \subset \mathfrak{A}$ (real in the actual) and $\nu_{jt} \in \mathfrak{G} \subset \mathfrak{U}$ (imagination in the potential), and the process of each categorial conversion is a problem–solution structure over the space of universal object set $((\mathfrak{H} \cup \mathfrak{G}) \subset \Omega = \mathfrak{U} \cup \mathfrak{A})$ such that there are transformation transactions between actual and potential varieties within the space of construction-destruction dualities with information-knowledge production which is also the variety production such that the following transformation equation holds:

$$\left\{\exists(\mathbf{C})\left(\nu_{it} \succsim \nu_{jt}\right) \Rightarrow \left(\nu_{jt} \succsim \nu_{it}\right), \forall \nu_{it} \in \mathfrak{A}, \& \forall \nu_{jt} \in \Omega\right\} \Rightarrow \left\{\mathbf{C}_{s}\left(\mathfrak{p}_{it} \rightarrow \mathfrak{s}_{jt}\right) \Rightarrow \left(\mathfrak{s}_{jt} \rightarrow \mathfrak{p}_{it}\right) \mid \nu_{it} = \mathfrak{p}_{it} \in \mathfrak{P}, \nu_{jt} = \mathfrak{s}_{jt} \in \mathfrak{S}\right\} \tag{6.3.1.3}$$

The symbol ($\exists$) means *there is* and the symbol ($\ni$) means *such that*, while the symbol ($\forall$) means *for all* and $\left(\succsim\right)$ means preferred or indifference to on the basis of real cost–benefit rationality Similarly, the symbol ($\rightarrow$) means from to, and ($\Rightarrow$) is an implication. The representation of Eq. (6.3.1.2) is such that the initial problem ($\mathfrak{p}_0$) is always one and the best solution ($\mathfrak{s}_0$) is also one. All new problem $\mathfrak{p}_i$ such that $i > 0$ emerging from either a solution ($\mathfrak{s}_0$) or ($\mathfrak{p}_{it}$) may be cost–benefit greater than or equal to one another. In other words, a solution to a problem may generate many and more difficult problems as a result. The number of the problems and the degree of difficulties will depend on the way in which the solutions are acquired. Let us keep in mind that the problem–solution dualities exist in the conflict zone between the problem dual and the solution dual in a game setting and that every variety problem is a duality and every variety solution is also a duality, where the assessments are in terms of real cost–benefit dispositions in the preference space of decision-choice activities. Let us also keep in mind that Eq. (6.3.1.2) may be viewed in terms of panel data where every time point represents a matrix of variety characteristic dispositions.

In the presented epistemic framework for knowing and information-knowing accumulation, the variety ignorance is the problem. It is also the variety real in the space of realities as the sub-space of the space of the actuals while the variety knowledge is the variety solution which is also the variety imagination in the sub-space of the space of the potentials as viewed from all varieties in the actual-potential space under neuro-decision-choice modules. Every process of variety knowing, whether classified as natural science, social science, physics, economics, mathematics or other sciences is a cognitive activity in the space of ignorance-knowledge dualities translated into the space of problem–solution polarities in a language that belongs to either the family of ordinary languages (FOL) or a family of abstract languages (FAL). Each variety of the ignorance-knowledge processes is a dualistic process with polar conditions, where there is always an irreducible minimal degree of ignorance corresponding to a maximum degree of knowledge to claim either a variety identification or transformation, as well as an irreducible minimal degree of knowledge corresponding to a maximum degree of ignorance to claim a maximum doubt on variety identification and transformation.

Each variety ignorance-knowledge duality in the process of knowing encounters the principle of acquaintance that generates experiential information or *knowledge by acquaintance* as a solution to the primary category of information-knowledge problem–solution dualities from which the *knowledge by description* is derived by cognitive agents with a paradigm of thought to establish the derived category of the information-knowledge problem–solution duality within the spaces of certainty-uncertainty dualities, doubt-surety dualities and decidability-undecidability dualities over the space of methodological constructionism-reductionism dualities. The point of special emphasis is that the stock-flow information-knowledge conditions are generated by conflicts within the principle of opposites where destruction-construction dualities are related to variety real cost–benefit dualities, variety progress-retrogressive dualities, failure-success dualities and variety ignorance-knowledge dualities. The emphasis on real cost–benefit duality as the foundation on decision-choice rationality composed of fuzziness, classical bounded logical constraint and others should not be underestimated. The dynamics of the world of knowing is nothing but identification-transformation processes defined in terms of real cost–benefit dynamics where a real cost is a real benefit and a real benefit is a real cost as seen in the space of opportunities contained in the space of imagination-reality dualities.

6.4.2 Paradigms of Thought and the Problem–Solution Sequential Structure

The paradigm of thought provides the algorithmic thinking process of knowing under the use of the toolbox of methodological constructionism-reductionism duality as an approach to reconcile the conflicts between the *degrees of ignorance* and the *degrees of knowledge* in the space of problem–solution dualities regarding variety identifications and transformations. An alternative analytical way of stating the same idea is that, in all areas of knowing, whether classified as science or non-science, we have the space of variety ignorance-knowledge dualities, where the variety ignorance-knowledge dualities are expressed in degrees of knowing and unknowing in the *space of relativity*. The degrees of knowing and unknowing are established by cognitive activities over the space of problem–solution dualities with the use of the paradigm of thought and are tested for acceptance within the same paradigm of thought under the principle of acquaintance for the development of knowledge by acquaintance or experiential information from variety signal dispositions, and then under the methodological constructionism-reductionism duality for the development of knowledge by description for all elements in the space of varieties where problem-solving is a variety destruction-replacement process in the space of actual-potential dualities. It is important to note that problems exist because variety preferences exist. Similarly, solutions exist because variety preferences exist as enablers to examine the commonness and similarities and differences of varieties which are translated into diversities and the unity of knowing, science and knowledge as input–output structures into the space of neuro-decision-choice processes guided by the elements from the space of cost–benefit dualities. The special importance of the paradigm of thought in information-knowledge investment and information-knowledge capital stocks must be clearly understood. It is this special importance of the role of paradigm in knowing and neuro-decision-choice actions in all areas of human activities over the action space that led Russell to write on the knowledge of the general principles of the classical paradigm [147, 149, 150] and the Russell-Brouwer debate on representation information [67, 74, 778].

In all areas of study and knowing, the nature of the paradigm that provides the framework of thinking and reasoning to generate and process information to arrive at conclusions is not discussed. It is always assumed that it is known or implicit in the teaching and learning processes. Similarly, the nature of the experiential information as input into knowledge production is not discussed. Likewise, the needs and foundation for the development of the paradigm in use with its strengths and weaknesses are also not make explicit, and hence the learners emerge as deformed intellectuals who have difficulties to relate their areas of learning and thinking to other areas of knowing. In this respect, the deformed intellectuals develop very little appreciation for unity of information-knowledge system and relational continuum of its diversities with give-and-take sharing mode that strengthen its disequilibrium growth and development. Furthermore, there is also a little development of critical understanding that the continual evolving structure of the information-knowledge system with additions of multiplicity of beliefs, culture misinformation, disinformation, propaganda and many others is always contained in the social philosophical consciencism that provides simplicity of neuro-decision-choice actions over the space of input–output dualities in relation to the space of imagination-reality dualities contained in the space of actual-potential dualities relative to the general space of ontological-epistemological dualities. Similarly, they do not develop appreciations for the relationship between cognitive capacity limitations over the thinking-reasoning activities relative to the elements in the space of imagination-reality dualities, where the need for risk–benefit calculations induce the continual developments of theories over the explanation-prescriptive dualities.

The analytical construct in this monograph about diversity and the unity of knowing and information -knowledge structure presents an epistemic structure, where, every variety presents itself at the space of acquaintance as a combination of degrees of ignorance and degrees of knowledge in relational proportionalities which sums up to unity. The distribution of the degrees of ignorance-knowledge dualities may be transformed into the distribution of degrees of certainty-uncertainty dualities to deal with information-knowledge-certainty and information-knowledge surety problems in all areas of knowing. The commonness of the space of variety ignorance-knowledge dualities and the space of certainty-uncertainty dualities to all areas of knowing enhance the space of the problem–solution dualities as the primary category of the principle of diversity and *the principle of unity in knowing* over all areas of the knowledge search including science, non-science, social management, engineering and planning, where knowing is an epistemic activity to generate intellectual investment flows as additions to information-knowledge capital stock which is continually being updated by the results of knowing.

This primary category of the principles of diversity of the unity of knowing, whether classified as science or non-science, is further amplified by the behaviors of the elements in the space of methodological constructionism-reductionism dualities for checks and balances in the creation and test of information-knowledge certainty through the derived conditions of *fuzzy-stochastic entropy*, where every area of knowledge search must provide solutions to *variety identification problems*, *variety transformation problems* and *variety information-knowledge-certainty problems* relative to spaces of variety relations and characteristic dispositions. As it has been explained, *categorial conversion* [669] initializes the present conditions of knowing and decision-choice necessity by helping to resolve the variety *identification problems*. The *philosophical Consciencism* [670] is used as a guide to neuro-decision-choice freedoms and actions for crafting variety categorial moments to induce problem–solution conversions, thus solving the variety transformation problems to move from variety ignorance to variety knowledge, from variety problem to variety solution, from uncertainty to certainty, from doubt to surety, from undecidability to decidability and many more in dualistic-polar conditions with fuzzy-stochastic conditionality where the system of intellectual investment and capital accumulation accounts for vagueness and volume constraints in terms of individual and social cognitive capacity limitations.

6.5 Human Life, Decision-Choice Actions, Problem–Solution Dualities and Social Change

Alternatively viewed, the primary category of the unity of knowing whether science or non-science finds expressions in *the spaces of variety identification problem-solution dualities, variety transformation problem-solution dualities and variety information-knowledge-certainty problem–solution dualities* under a matter-energy equilibrium state. A question the arises regarding what the relationship between the problem–solution duality and the meaning and essence of human life. In other words what are the factors that define the meaning and essence of human life and its existence? Every living being seeks conditions of protection of life and existence. In the case of human beings these conditions define the meaning and essence of life over the space of variety problem–solution dualities, the space of problem–solution dualities is contained in the space of imagination-reality dualities. There are many literatures that have examine the meaning of life. The approach here is based on what defines humanness (*onipa*), the will to live, the fear to die and the drive for increasing progress defined in terms of more comfort and less suffering. All these involve neuro-decision-choice activities over the action space as the space of variety production-consumption dualities which is established over the space of problem–solution dualities, where the flows of services from the information-knowledge accumulation and intellectual capital stocks are the inputs under philosophical consciencism that have taken hold of the social consciousness which may establish social behavioral regularity, commonness and implicit institutional discrimination and bias from the politico-legal structure in the process of creating social cohesion and stability for collective preference from conflicting individual preferences over the space of collective decision-choice actions..

It is the cognitive activities over the space of problem–solution dualities in the space of transformation-substitution dualities over the variety space that give a *meaning to life* and human negative and positive progress in relation to things seen as varieties. It is also the way that problem–solution replacements are undertaken by decision-choice actions in substituting variety solutions for variety problems that defines the *essence of life* in relation to the rise of new problems and the direction of human progress in the space of negative–positive dualities under the cost–benefit conditions in a relative structure over the space of varieties. The cost–benefit relativity conditions are recognition that a variety solution may produce more undesirable replacements than the variety problem. Every problem identification is an information-knowledge creation that provides initial conditions for neuro-decision-choice activities to transform the existing social conditions. Similarly, the progress of human existence, as expressed over the space of fear-hope dualities or joy-sorrow dualities with relational continuum and unity, finds meaning and essence over the space of problem–solution dualities under the *general principle of opposites* with relational continua and unity which provide explanatory and prescriptive structures for the existence of the spaces of peace-war dualities, conflict-harmony dualities, love-hate dualities, positive–negative dualities, success-failure dualities and the placement of cognitive agents in the zones of irreducible ignorance, imbecility, irrationality and the continual search for information-knowledge elements to deal with the never-ending dynamics of problem–solution processes in the space of parent–offspring or predecessor-successor disequilibrium dynamics to provide a continuity among the telescopic past, the telescopic present and the telescopic future over the spaces of beginning-end, and nothingness-somethingness dualities.

The set of the conditions of the present time is what the individual and collective information-decision-choice-interactive processes have produced from the individual and collective cognitive activities over the space of problem–solution dualities on varieties from the past. The set of conditions of the future will be the outcomes of the present-time cognitive activities over the space of the problem–solution dualities, the efficient use of the conditions of the past and the reflections of the telescopic present. In other words, the history of negative or positive human progress and achievements is an enveloping of neuro-decision-choice dynamics over the space of success-failure dualities relative to the elements of the space of the problem–solution dualities as the unifying force of activities of knowing as intellectual investments and information-knowledge accumulation as intellectual capital stocks in all areas of human endeavor under the principles of diversity and unity in the universal system of oneness over the space of static-dynamic dualities. This negative or positive human progress may also be viewed in terms of stock-flow of information-knowledge disequilibrium processes under neuro-decision-choice actions where information-knowledge input–output is the lifeblood of the neuro-decision-choice actions over the space of general input–output dualities for production-consumption disequilibrium dynamics.

In this respect, any human civilization may be conceptualized as an enveloping of success-failure conditions of human cognitive activities over the space of problem–solution dualities, where the success-failure enveloping establishes the past-present-future (Sankofa principle) history of individual and collective neuro-decision-choice activities under cognitive capacity limitations. The cognitive capacity limitations generate qualitative and quantitative uncertainties of outcomes and blind forces of telescopic tomorrow over the space of necessity-freedom dualities. The space of necessity-freedom dualities may be conceptualized from the space of possibility-probability dualities that relates human cognitive actions over the space of imagination-reality as a sub-space of actual-potential dualities in a relational continuum and unity, where actual varieties exit from the space of actual varieties and enter into the space of potential varieties and potential varieties exist from the space of the potential varieties and enter into the space of actual varieties under the destruction-replacement process in disequilibrium states over the space of imagination-reality dualities.

The qualitative and quantitative uncertainties, amplified by blind forces of telescopic tomorrow, generate the spaces of decidability-undecidability and doubt-surety dualistic-polar problems in information-knowledge constructs with negative or positive unexpected outcomes in the neuro-decision-choice space relative to preference ordering over the space of variety real cost–benefit dualities. Some unwanted paradigmatic outcomes generate conditions of *decision-choice paradoxes* and unintended variety outcomes, where a collective action for social welfare-improvements leads to social welfare reduction with an increasing suffering and dissatisfaction requiring neuro-decision-choice action over the space of actual-potential dualities as demanded by the meaning and essence of human life. It is within this analytical framework that social histories, defined as enveloping of human success-failure decision-choice processes over the space of problem–solution dualities, find epistemic stands within the spaces of doubt-credulity and hope-despair dualities in the knowing process, where doubt-credulity dualities find expressions in individual and social preferences with hope-despair constraints over the space of goals and objectives, and where good-bad decisions potentialize the existing varieties and actualize potential varieties under real cost–benefit rationality of the *asantrofi-anoma principle* which is constrained by fuzzy-stochastic conditionality that deals with cognitive capacity limitations of all forms and over all areas of knowing.

It is here, that nominal and real cost–benefit analyses act as a toolbox for interrogating individual and collective preference schemes over time, where both the individual and collective preference schemes change over generations as well as with the discovery of new information-knowledge constructs that reveal conditions in the space of ontological-epistemological dualities to affect belief systems and cultural dynamics with the principles enshrined in the ruling Philosophical Consciencism that has taken hold of the society. The changing individual and collective preference schemes shift the enveloping curve of social history upwards or downwards as one journeys through different generations and cultural times, where the cultural times and preferences are mutually determined with differentially mutual impact on the social collectivity and the concepts of diversity and unity of knowing and unity of science. It may be kept in mind that human problem solving involves weaponizing all the gains of all or some areas of knowing into unity, pulling together leaning protocols to penetrate the complexities of the space of problem–solution dualities and variety interdependence in other to set the solution varieties against the problem varieties as assessed by cognitive agents under individual and collective preference schemes within the space of real and nominal cost–benefit dualities, either preserving existing varieties, or destroying existing varieties and creating new varieties in continual information-knowledge disequilibrium dynamics.

In this respect, human life as we know it, finds meaning in the space of information-decision-choice-interactive processes which then finds meaning in the space of problem–solution dualities that engenders either negative or positive social change with the essence of life which then finds meaning in the manner of the social problem–solution replacement process. The progress of different information-knowledge structures is interdependent which also depends on the social organization and its institutional arrangements and management which then unite them into the space of problem–solution dynamics for knowing, covering all areas of human endeavor including science, technology, engineering and mathematics (STEM) which inter-disciplinarily empower cognitive agents to enhance the development of the applied knowledge systems for decision-choice actions over the space of applied problem–solution dualities toward the creation of new social varieties and forms.

Chapter 7
The Theory of Relational Problem–Solution Polarities, Diversity and the Unity of Knowing and Science

Abstract The chapter contains discussions on the concepts and phenomena of problem–solution polarities and the role they play in the understanding of knowing as intellectual investment flows and information-knowledge accumulation as intellectual capital stocks in order to establish the principles of diversity, relational continuum and unity that they entail. The relational structure between duality and polarity is made explicit in an epistemic organicity, where the diversities are in a multiplicity of give-and-take sharing mode. The poles of the polarities are established by the residing dualities such that the negative (positive) duality defines the negative (positive) pole while the dominance of the negative or the positive pole reveals the identity of the polarity in variety transformations. The categorial conversion establishes the external conditions for dualistic-polar change while the philosophical consciencism establishes the internal conditions for change over the space of variety quantitative–qualitative dualities relative to the space of statics-dynamics dualities with extensive discussions on the variety dynamics in knowing and information-knowing development.

7.1 Relational Introduction

The progress of human existence is info-generational and info-generating under continual variety transformations at both the levels of ontology and epistemology which are mutually interdependent over the space of problem–solution dualities where the neuro-decision-choice system acts as a connector between the ontological space and the epistemological space. There is no epistemology without ontology which exists independently of epistemology. There is no epistemological space without the concept of ontological space. The meaning of social progress is found in decision-choice actions and the essence of this progress is found in the continual search of destiny control and vision setting through continual decision-choice actions over the space of problem–solution dualities under the stock-flow disequilibrium

K. K. Dompere, *The Theory of Problem-Solution Dualities and Polarities*,
Studies in Systems, Decision and Control 405,
https://doi.org/10.1007/978-3-030-90279-7_7

dynamics of information-knowledge systems as input–output structures into individual and collective decision-choice processes. Information, knowledge and neuro-decision-choice actions have always been central to the existence and progress of cognitive agents with a continual development of culture and civilization. The contemporary claims of an information-knowledge economy are simply emphasis on the explicit acknowledgement of the information-decision-choice-knowledge interactive systems of general human activities with differential systems of source–destination processes relative to the space of production-consumption dualistic-polar structures in terms of dualities and polarities. At the level of ontology, the transformations are under natural forces through the *natural decision-choice processes* to change the relationships between the *actual natural varieties* and *potential natural varieties* to maintain the bio-ecological diversities and stabilities within the natural environment and as the environment alters in response to variety activities within the conditions of matter-energy equilibrium. The matter-energy equilibrium and the disequilibrium dynamics of varieties and the information-knowledge stock-flow system will be defined, explicated and analyzed in the chapters to follow. The current chapter is about the role of polarity, problem–solution polarities and their relationship to duality and problem–solution dualities.

At the level of epistemology, the transformations are under social forces through *individual* and *social neuro-decision-choice processes* to alter the relationships between the *actual social varieties* and *potential social varieties* where the alterations leads to the creation of the elements of reality and the space of realities as a sub-space of the space of actuals. We must note that the space of actuals is independent of cognitive agents while the space of realities cognitive-agent dependent in such a way that realities belong to the space of acquaintance-description dualities with fuzzy-stochastic conditionality. These variety transformations find ultimate expressions in problem–solution processes under the internal activities of varieties and categorial varieties to generate *actual-potential replacement dynamics* through parent–offspring dynamics within the space of imagination-reality dynamics. Both the natural and social transformations are induced by conditions of opposites in terms of conflicts, tensions, forces and motions (qualitative and quantitative) in varieties and categorial varieties with different actors under matter-energy conditions. Our main concern here is on *social actual-potential replacements* as problem-solving activities that generate knowing as an intellectual investment flows which go to update intellectual information-knowledge capital stocks. The incentive structure of natural transformation is from the space of natural environmental balances to maintain the matter-energy equilibrium dynamics. The incentive structure of social transformation is from the space of cost–benefit environmental balances to maintain information-knowledge disequilibrium dynamics in the dynamic space of generational and intergenerational preferences. In other words, every correction process in the space of static-dynamic dualities relative to the space of identification-transformation dualities involving the varieties in the spaces of destruction-construction dualities and destruction-replacement dualities has internal or external incentive structures or both, where the external incentives indicate a necessity for change and internal incentives indicate the freedom and direction for change.

The actual-potential replacement dynamics may be explained through the understanding of the spaces of the problem–solution polarities and the problem–solution dualities as unified components of the principle of opposites with relational continua and unity to give rise to the space of realities within the space of acquaintance-description dualities. The nature of the space of the problem–solution dualities as it relates to the space of identification problems to generate the knowledge of *what there was* and *what there is* has been discussed in chapters two and three of this monograph in terms of intimate relational dualistic-polar dynamics. The problem–solution process is composed of two inter-dependent processes of *problem–solution duality* and the *problem–solution polarity* which together constitute the essential components of the principle of the opposites and the dynamics of internal transformations. Chapters two and three are devoted to the discussions on the dualistic components of the problem–solution processes. This chapter is devoted to the discussions on the polar components of the problem–solution process under cost–benefit rationality and the individual and collective preferences over the elements of the goal-objective space.

We shall now turn our attention to examine the nature of the space of problem–solution polarities and its relationship to the space of problem–solution dualities as a continuity to integrate the problem-solving process to construct the derived category of diversity and unity of knowing and then examine the principles of diversity and the-unity of science. The thesis put forward here is that the general dynamics of knowing find meaning and essence over the space of problem–solution dualities under the general principle of opposites with relational continua and unity which provides explanatory and prescriptive structures for the transformative behaviors of varieties under cost–benefit conditions at the level of epistemology. The space of the problem–solution dualities links all areas of knowing together whether classified as science or non-science. The problem–solution dualities in the process of knowing take place at both the level of static states in relation to variety identifications and at the level of dynamic states in relation to variety transformations and changing variety identities in reference to variety relation and variety properties.

We shall first discuss the notion of the relational structures of the relevant conceptual elements and then link them to the general space of problem–solution activities. It must be kept in mind that without variety identifications there will be no goal-objective element, no problem–solution duality, concept of conflict, individual and collective preferences, decision-choice action, static-dynamic duality and many more of dualistic-polar structures. The cognitive information-knowledge structure is about variety identification that initializes the static states of being and transformation that involves variety behavior at the dynamic states of variety identity in relation to variety problem–solution dualities and change of properties. Here, information and knowledge reside in a conceptual and practical unity, while knowledge and ignorance reside in inseparable duality in a relational unity of a give-and-take sharing structure that maintains their mutual existence and destruction. There is no epistemic paradox under the general principle of opposites composed of dualities and polarities with relational continua and unity where the dualities are connected, the poles are relationally linked, and the thinking and reasoning are under the fuzzy paradigm of thought over decidability-undecidability duality with epistemic conditionality.

An epistemic paradox and conditions of ill-problem representation may arise when the duality and polarity are set in mutual separation with the-principle of the-excluded middle, where the duals are relationally unconnected, and the poles are relationally unlinked and thinking, and reasoning are under the classical paradigm of thought with non-acceptance of contradictions and vagueness. Chapter two and three are used to discuss transformation conditions over the space of problem-answer dualities relative to the role of analytical duality in diversity and the unity of science through the theory of knowing, information-knowledge development and decision-choice actions over the space of problem–solution dualities. This chapter is used to discuss the conditions of neuro-decision-choice actions over the role of problem–solution polarity and the space of problem–solution polarities as complements to the problem–solution duality and the space of problem–solution dualities in understanding knowing as an intellectual investment and information-knowledge accumulation as an intellectual capital stock as an input–output system of continual economic production under neuro-decision-choice actions in disequilibrium dynamics. Let us keep in mind that there are many interdependent steps defining productions processes within the space of input–output dualities for variety generation in destruction-construction dualities, where destructions are inputs to processes and constructions are outputs in the chain of continual transformations in the space of telescopic past-present, present-future and past-future dualities to establish under the Sankofa-anoma traditions in the forecasting-discounting dualities and polarities.

7.2 The Principle of Opposites, Polarity, Problem–Solution Polarities and the Problem–Solution Dualities

In variety transformation dynamics under categorial conversion, the natural transformations, viewed as problem–solution phenomena, take place over the ontological space in a continual non-stopping mode over the space of actual-potential replacement processes, where the internal and external relational forces of varieties destroy and create in the socio-natural negative and positive transformations under no goal-constraint conditions as may be abstracted. The human problem–solution dynamics under categorial conversion, however, relate to neuro-decision-choice processes over the epistemological space of varieties under goal-constraint conditions in the epistemological space where the goal-constraint conditions are under the general utility principle of non-satiation and matter-energy equilibrium states. There are important theoretical conditions that must be considered at the ontological space as we look at variety transformations and information production at the level of nature where natural information and knowledge are the same and thus inter-changeable with zero uncertainty, doubt, undesirability and risk without connatural knowledge. Should we accept or reject the claimed idea in the physical sciences that matter, and energy are in stock-flow equilibrium which simply implies that matter-energy structure has zero flows and hence stocks cannot be updated, where the matter-energy equilibrium imposes conditions of input–output scarcity leading to restrictive conditions on general freedom and necessity in the space of possibility-probability dualities relative to the space of decision-choice processes? It may be noted that the economic theories of scarcity and cost–benefit analysis derive their foundations from conditions of matter-energy equilibrium in the variety transformations over the spaces of production-consumption dualities and input–output dualities.

The acceptance of matter-energy equilibrium over the universal existence implies that no new matter with the corresponding energy is being created in the sense that there is no new matter-energy investment and that all new varieties of matter and energy are transformations from the existing matter-energy natural capital stocks. The acceptance of matter-energy stock-flow equilibrium and the concept of universal infinity restrict the understanding of the claim that the universe is expanding, where the nature of expansion must be related to matter-energy stocks of nature. The rejection of the matter-energy equilibrium and imposition of matter-energy disequilibrium over the universal existence implying that new matter-energy forms are being created not from the existing mater-energy stock, but from a source that must be indicated that all new varieties of the matter-energy form may be transformations from either the existing matter or from new matter-energy forms and that our concept of infinity requires some critical reflections, the sources of which must be indicated and explained. In all critical reflections of equilibrium-disequilibrium dualities over the space of static-dynamic dualities, the economic theory of scarcity must be reexamined over the space of input–output processes. Given the matter-energy equilibrium and continual variety transformation, the economic scarcity is conditional on the cognitive capacity limitations relative to the space of process-technology duality and the evolving information-knowledge disequilibrium dynamics. The increasing knowing about processes and technologies relaxes the scarcity constraints on the input–output transformations where same varieties may serve as input–output elements in the understanding of variety recycling in nature.

The rejection of matter-energy stock-flow equilibrium promotes the understanding of the epistemic claim that the universe is expanding where new matter-energy forms are in creation, and hence, matter-energy is in stock-flow disequilibrium dynamics with investment flows. In this respect, the concept of infinity in our knowing and information-knowledge processes about the universe must be re-examined and redefined [672]. In the case of matter-energy equilibrium, increasing varieties are internally induced from transformations of existing varieties. In the matter-energy disequilibrium, increasing varieties and categories may be due to internal and external transformations under the dualistic-polar game principle of opposites with relational continua and universal unity. The internal variety transformations may be explained by the logic of the principle of opposites in a gamified process supported by the fuzzy paradigm of thought and category theory. The explanatory process of external transformation must be sought to define functions for the prescriptive process for variety transformations under the neuro-decision-choice action of cognitive agents relative to the space of cost–benefit dualities that defines the ranking and decision-choice rationality over the space of variety destruction-replacement processes through the space of problem–solution dualities and polarities within the space of actual-potential dualities. The whole of knowing and information-knowledge activities is to provide ingredients into connectivity among neuro-decision-choice actions and problem–solution processes over the space of decidability-undecidability dualities that connect statics to dynamics. Our understanding of the relational structure information-decision-choice-knowledge processes over the space of problem–solution dualities is made possible by conditions of the principle of opposites as providing a framework of the internal transformation of diversities in unity. Let us attend to the principle of opposite.

7.2.1 *Representation Analytics of the Principle of Opposites Under Polarity*

The decision-choice processes relate to the principle of opposites composed of the space of polarities ($\mathbb{P}$) and a space of dualities ($\mathbb{D}$), where the system of technologies of the transformation dynamics ($\mathfrak{T}$) is generated by a system of information-decision-choice processes ($\mathfrak{D}$). The neuro-decision-choice space relates to the space of varieties ($\nu \in \mathbb{V}$) which is identified by the space of information structures of the form ($\mathbb{Z} = \mathbb{X} \otimes \mathbb{S}$) where ($\mathbb{X}$) is the general characteristic disposition representing the contents of the information, and $\mathbb{S}$ is the general signal disposition sent by the content disposition to reveal the contents of the identities [671]. Every variety ($\nu \in \mathbb{V}$) has supporting information, defined as a characteristic-signal disposition ($\mathbb{Z}_\nu = \mathbb{X}_\nu \otimes \mathbb{S}_\nu$) for its identity and signal at the epistemological space. The characteristic disposition $\mathbb{X}_\nu$ of any variety (ν) is a union of a negative sub-set $\left(\mathbb{Z}_\nu^N = \mathbb{X}_\nu^N \otimes \mathbb{S}_\nu^N\right)$ and a positive sub-set of the form $\left(\mathbb{Z}_\nu^P = \mathbb{X}_\nu^P \otimes \mathbb{S}_\nu^P\right)$ such that $(\mathbb{Z}_\nu = \mathbb{X}_\nu \otimes \mathbb{S}_\nu) = \left(\mathbb{Z}_\nu^N = \mathbb{X}_\nu^N \otimes \mathbb{S}_\nu^N\right) \cup \left(\mathbb{Z}_\nu^P = \mathbb{X}_\nu^P \otimes \mathbb{S}_\nu^P\right)$ defines the internal negative and positive duals respectively to make up of the *internal problem–solution duality* with a fuzzy conditionality of the form: $\left\{\left(\mathbb{Z}_\nu^N = \mathbb{X}_\nu^N \otimes \mathbb{S}_\nu^N\right) \cap \left(\mathbb{Z}_\nu^P = \mathbb{X}_\nu^P \otimes \mathbb{S}_\nu^P\right) \neq \emptyset\right\}$.

The concept of internal duality (as has been defined and explained in chapter two and three of this monograph) is used to represent the notion that each variety in a separate existence in the space of varieties also projects *internal cost–benefit duality* and that the variety is simultaneously desirable and undesirable in a problem–solution process under the *asantrofi-anoma principle* and *rationality*.

Definition 7.1.1.1: The Asantrofi-anoma Principle and Rationality The *asantrofi-anoma principle* simply states that every variety is simultaneously attractive and repulsive where the attractiveness relates to its benefit characteristic disposition which is associated with both duals of the negative–positive duality, and where the repulsiveness is related to its cost characteristic disposition which is also associated with the both the duals of the negative–positive duality. The *asantrofi-anoma rationality* simply states that the decision-choice action must balance cost–benefit dispositions and select the variety with real benefit dominance. Every decision-choice action confronted with simultaneous existence of attractiveness and repulsiveness is called the *asantrofi-anoma problem* and the resolution of the variety internal and external cost–benefit conflicts under real benefit dominance is the *asantrofi-anoma rationality*.

Note 7.1.1.1: Asantrofi-Anoma Conditions The *asantrofi-anoma principle* defines the general conditions of variety existence and identity in the decision-choice space in relation to the space of imagination-reality dualities. The *asantrofi-anoma problem* defines the general conditions of decision-choice problem of variety selection over the variety space. The *asantrofi-anoma rationality* defines the general conditions for decision-choice resolution of multi-variety decision-choice conflicts in variety destruction and replacement. The *asantrofi-anoma principle, problem and rationality* help to explain the possible emergence of decision-choice paradox under the classical system of thought where exactness and principle of the excluded middle are incompatible with vagueness, quality and subjectivity of information-knowledge input in neuro-decision-choice actions. The asantrofi-anoma principle affirms conditions surrounding a decision-choice process where every variety for selection has a cost characteristic disposition (dual) and benefit characteristic disposition (dual), and that it is not possible for any neuro-decision-choice agent to select the real benefit dual and leave the real cost dual of the same variety. The real benefit information-knowledge conditions are constraints on real cost information-knowledge conditions and the real cost information-knowledge conditions are constraints on real benefit information-knowledge conditions in the decision-choice calculus where the real values may or may not be translated into nominal values [688]. The constraint conditions create internal and external conflicts of the dualistic-polar structure in the decision-choice process over the space of varieties where the rules variety attractions and repulsions operate in the space of decidability-undecidability dualities [667].

These dualistic-polar conflicts in varieties are the basis of cost–benefit analysis as general decision-choice analytics that affect all elements in the space of neuro-decision-choice actions under the principle of opposites and the fuzzy paradigm of thought. These complex conditions led me to the discussion of the economics, mathematics and the implied paradigm of thought in a twin monograph on cost–benefit analysis as a general approach to the development of decision-choice theory composed of sub-theories of cost–benefit measurements and fuzzy value analysis and their extensions to optimal control in every area over the space of cost–benefit dualities [218, 219]. The cost–benefit conditions also form the basis of the fuzzy and non-fuzzy optimization theory with real cost–benefit mutual constraints leading to the duality theory in optimization, in that the real cost dual constrains the behavior of the real benefit dual in maximizing conditions, where the real benefit dual also constrains the behavior of the real cost dual in minimizing conditions in the space of computable-incomputable dualities. Furthermore, the cost–benefit duality in the general framework of the principle of opposites provides an interesting understanding of the game theory and the role of information-knowledge and information-knowledge conceal ment and security classifications as constraints in all game decisions contained in the space of imagination-reality dualities, where information concealment and security conditions for an ignorance and decision-choice surprise create net cost–benefit advantage and disadvantage. The complexity of neuro-decision-choice systems is such that dualistic-polar give-and-take sharing relations is such that a variety may represent a problem at some conditions as well as representing a solution at some other conditions relative to the space of the goal-objective elements. In other words, the variety problem and variety solution are defined by the nature of goal-objective elements that may demand changes in cost–benefit conditions through the relational structure over the space of internal and external dualities. The relative relational structure of the internal duality provides us with an analytical way to introduce the relationality of concepts of external problem–solution duality and problem–solution polarity, and to abstract the differences and similarities between the spaces of problem–solution dualities and problem–solution polarities.

Let us keep in mind that each negative–positive duality also resides in a variety where the negative or positive characteristic disposition may be either real benefit or real cost under a defined goal-objective element. The negative–positive and cost–benefit interdependent conditions of dualities in a give-and-take sharing relational mode amplify the explanation of the existence of social conflicts that generate the energy for resistance to the general acceptance of new information-knowledge ideas as well as social change and at the same time generate forces for negative and positive social changes in the dualistic-polar game process. The resistances or non-resistances to the acceptances of new information-knowledge ideas and social change are explainable by the dynamics of variety *internal problem–solution duality* with corresponding internal cost–benefit duality, while the acceptances and non-acceptances of new ideas and social changes are explainable through the dynamics of variety *external problem–solution dualities* with corresponding external cost–benefit dualities (the concept of internal and external dualities have been defined and explicated in chapters two and three of this monograph) and their influences on the poles of any problem–solution polarity in the social set-up.

The conflicts within social set-ups, social changes, the resistance to social changes and variety transformations are also explainable by the distributions of real cost–benefit conditions within the space of actual-potential dualities in relation to destruction-replacement processes that generate new information-knowledge elements and changes in the real social cost–benefit distribution. Let us keep in mind that every decision-choice action is a change in either relation or property which is a destruction of the corresponding real cost–benefit duality and its social distribution in the space of input–output dualities. The same decision-choice action is a change in relation, property or both which is simultaneously a creation of the corresponding real social cost–benefit duality and its social distribution in the spaces of input–output dualities and production-consumption dualities where the anticipated distributional disparities, as seen in information-knowledge structures, influence elements in the space of acceptance-rejection dualities neurocognitive responses to variety behavior over the space of static-dynamic dualities. Let us examine the similarities and differences between the external problem–solution duality and problem–solution polarity.

7.2.2 External Problem–Solution Duality and the Space of Polarities

The negative external duality is identified by the negative internal duality where the negative dual dominates the positive dual. Similarly, the positive external duality is identified by the positive internal duality where the positive dual dominates the negative dual. The negative external duality identifies the negative pole while the positive external duality identifies the positive pole in all polarities within the transformation games in destruction-replacement dynamics under the principle of opposites in relational continuum and unity. The variety problem–solution processes over the epistemological space present themselves in a way where there are a system of *problem–solution polarities* with a *system of solution poles* ($\mathbf{P}_{\mathfrak{S}}$) and a *system of problem poles* $\left(\mathbf{P}_{\mathfrak{P}}\right)$ with corresponding relationality of continuum and unity such that $\left[\left(\mathbf{P} = \left(\mathbf{P}_{\mathfrak{P}} \otimes \mathbf{P}_{\mathfrak{S}}\right)\right) = \left\{\mathbf{p} = \left(\mathbf{p}_{\mathrm{p}}, \mathbf{p}_{\mathfrak{s}}\right) | \mathbf{p}_{\mathrm{p}} \in \mathbf{P}_{\mathfrak{P}}, \mathbf{p}_{\mathfrak{s}} \in \mathbf{P}_{\mathfrak{S}}\right\}\right]$. In the case of polar structure, each variety problem pole has a corresponding variety solution pole under a give-and-take sharing relation that maintains their mutual existence. There is also a system of problem–solution dualities ($\mathbf{D}$) with a corresponding *system of solution duals* ($\mathbf{D}_{\mathfrak{S}}$) and a *system of problem duals* $\left(\mathbf{D}_{\mathfrak{P}}\right)$ such that $\left[\left(\mathbf{D} = \left(\mathbf{D}_{\mathfrak{P}} \otimes \mathbf{D}_{\mathfrak{S}}\right)\right) = \left\{\mathbf{d} = \left(\mathbf{d}_{\mathrm{p}}, \mathbf{d}_{\mathfrak{s}}\right) | \mathbf{d}_{\mathrm{p}} \in \mathbf{D}_{\mathfrak{P}}, \mathbf{d}_{\mathfrak{s}} \in \mathbf{D}_{\mathfrak{S}}\right\}\right]$. In the case of polar structure, each variety problem dual has a corresponding variety solution dual under a give-and-take sharing relation that maintains their mutual existence and connect then to the system of polarities. All these are established by their corresponding characteristic-signal disposition ($\mathbb{Z} = \mathbb{X} \otimes \mathbb{S}$) as the information structure for identities, information-knowledge connectivity between the space of knowledge by acquaintance and the space of knowledge by description.

Let always keep in mind the concepts of primary-derived dualities and the relational structure of socio-natural spaces of dualities and polarities and the role they play in in the space of variety identification-transformation dualities. We also must keep in focus the space of natural activities, where there is information-knowledge equality with complete certainty and no risk; and the space of social activities, where there is information-knowledge inequality establishing the space of certainty-uncertainty dualities with the space of risk-riskless dualities, where the space of problem–solution dualities find definitions and meaning. All neuro-decision-choice actions take place over the space of imagination-reality dualities as a subspace of actual-potential dualities. All-natural actions take place over the actual-potential dualities.

The epistemic system for knowing as an intellectual investment flows and information-knowledge accumulation as intellectual stocks is such that for every variety problem in the space of actuals, there are corresponding variety solutions in the potential space, and for every variety solution there are corresponding variety actual problems that may be replaced under general cost–benefit in the space of knowing, information-knowledge accumulation and problem–solution dualities. The neuro-decision-choice processes are defined in the space of actual-potential dualities with a system of dualistic-polar games for continual variety transformation between the actual variety problems and potential variety solutions where the rewards are structured in the space of cost–benefit dualities establishing interdependencies in the space of actual-potential dualities. These actual-potential interdependencies are such that every selected variety solution generates a corresponding variety problem or problems where the variety solution may come with positive or negative externalities creating distortions in the social cost–benefit distributions. The analytical reasoning here is such that every variety problem has multiple variety solutions and an optimal variety solution where the multiple solutions reside in the space of potential varieties of $(\nu_{\mathfrak{u}} \in \mathfrak{U})$ as possible replacements of a problem variety $(\nu_{\mathfrak{a}} \in \mathfrak{A})$, and where such possible replacements may be ranked in the order of the best in accordance with both individual and collective preferences over the space of cost–benefit dualities of any decision-choice setup. The analytical construct also applies to theory of knowing and the unity of knowing and science. It also applies to problem-selection decisions for investigation by researchers as well as provides conditions for the development of the theory of optimization to abstract the maximal variety benefit under variety a cost constraint or to abstract the minimal variety cost under a variety benefit constraint.

The epistemic structure of the application of the principle of opposites in understanding the problem-solving process is such that the information-knowledge development is seen in terms of dynamic conditions of integrated problem–solution dualities mapped into the spaces of certainty-uncertainty, doubt-surety and decidability-undecidability dualities over the epistemological space, how the conditions of problem–solution dualities relate to the transformation-knowledge conditions of problem–solution polarities to bring about a variety solution to a variety problem, and for the variety solution to generate new variety problems as has been discussed in chapter two and three of this monograph. The explanatory understanding of the problem–solution process is to relate the activities over the space of problem–solution dualities to the elements in the decision-choice space and implementations. The activities of the internal dualities create the forces for the actions of the external dualities represented by the poles of the polarity to bring about, as well as sustain, the continual actual-potential destruction-replacement dynamics. The internal cost–benefit dualities indicate the conditions of change while the external cost–benefit dualities indicate the process of change. Let us keep in mind that problems and solutions relate to varieties where a variety problem is an actual, a variety solution is a potential and the variety solution may be considered as the optimal from many variety solutions. A solution may be an intermediate goal in a sequence of intermediate goals to a terminal goal such as a vision with dualistic-polar conditions.

All variety problems belong to the space of the realities contained in the space of the actuals, while all variety solutions belong to the space of imagination contained in the space the potentials, where the cost–benefit disparities between variety reality and variety imagination create conflicts in preferences to bring about transformation decisions in the destruction-replacement dynamics. How are the variety problems and variety solutions specified in terms of the principle of opposites? A solution variety is nothing more than a transformation of the variety problem with a replacement of a variety solution as a new reality with a new information structure for its identity while the variety problem is sent into the potential space as new imagination that may turn out to be a variety solution in the time processes with its information structure left as part of the information-knowledge accumulation in ***the*** historic sense. It is useful to keep in mind that an acquisition of knowledge about any variety is a successful cognitive operation over the space of identification-transformation dualities. The analytical process is to see human information-knowledge processes and advances as an integrated system of problem–solution dualities and polarities under decision-choice actions within the space of diversity-unity dualities in relation to either variety identification, variety transformation or both.

In terms of dualistic-polar analytics under the principle of opposites, every variety exists simultaneously as a problem–solution polarity with a supporting system of dualities defined over both the spaces of cost–benefit dualities and negative–positive dualities that find expressions, identity and meanings over the space of characteristic dispositions. Every variety also exists simultaneously as internal duality and external duality with relational continua and unity. The relational nature of the duals of the internal duality presents the character of the duality, where the internal duality is said to be negative (positive) if the negative (positive) internal dual dominates the positive (negative) internal dual in dispositional effects. The information-knowledge conditions of the interplays of the internal duality and external dualities allow variety solution to be abstracted through decision-choice actions acting as a *categorial convertor* creating conditions of technological know-hows in all areas of knowing, where knowing may be seen in terms of either explanatory or prescriptive sciences.

The explanatory science includes the theories and practice of social sciences and non-social sciences to solve the identification problem in knowing. It involves in establishing variety identifications and finding explanations to variety behaviors over the spaces of problem–solution dualities and polarities. The prescriptive science includes the theories and practice of engineering science and all forms of planning science to solve the transformation problems over the space of problem–solution dualities in terms of information and knowledge in variety dynamics. It also involves the processes of establishing the conditions of variety transformations and finding the understanding of variety dynamics over the spaces of problem–solution dualities and polarities [671, 672, 685, 700]. The identification and transformation conditions of the varieties of the space of problem–solution dualities are central to the establishment of the principles of the unity of knowing, the unity of science and unity of non-science through the principle of diversity and dividedness. The methods and techniques for establishing the problem–solution replacement processes may vary over categorial varieties but are interdependent in knowing and utilization under a general methodology where we must deal with uncertainty-certainty, doubt-surety, undecidability-decidability and risk-riskless dualities in knowing and information-knowledge accumulation. The set of methods and techniques in a methodological toolbox for dealing with the problem–solution dualities of one area of knowing may not only help in discovering other methods and techniques relevant in dealing with problem–solution dualities of other areas of knowing but may be transferable with or without modifications. The efficiencies of all areas of knowing are united by the efficiency of gains and uses of the information-knowledge structure of social science which also induce unity among the acts of thinking, reasoning, organizing and learning of cognitive agents.

To understand the dynamics of the space of problem–solution dualities, and how problems to solutions are internally or externally achieved through the interplay of analytical dualities and polarities under the principle of opposites within the space of decision-choice actions, the following working definitions will be useful. Let us keep in mind that a definition, whether real or nominal, is a *definitional constraint* restricting the allowable area of meaning, use and understanding. Similarly, explication is an *explicative constraint* with further restriction on the domain of the meaning, use and scientific understanding of linguistic varieties. The uses of all these elements of definitional constraints and explicative constraints in communications and discussions of problem–solution dualities are not restricted to one area of knowing such as science or non-science. They are part of methods of source–destination relations. All the concepts and restrictions that they impose are also related to the principle of opposites and the developments of linguistic varieties, variety identification, variety transformation and neuro-decision-choice actions [87, 501]. We have worked with the principle of opposites. Let us now provide a working definition and how it relates to cognition and neuro-decision-choice processes of general human actions over the space of imagination-reality dualities as a sub-space of the space of actual-potential dualities.

Definition 7.1.2.1A: The Principle of Opposites (verbal) The principle of opposites is an analytical approach that states that the universe is composed of a system of matter-energy actual-potential varieties with a system of characteristic dispositions that places distinctions among the varieties for identification, where the system of matter-energy varieties exists as a system of dualities in variety identification processes with actual duals and corresponding potential duals, and as a system of polarities in transformation processes with actual poles and corresponding potential poles, where each pole has a residing duality and each duality is composed of a positive dual defined by a positive characteristic sub-set and a negative dual defined by a negative characteristic sub-set such that the union of the negative and positive characteristic sub-sets and their relative combinations define the identity of each variety in the universal existence, while the union of the negative and positive poles with their residing dualities defines the transformation conditions of varieties to maintain the system of continual variety transformations under the *law of opportunity cost* in diversities, unity, maintenance of matter-energy equilibrium states and information-knowledge disequilibrium dynamics.

Definition 7.1.2.1B: The Principle of Opposites (Algebraic) The principle of opposites is an analytical approach that states that the universe (Ω) with a generic element ($\omega \in \Omega$), composed of the space of the actuals ($\mathfrak{A}$) with a generic element ($\mathfrak{a} \in \mathfrak{A}$) and the space of the potentials ($\mathfrak{U}$) with a generic element ($\mathfrak{u} \in \mathfrak{U}$), is made up of a system of matter-energy varieties ($\mathbb{V}$) with a generic element ($\nu \in \mathbb{V}$) and a defining system of characteristic dispositions ($\mathbb{X}$) partitioned by ($\nu \in \mathbb{V}$) in the form $(\mathbb{X}_\nu)$ with $\left(\mathbb{X} = \bigcup_{(\nu \in \mathbb{V})} \mathbb{X}_\nu\right)$ such that distinctions are placed among the varieties $(\nu \in \mathbb{V})$ for identification, and where the system of mater-energy varieties exists as a *system of dualities* $(\mathbf{D})$ with a generic element $(\mathbf{d} \in \mathbf{D})$ in establishing a *system of solutions* $(\mathfrak{S}_\mathbf{I})$ to the *system of variety identification* $(\mathbf{I})$ *problems* $(\mathfrak{P}_\mathbf{I} \subset \mathfrak{H} \subset \mathfrak{A})$ and as a *system* $(\mathbf{P})$ *of polarities* with a generic element $(\mathbf{p} \in \mathbf{P})$ in the establishment of the *system of solutions* $(\mathfrak{S}_\mathbf{T})$ with a generic element $(\mathfrak{s} \in \mathfrak{S}_\mathbf{T} \subset \mathfrak{G} \subset \mathfrak{U})$ to the system of *variety transformation* $(\mathbf{T})$ *problems* $.(\mathfrak{P}_\mathbf{T})$. with a generic element $(\mathfrak{s} \in \mathfrak{S}_\mathbf{T} \subset \mathfrak{G} \subset \mathfrak{A})$, a corresponding system *of actual poles* $(\mathbf{P}_\mathfrak{A})$ and a corresponding system of *potential poles* $(\mathbf{P}_\mathfrak{U})$, where each pole has a residing duality $(\mathbf{d} \in \mathbf{D})$ and each duality is composed of positive dual $\left(\mathbf{d}_p = \mathbb{X}^P \in \mathbf{D}_\mathfrak{A} \subset \mathbf{P}_\mathfrak{A}\right)$ or $\left(\mathbf{d}_\mathfrak{p} = \mathbb{X}^P \in \mathbf{D}_\mathfrak{U} \subset \mathbf{P}_\mathfrak{U}\right)$ and negative duals $\left(\mathbf{d}_\mathrm{N} = \mathbb{X}^\mathrm{N} \in \mathbf{D}_\mathfrak{A} \subset \mathbf{P}\right)$ or $\left(\mathbf{d}_\mathrm{N} = \mathbb{X}^\mathrm{N} \in \mathbf{D}_\mathfrak{U} \subset \mathbf{P}\right)$ in different relative proportions to form the systems of dualities and polarities to establish the infinite system of opposites $\mathbb{Q}$ with a generic element $(\mathfrak{q} \in \mathbb{Q})$, an *infinite index set* $\left(\mathbb{I}_\mathbb{Q}^\infty\right)$ and a condition $\Omega = (\mathfrak{A} \cup \mathfrak{U})$.

Note 7.2.1: On the Principle of Opposites The principle of opposites is a general characterization of variety identities in dualistic-polar conditions within the space of variety identification-transformation dualities in socio-natural static-dynamic conditions in relation to the elements in the space of input–output dualities where every variety exists as a union of opposite dispositions in relational continuum and unity with continual struggle for identity dominance in dualistic-polar game dynamics. The principle of opposites provides us the conditions for the understanding of all internal transformations including all forms of socio-natural systems of varieties in the space of actual-potential dualities which contains the space of imagination-reality dualities, where changes in polarity require changes and restructures of residing dualities. The understanding of the conditions of the principle of opposites will help decision-choice agents in engineering rational variety transformations over the spaces of destruction-replacement dualities with efficient neuro-decision-choice actions over the space of input–output dualities contained in the space of imagination-reality dualities. Additionally, the understanding of the conditions of the principle of opposites reveals the forces of transformation in the space of internal–external dualities and thus helps cognitive agents to understand the permanent conflicts in social systems and use this understanding to manage social transformations and stability, where there is always balancing actions among variety opposites from the knowing and information-knowledge stock-flow disequilibrium dynamics, where the negative opposite is set against the positive opposite and vice versa, and the benefit opposite is also set against the cost opposite and vice versa within the spaces of problem–solution dualities and ignorance-knowledge dualities. The knowing and information-knowledge outcomes over the space of equilibrium-disequilibrium dualities become inputs for neuro-decision-choice actions over the pace of imagination-reality dualities as a sub-space of the space of the actual-potential dualities. The principle of opposite with relational continuum and unity allows us to understand it's the relationship to the nature of fuzzy paradigm of thought that accepts contradictions in thinking and reasoning, where the principle of opposites creates conflicts in preferences and neuro-decision-choice actions over the space of individual-collective dualities in all aspects of human interactions and endeavors.

7.3 Knowing, Problem–solution Polarity and Cost–benefit-Analysis

The principle of opposites is established by the space of opposites with its static and dynamic behaviors in terms of the problem–solution destruction-solution-replacement process. All the opposites may be taken in relational continua and in an organic unity under the *give-and-take-sharing principle* in relational continuum and unity for their mutual natural existence in the space of relativity. They may also be taken in relational separation with excluded middle and with assumed relational disunity under individual existence with *no-relational give-and-take sharing principle* in the space of absoluteness. The internal and external dualistic-polar relational continua generate conflicts, friction and contradictions setting the opposites to produce internal energies for internal self-transformation with the backward-forward dualities of the *telescopic processes* defining the variety paths of past history and future history around the present at any time point to establish an indestructible unity of universal existence and unity in time, for the processes of knowing to generate diversities and unity of knowledge, where old socio-natural varieties are destroyed to give way to new socio-natural varieties as solutions in the *destruction-replacement dynamics* with the creation of new varieties information-knowledge conditions on the opportunity cost of real variety problem with the understanding that the space of realities finds expression from the space of the actuals. Information-knowledge items are created over the space of success-failure dualities leading to the development of error-correction processes as an integral part of the disequilibrium dynamics of information-knowledge system in continual improvements within the conditions of give-and-take sharing mode.

7.3.1 Cost–Benefit Analytics, Dualistic-Polar Analytics and Problem-Solution Analytics

There is a cost–benefit distributional complexity defined over the space of varieties ($\upsilon \in \mathbb{V}$) that must be distinguished and understood in the information-knowledge process as neuro-decision-choice activities undertaken by cognitive agents over the space of problem–solution dualities. The destroyed variety problem is the real opportunity cost for bringing into its place the variety solution which is also the real opportunity cost for the destroyed variety problem. In other words, every variety in both the space of actual and the space of potentials is an opportunity cost in the destruction-replacement dynamics such that variety ignorance is destroyed and transformed into variety knowledge as its replacement in the space of ignorance-knowledge dualities (as represented by adinkra symbol of, *onim-sua-ohu*). Additionally, every variety exists as an internal negative -positive duality with a relational continuum and unity in identification, and as an external pole of a polarity in a relational continuum with a potential pole for variety transformation. The internal negative–positive dualities also exists in an interdependent system of real cost–benefit dualities, where the negative dual may be either real cost or real benefit and the positive dual may be either real cost or real benefit depending on the goal-objective conditions from a given preference order over the space of visions. The structure of any problem–solution polarity may be illustrated with a cognitive geometry of the form in Fig. 7.1 to show the distribution of the negative and positive duals, and the cost and benefit duals in relation to the poles of polarity and how they define and affect the positive and negative poles and the polarities, and their relational structures to cost–benefit conditions and neuro-decision-choice processes. The cognitive geometry may also be viewed as a paremiological geometry of thought that shows the structure of multiplicity of the relational connectivity of concepts and meanings of terms within the space of static-dynamic dualities requiring increasing epistemic falling in linguistic system for continual development of intellectual investment and information-knowledge accumulation.

A variety solution to a variety problem means the destruction of a actual variety in existence and the corresponding real cost–benefit duality to create a vacuum for a new variety replacement with the corresponding real cost–benefit duality from the potential space where the actual variety acts as an opportunity cost within the real cost–benefit duality to obtain the real net benefit of the corresponding cost–benefit duality of the replacement variety. Problem variety implies a preference dissatisfaction of the existing variety's internal real cost–benefit balance and a search for a potential variety with better real cost–benefit balance as a variety replacement in terms of a variety solution where both destruction and replacement create new information-knowledge items. In terms of problem–solution interactive processes, there are two socio-natural decision-choice processes, where all social decision-choice processes are about cognitive behaviors over the space of problem–solution dualities as established over the space of varieties for knowing, relative to variety identification, transformation or both. Every variety in the universal space of varieties resides in dualistic-polar conditions where the variety is simultaneously a problem and a solution relative to preferential conditions as they relate to goals, objectives and vision. A variety solution is nothing more than a transformation of a variety problem with lower net cost–benefit disposition for a variety with higher net cost–benefit disposition under a cost–benefit relationality within the space of actual-potential dualities. Similarly, a variety problem is nothing more than a transformation of a variety solution with higher real net cost–benefit conditions for lower real net cost–benefit conditions under a real cost–benefit rationality within the same space of actual-potential varieties where goals, objectives and vision are also defined in

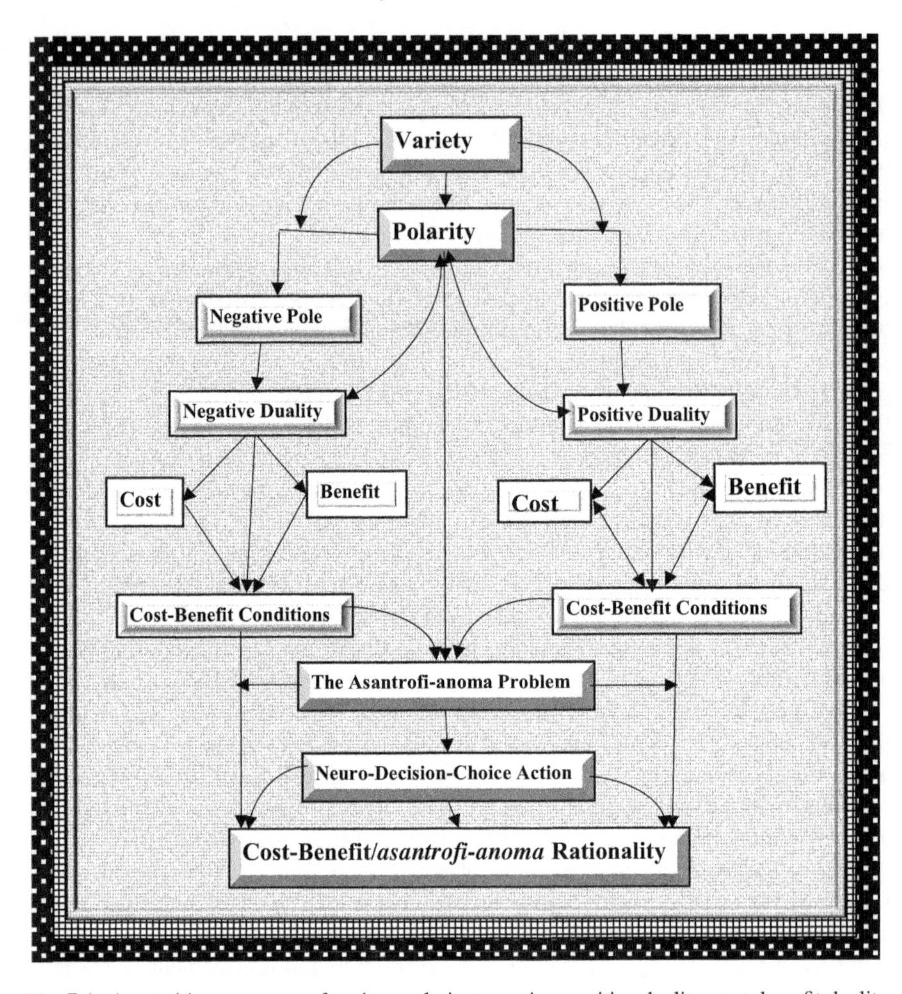

Fig. 7.1 A cognitive geometry of variety polarity, negative–positive duality, cost–benefit duality and the dynamics of problem–solution duality with categorial convertors as technologies induced by decision-choice activities over the action space

the same space of varieties under information-knowledge conditions of stock-flow dynamics.

The epistemic structure and the analytical principle of opposites with sub-principles of duality and polarity that provide the explanatory process of the general variety destruction-replacement dynamics of the problem–solution duality may be used as a predictive framework and prescriptive process for the development of social policy, science policy, economic policy, medical policy, educational policy, technological policy and engineering policy in all areas of knowing that may be followed by any society in creating the social information-knowledge bases as inputs into individual and collective neuro-decision-choice actions at the level of theory and practice.

It may also be used to construct a theory of the development of economics and social change where categories of economic development stages or stages of social change are seen in terms of problem–solution dualities, and where socio-economic progress and retrogress are viewed as the destructions of the old forms with lower real net cost–benefit conditions and the replacement of the new forms with higher real net cost–benefit conditions as governed by continual decision-choice dynamics of the general information-knowledge structure.

The nature and the case of constructing an incisive theory of economic development based on decision-choice activities over the space of problem–solution duality are the driving forces for my development of theories on information, decision and choice in that socioeconomic development is a work of information-decision-interactive processes with increasing neuro-preferences for better conditions of life no matter how they are measured where stages of development may be seen as varieties with defined characteristic dispositions. Hopefully, time and place will allow me to construct this theory of economic development based on the conditions of problem–solution duality within the unity of the theory of knowing, the information-knowledge system and decision-choice technologies (*categorial convertors*) governed by social philosophical Consciencism as is indicated in the monograph of epistemics of development economics [668] and the theory of Philosophical Consciencism [670], the theory of info-statics [671] and the theory of info-dynamics [672] on the basis of the asantrofi-anoma principle and rationality with conditions over the space of cost–benefit dualities in relation to market and non-market institutions of production where the cost–benefit dualities may or may not be related profit or non-profit conditions in variety transformations [218, 219].

In terms of problem–solution analytics in support of the unity of science through the unity of knowing, knowing is a variety solution to a variety problem of ignorance where ignorance and knowledge create the dualistic-polar conditions for cognition in the continual process of matter-energy variety dynamics in disequilibrium states just like economic development resides in states of disequilibrium dynamics. The important thing is that the problem–solution analytics connect all areas of knowing, and the networks of problem–solution connectedness are the ultimate strength in the progress of science, non-science, exact science, inexact science, and the organization of human society in relation to productive activities, culture, success, failure and progress in their mutual existence that defines the general disequilibrium dynamics of the-economic development process. It is the disequilibrium dynamics of economic development and its input factors that form the foundation for the studies of economics of health, education, labor, science, technology, law, politics, agriculture, information and many others since all these are economic production in the management of social organization and its transformations in the space of imagination-reality dualities as a sub-space of actual-potential dualities, where the source of actual-potential dualities constitutes the ontological space as the knowing identity which is linked to the epistemological space as a derived identity through a neuro-decision-choice system connected to the space of imagination-reality which then generates goals, objectives and visions under principle of non-satiation to create the spaces of certainty-uncertainty dualities, doubt-surety and fulfilment-disappointment dualities.

The nature of the actual-potential replacement processes requires us to search for the best replacement from the potential space through decision-choice actions and information-knowledge processes in the space of imagination since there are many potential variety replacements that may be conceptualized through imagination. The variety search over the potential space acknowledges the *Euler mini-max principle* that nothing happens in this world without a minimum or maximum and the justification of the theory of optimization to abstract the optimum in terms of minimum or maximum of variety cost or benefit where costs constrain benefits and benefits constrain cost under the *asantrofi-anaoma* principle and rationality.

Over the space of problem–solution dualities, there are many potential variety solutions that may be coordinated with the existing variety problem under cost–benefit conditions. These many potential variety solutions may be ranked in increasing or decreasing degrees of replacement efficiency defined in terms of cost–benefit relative values under resource constraint. This requires a search for an efficient variety replacement in terms of the best among the potential variety solutions relative to a variety problem. The search for the best variety replacement requires the development of a process to represent the structure of the decision-choice actions that are in the space of computable-incomputable dualities and algorithms to abstract an optimal replacement in terms of an optimal or efficient variety solution with necessary and sufficient conditions to the variety problem. This is since for any actual variety problem there are many potential variety solutions, where the attainment of each variety solution is constrained by the available resource to bring it into action. The best unconstrained solution is never available in the space of relativity and hence unattainable in a resource-constrained environment within the matter-energy equilibrium. The relational structure of the internal duality, the external dual and the actual pole of the actual-potential polarity in the solution process is shown as an epistemic geometry in Fig. 7.2.

Let us keep in mind that the search for optimal problem–solution replacements and optimal replacement path is in relation to the search for an optimal strategy for the optimal path of ignorance-knowledge polarity to establish the path of information-knowledge stock-flow disequilibrium dynamics in self-correcting neuro-decision-choice system. This search for the optimal path of knowing is not restricted to one area of cognitive activities. Not only does it involve all areas cognitive activities in the enterprise of knowing but that the cognitive activities in different categories of the search are interdependent to optimize the cumulative results of the totality of knowing even though the path to each variety knowledge may be different from others. We have already defined the problem and solution dualities. We now need analytical definitions and explications for *problem polarity* and *problem pole* and to show their relationality and continuity for the establishment of diversity and the unity principles of knowing. It must always be kept in mind that knowing is an activity of intellectual investment, while the information-knowledge system is an intellectual capital stock that is continually being updated by the enterprise of knowing where the acceptable investment is always equal to intellectual savings.

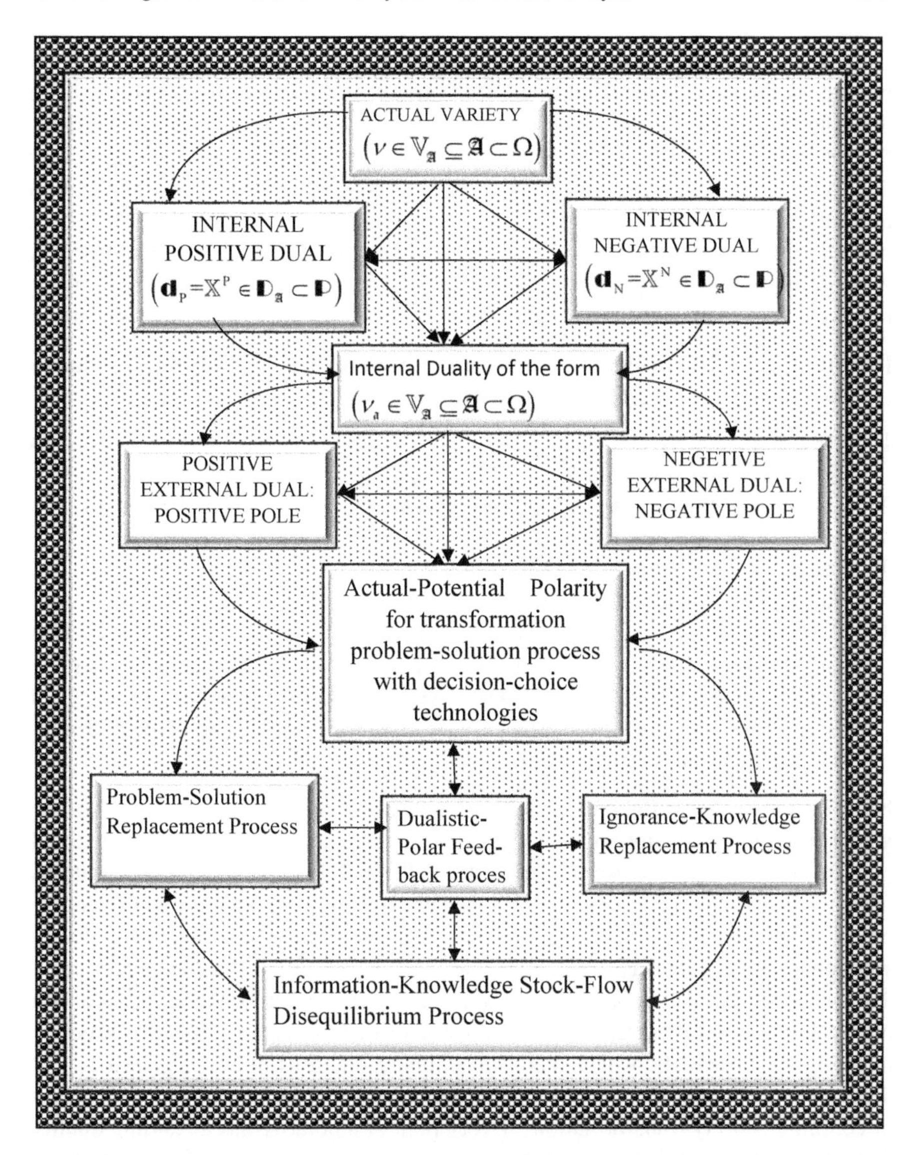

Fig. 7.2 Paremiological geometry of the relational connectivity of problem–solution duals, poles, internal problem–solution duality, external problem–solution duality and problem–solution polarity over the decision-choice space: the case of actual poles and polarity which may also be viewed as the case of reality poles and imagination poles under dualistic-polar conditions

Definition 7.2.1.1A: Problem Polarity and Problem Pole (verbal) A problem polarity is dualistic-polar conditions of a variety in the space of relativity, where the polarity belongs to the space of actual polarities and problem pole dominates the solution pole with the problem external duality dominating the solution external duality, and where the problem dual of the internal duality dominates the solution dual of the same internal duality of the variety. An actual variety is said to be a problem if it is a problem pole with real cost disposition dominating the real benefit disposition relative to a potential pole as assessed by a neuro-decision-choice agent relative to either a goal-objective element as viewed under cost–benefit conditions.

Definition 7.2.1.1B: Problem Polarity and Problem Pole (algebraic) A polarity $(\mathbf{p} \in \mathbb{P})$ is said to be a problem polarity if.

1. $(\mathbf{p} \in \mathbb{P}_{\mathfrak{A}})$, and $\mathbf{p} \in (\mathfrak{P}_{\mathbf{I}} \cup \mathfrak{P}_{\mathbf{T}})$, where $(\mathbf{I})$ represents identification and $(\mathbf{T})$ represents transformation.
2. $\left(\mathbf{p}_{\mathfrak{P}} \in \mathbb{P}_{\mathfrak{A}}, \mathbf{p}_{\mathfrak{P}} \ggg \mathbf{p}_{\mathfrak{S}} \in \mathbb{P}_{\mathfrak{A}}\right)$ with $\left(\mathbf{p}_{\mathfrak{P}} \cup \mathbf{p}_{\mathfrak{S}} = \mathbf{p} \in \mathbb{P}_{\mathfrak{A}}\right)$ where $(\ggg) \Rightarrow$ substentially greater than
3. $\left(\mathbf{d}_{\mathfrak{P}} \in \mathbb{D}_{\mathfrak{A}}, \mathbf{d}_{\mathfrak{P}} \ggg \mathbf{d}_{\mathfrak{S}} \in \mathbb{P}_{\mathfrak{A}}\right)$ with $\left(\mathbf{d}_{\mathfrak{P}} \cup \mathbf{d}_{\mathfrak{S}} = \mathbf{d} \in \mathbb{D}_{\mathfrak{A}}\right)$ where $(\ggg) \Rightarrow$ substentially greater than
4. A variety $(\nu \in \mathbb{V})$ is a problem if it is of the form $(\nu_{\mathfrak{a}} \in \mathbb{V}_{\mathfrak{A}} \subseteq \mathfrak{A} \subset \Omega = \mathfrak{A} \cup \mathfrak{U})$ and satisfies conditions 1–3 where $\mathfrak{P}_{\mathbf{I}} =$ identification problem, $\mathfrak{P}_{\mathbf{T}} =$ transformation problem , $\mathbf{p} = \left(\mathbf{p}_{\mathfrak{P}} \cup \mathbf{p}_{\mathfrak{S}}\right)$ is a problem–solution polarity and $\mathbf{d} = \left(\mathbf{d}_{\mathfrak{P}} \cup \mathbf{d}_{\mathfrak{S}}\right)$ is a problem–solution duality as established by cost–benefit conditions in the space of relativity.

Note 7.2.1.1 Problem Polarity The relational parts of the internal conditions of the actual problem variety, internal duality, external dual and the pole of actual-potential polarity are shown as an epistemic geometry in Fig. 7.1. In terms of human conditions in neuro-decision-choice actions in the space of problem–solution dualities, every problem presents itself as an actual variety that belongs to the actual space. Every solution presents itself as a potential variety that belongs to the potential space. The concepts of dualistic-polar conditions imply the simultaneous existence of a problem–solution structure, a cost–benefit structure and a negative–positive structure in the space of actuals, where there are intimate relations between polarities and dualities in the space of opposites in relational give-and-take sharing modes in terms of continual variety transformations. Every pole of a polarity has a residing duality that defines its essence and transformation behavior. Every duality is composed of a negative dual and a positive dual. Every dual in the negative–positive duality is viewed as a cost–benefit duality the structure of which rests on preferences over the elements of the goal-objective space. Every variety problem resides in the space of cost–benefit dualities, where the cost dual is substantially greater than the benefit dual and every variety solution also resides in the space of cost–benefit dualities, where the benefit dual is substantially greater than the cost dual relative to a goal-objective element from the goal-objective space.

It must be understood that every problem is defined over the space of realities which is a sub-space of the space of the actuals. The important reflection in this regard is that the elements in the space of reality depend on the approvals of neuro-decision-choice system such that a problem exists if it is claimed by neuro-decision-choice action. In this respect, a problem may be identified in the space of realities without belonging to the space of the actuals creating a phantom problem. The existence of the space of realities depends on the existence of cognitive agents. Similarly, the existence of the space of actuals is independent of the existence of neuro-decision-choice agents. Translated in another way, the existence of ontological space and its varieties are independent of the existences of ontological varieties, however, the existence of epistemological space depend on the existence of neuro-cognitive agents. The socio-natural processes, varieties, states in the space of ignorance-knowledge dualities help to establish the space of ontological-epistemological dualities that gives meaning to knowing as intellectual investment flows and information-knowledge accumulation as intellectual stocks in relational diversities, continuum and unity to establish a system of intellectual input–output stock-flow for continual epistemological transformations.

Definition 7.2.1.2A: Solution Polarity, Solution Duality and Solution Pole (Verbal) A solution polarity is a set of dualistic-polar conditions of a variety, where the variety solution pole dominates the variety problem pole with the *solution external dual* dominating the *problem external dual,* and where the *solution dual* of the *internal duality* dominates the *problem dual* of the same internal duality of a variety in the space of relativity. A potential variety is said to be a solution to a problem variety if it is a solution polarity with an assessed *benefit disposition* dominating an assessed cost disposition relative to the cost–benefit conditions of the actual problem polarity.

Definition 7.2.1.2B: Solution Polarity and Solution Duality (Algebraic) A polarity ($\mathbf{p} \in \mathbb{P}$) is said to be a solution polarity if:

1. ($\mathbf{p} \in \mathbb{P}_{\mathfrak{U}}$), and $\mathbf{p} \in (\mathfrak{S}_{\mathbf{I}} \cup \mathfrak{S}_{\mathbf{T}})$, where ($\mathbf{I}$) represents identification and ($\mathbf{T}$) represents transformation
2. $\left(\mathbf{p}_{\mathfrak{S}} \in \mathbb{P}_{\mathfrak{U}}, \mathbf{p}_{\mathfrak{S}} \ggg \mathbf{p}_{\mathfrak{P}} \in \mathbb{P}_{\mathfrak{U}}\right)$ with $\left(\mathbf{p}_{\mathfrak{P}} \cup \mathbf{p}_{\mathfrak{S}} = \mathbf{p} \in \mathbb{P}_{\mathfrak{U}}\right)$ where $\ggg \Rightarrow$ substantially greater than
3. $\left(\mathbf{d}_{\mathfrak{S}} \in \mathbb{D}_{\mathfrak{U}}, \mathbf{d}_{\mathfrak{S}} \ggg \mathbf{d}_{\mathfrak{P}} \in \mathbb{D}_{\mathfrak{U}}\right)$ with $\left(\mathbf{d}_{\mathfrak{P}} \cup \mathbf{d}_{\mathfrak{S}} = \mathbf{d} \in \mathbb{D}_{\mathfrak{U}}\right)$ where $\ggg \Rightarrow$ substantially greater than
4. An element ($\nu \in \mathbb{V}$) is a solution if it is of the form ($\nu_{\mathfrak{u}} \in \mathbb{V}_{\mathfrak{U}} \subseteq \mathfrak{U} \subset \Omega = \mathfrak{A} \cup \mathfrak{U}$) as well as satisfies conditions 1–3 where $\mathfrak{S}_{\mathbf{I}} =$ identification solution, $\mathfrak{S}_{\mathbf{T}} =$ transformation solution, $\mathbf{p} = \left(\mathbf{p}_{\mathfrak{P}} \cup \mathbf{p}_{\mathfrak{S}}\right)$ is a problem–solution polarity and $\mathbf{d} = \left(\mathbf{d}_{\mathfrak{P}} \cup \mathbf{d}_{\mathfrak{S}}\right)$sis a problem–solution duality as established by cost–benefit conditions in the space of relativity.

The epistemic geometry of the solution polarity is given below in Fig. 7.3.

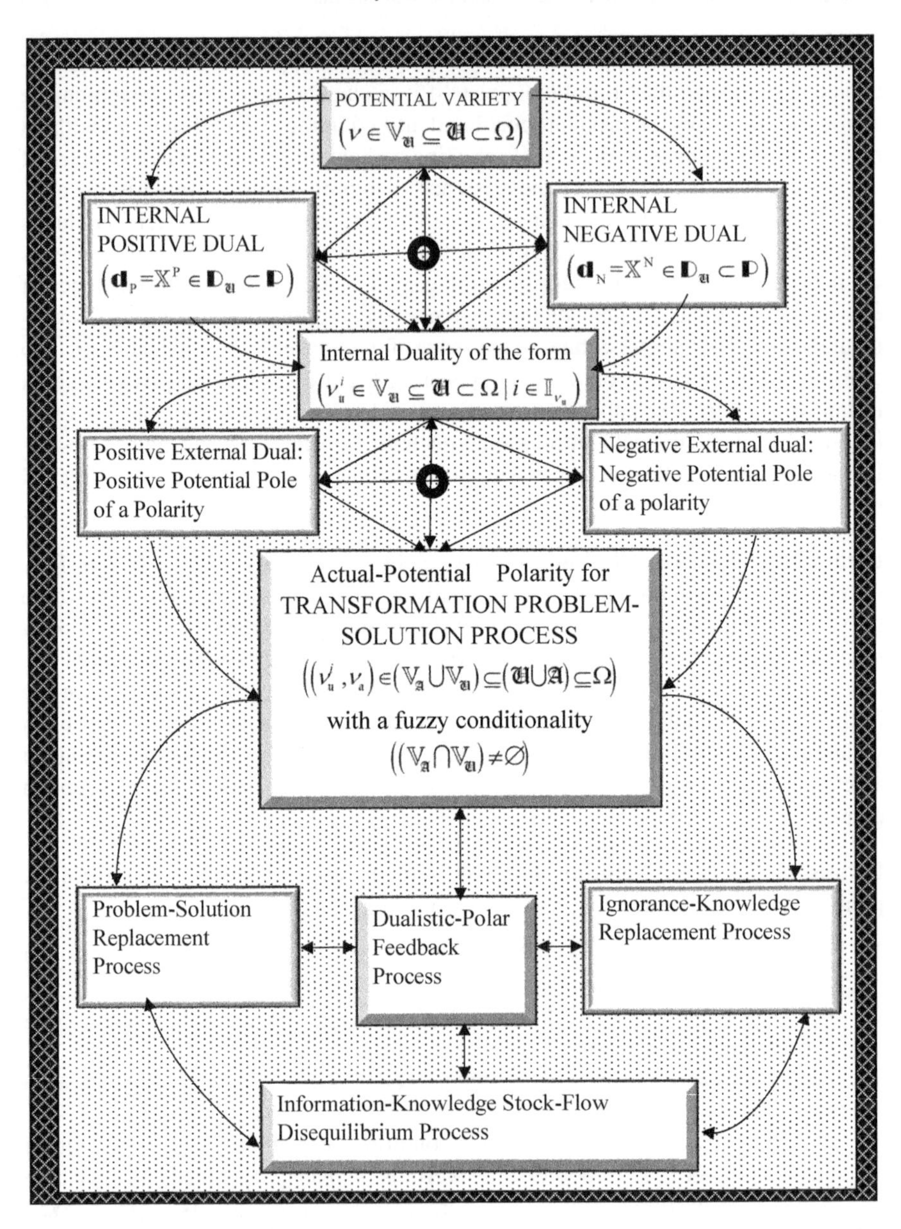

Fig. 7.3 An epistemic geometry of the relational connectivity of problem–solution duals, poles, internal problem–solution duality, external problem–solution duality and the problem–solution Polarity over the decision-choice space: the case of potential poles and polarity

7.3.2 Paremiologically Geometric Analytics of Dualistic-Polar Relational Connectivity

The epistemic geometry of solution polarity is given in Fig. 7.3. The problem–solution process as presented and illustrated in Figs. 7.1, 7.2 and 7.3 are in relational connectivity. By analytical aggregation these geometries are equivalent to the epistemic geometry of Fig. 7.4. All of them must be translated into real cost–benefit characteristic dispositions, where variety problems are viewed in terms of dominant cost characteristic dispositions and variety solutions are viewed in terms of dominant characteristic benefit dispositions under a given preference scheme of either an individual or a collective. The relational geometry of aggregation must connect internal and external dualities to actual and potential poles to present actual problem pole with

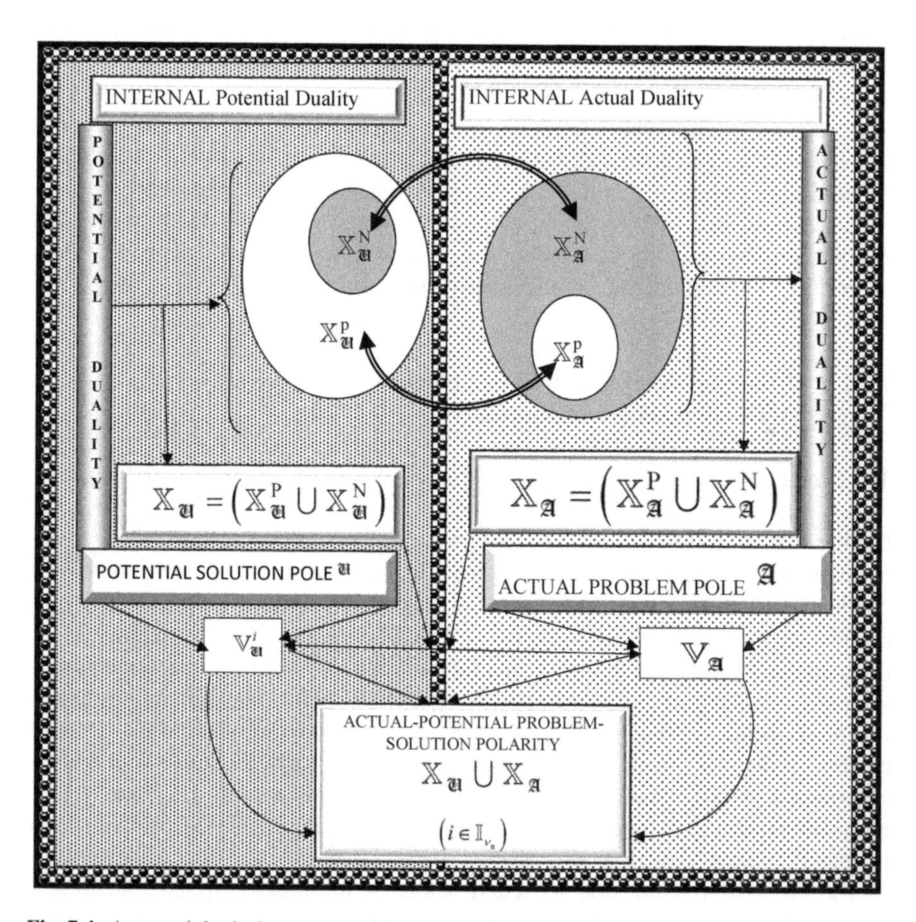

Fig. 7.4 A paremiological geometry of the relational structure of polarity, duality, continuum and unity in the categorial-conversion process of a problem to a solution and a solution to a problem with decision-choice actions as the transformations technologies

residing problem duality, and potential solution pole with residing solution duality to arrive at a problem–solution polarity. We must always keep in mind the important role that real cost–benefit duality plays in the definitions of problem–solution duality and its extensions to problem–solution polarity as well as the relativity cost–benefit conditions in defining variety problem and variety solution and by extension problem duality and solution duality in the problem–solution analytics including conditions of knowing as intellectual investment flows and information-knowledge system as intellectual capital stock.

In this complex epistemic system in multilateral levels and multi-dimensional connectedness in sub-systems of dynamic matrixes, the space of the actuals may be replaced with the space of realities which is contained in the space of the actuals, the space of variety potentials may be replaced by the space of variety imaginations which is contained in the space of variety potentials, and the space of variety actual-potential dualistic-polar structure may be replaced by the space of variety imagination-reality dualistic-polar structures which is contained the space of the actual-potential dualistic-polar structures. All human actions take place over the space of imagination-reality dualities which houses all spaces of human actions in relation to hope, visions, objectives goals, aspiration anticipations, expectation, beliefs and many others that drive thinking, reason and actions. The external conditions define the necessity to change and prompt the need of the pole to change. The internal conditions define the freedom and the actions to chance which proceed from the residing internal dualities of the polarity as illustrated in Fig. 7.4.

7.3.2.1 Efficiency Analytics of Paremiological Geometry of Dualistic-Polar Connectivity

The complexity of dualistic-polar conditions and the variety identity and transformation may be represented as an analytical geometry in the form of an efficiency frontier that shows the collection of internal dualistic-polar possibilities of variety representations in terms of dualistic and polar characteristic dispositions where each combination may represent a point of transformation. The analytical geometry is in support of Secs. (5.2.1.4, 5.1.2.3 and 5.1.2.2). The dualistic-polar efficiency curve is called the *dualistic-polar transformation possibility frontier* for duals of dualities and poles of polarities where every pole has a residing duality that defines its identity.

The epistemic geometry of the *dualistic-polar transformation possibility* frontier of Fig. 7.5 *may be* related to the resolution of the conflict of the dualistic membership game of transformations and dominance from a reduction of the characteristic disposition of Dual (1) (negative dual) to the increase in the characteristic disposition of Dual (2) (positive dual) as is presented in Fig. 7.6.

The movements along the AB curve which is the dualistic-polar transformation possibility frontier (DTPF) represent variety dualistic-polar games of categorial conversions where the acquisition of power-dominance in defining the variety identity and establishing the polar conditions are the reward. The degree of the power dominance depends on the *characteristic disposition load* of a dual relative to the

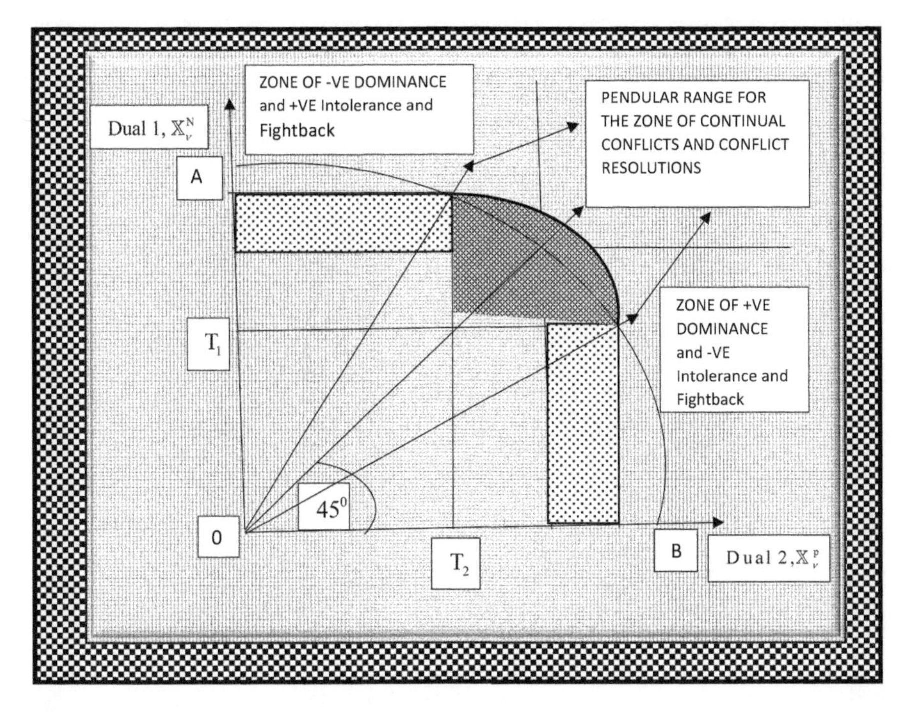

Fig. 7.5 Dualistic-polar transformation possibility frontier of AB curve between the positive dual and the negative dual of a duality

corresponding dual and to the total variety characteristic disposition of the duality. The internal variety dualistic-polar transformation process is always zero-some game in the process of acquiring power dominance by any dual to establish the identity of a pole of polarity for mutual destruction and existence in the sense of what is gained by one dual is lost by the other dual of the duality. There is no dualistic-polar negative or positive sum game in the variety principle of opposites. In this respect, every variety transformation possibility frontier in the game process may be represented as:

$$G\left(\mathbb{X}_\nu^P, \mathbb{X}_\nu^N\right) = 0 \Rightarrow \mathbb{X}_\nu^P = \mathbb{X}_\nu - T_\nu^P\left(\mathbb{X}_\nu^N\right) \text{for transformation of negative to increase the positive} \quad (7.2.1.1)$$

$$G\left(\mathbb{X}_\nu^P, \mathbb{X}_\nu^N\right) = 0 \Rightarrow \mathbb{X}_\nu^N = \mathbb{X}_\nu - T_\nu^N\left(\mathbb{X}_\nu^P\right) \text{for transformation of positive to increase the negative} \quad (7.2.1.2)$$

The dualistic polar game is between the positive and negative transformation functions of the form T_ν^P and T_ν^N creating conflict, tension, energy and force in implicit functions for categorial-conversion dominance and variety conversions where tension

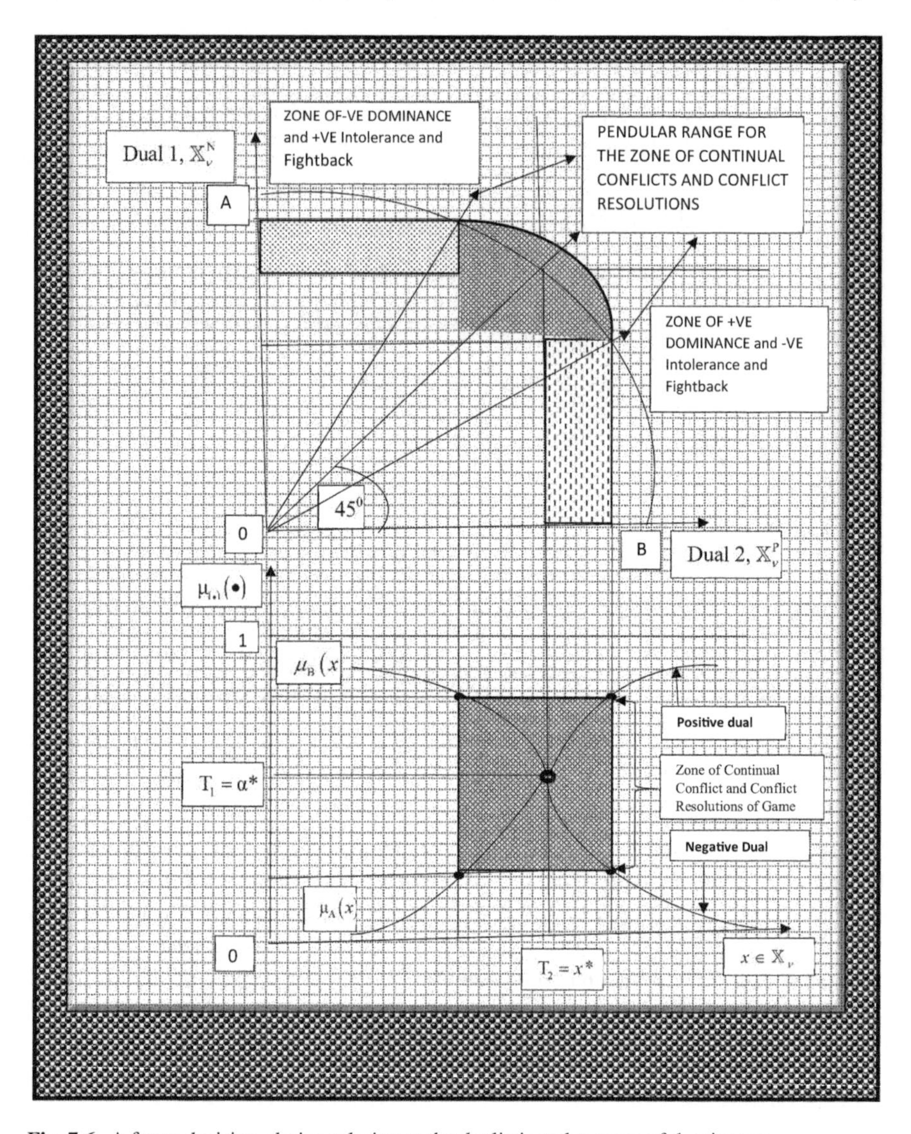

Fig. 7.6 A fuzzy decision-choice solution to the dualistic-polar game of dominance

generates conflict $\mathfrak{C} = \mathcal{F}_{\mathfrak{C}}\left(T_\nu^P, T_\nu^N\right)$, and where the conflict degrees ($\mathfrak{C}$) depend on the behavior of $\mathcal{F}_{\mathfrak{C}}(.)$ which then generates energy ($\mathfrak{E}$) in the form $\mathfrak{E} = \mathcal{F}_{\mathfrak{E}}\left(\mathcal{F}_{\mathfrak{C}}\left(T_\nu^P, T_\nu^N\right)\right)$ and a conversional force, ($\mathfrak{T}$) in the form $\mathfrak{T} = \mathcal{F}_{\mathfrak{T}}\left(\mathcal{F}_{\mathfrak{E}}\left(\mathcal{F}_{\mathfrak{C}}\left(T_\nu^P, T_\nu^N\right)\right)\right)$. The natural decision-choice activities over the ontological space, and the neuro-decision-choice activities over the space of identification-transformation dualities and polarities over the space of acquaintance-description dualities and polarities are to reconcile conditions of the information trinity of *what there was*, *what there is* and *what there would*

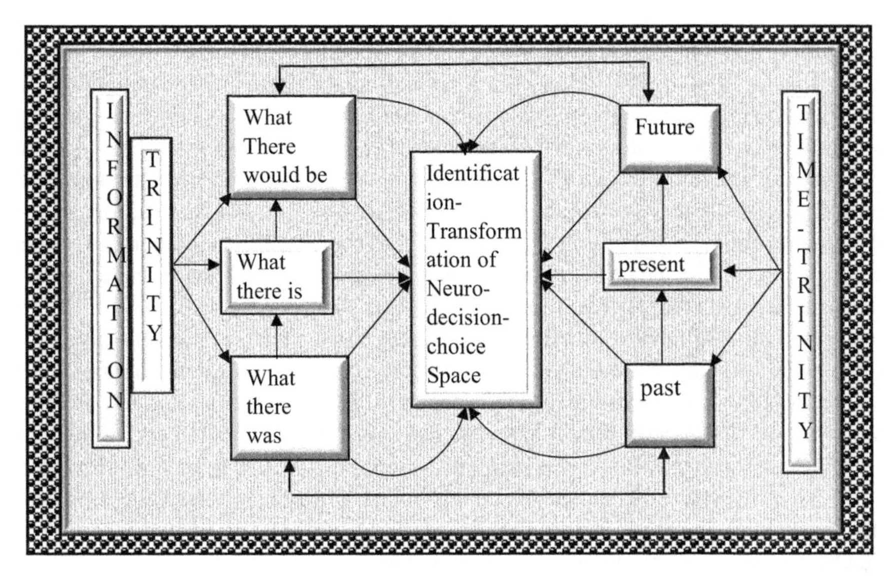

Fig. 7.7 Information-time trinities in knowing and information-knowledge production system

be and conditions of the time trinity of *past*, *present* and *future* in the intellectual investment and intellectual capital accumulation as may be represented in Fig. 7.7, where what there was relates to the past, what there is relates to the present and what there would be relates to the future in Sankofa-anoma tradition of time processes.

The function $T_{\upsilon}^{P}(\cdot)$ is a technology of a process that acts on the negative characteristics of the negative dual and transforms them into the positive characteristics to empower the positive dual for positive dualistic dominance and pole identification. Similarly, the function $T_{\upsilon}^{N}(\cdot)$ is a technology of a process that acts on the positive characteristics of the positive dual and transforms them into negative characteristics to empower the negative dual for negative dualistic dominance and pole identification. It is important to note that positive (negative) transformation success generates a reversal process within a defined pendular range of the game of continual conflicts and conflict resolutions changing the relational structures in the positive–negative duality of the variety. The positive (negative) dualistic-polar dominance leads to a critical re-examination of the power positions of the negative (positive) dualistic-polar characteristic disposition, design of new tactics, strategies and change of response in terms of neuro-decision-choice actions over the problem–solution dualities, where every change in relation presents simultaneous conditions of a variety solution to one of the duals and a variety problem to the other dual of the variety duality. The neuro-decision-choice actions change the relative strengths in the system of transformational technologies to affect the processes of knowing and the information-knowledge stock-flow disequilibrium dynamics without affecting the matter-energy equilibrium stock except the number of its varieties in the space of quality-quantity dualities.

Each round of the dualistic-polar game, therefore, generates information-knowledge input–output relations from the continual dynamics of the elements in the space of problem–solution dualities and polarities, where a variety problem is replaced by a variety solution which then generates another variety problem within the pendular range for a search of a new variety solution. Let us keep in mind that problems arise due to the distribution of cost–benefit disparities in the space of actual-potential dualities over the variety space relative to neural preferences that also create conflicts over the space of the collective neuro-decision-choice space and aggregation conflicts over the space of problem–solution dualities and polarities. Change requires identification of *what there is* and preference of *what is not* over the present-future dualities relative to the elements in the space of real cost–benefit dualities. The automatic self-correction information-knowledge disequilibrium system through the automatic self-correction neuro-decision-choice system requires the correct understanding with epistemic honesty and integrity of the past-present telescopic history of socio-natural events without self-deception, social propaganda misinformation and disinformation for neuro-decision-choice activities in the telescopic present-future dualistic-polar varieties in the space of imagination-reality dualities.

7.4 The Categorial Conversion Process and Categorial Analytics

To follow the problem–solution dynamics, it is the case that the *categorial conversion process* must indicate the identity of every problem–solution duality $\mathbf{D}$ which may be defined as a function of the negative and positive relational characteristic dispositions at any point of time as well as the transformation dynamics of their characteristic dispositions. In this respect, we may specify the potential problem–solution duality $\left(\mathbf{D}_{\mathfrak{U}}\right)$ that resides in the potential solution pole $\left(\mathbf{P}_{\mathfrak{U}_{\mathfrak{S}}}\right)$ and the actual problem–solution duality $\left(\mathbf{D}_{\mathfrak{A}}\right)$ that resides in the actual problem pole $\left(\mathbf{P}_{\mathfrak{A}_{\mathfrak{P}}}\right)$ as a transformation functional relation that maps the sub-set of potential problem–solution dualities onto the sub-set of the actual variety problem–solution dualities on the basis of variety cost–benefit characteristic dispositions. It is a set-to-point mapping where an optimal potential variety solution is mapped onto the actual variety problem, where the categorial conversion process is an information-decision-choice action generated by constructing an appropriate *categorial moment* as guided by a social Philosophical Consciencism to affect a variety solution under cost–benefit rationality and the system's preferences that guide the preferred disequilibrium path but not the path of actual outcomes which may deviate from the preferred.

It must be noted that the system that is set up for the understanding of knowing as intellectual investment flows and information-knowledge accumulation as intellectual capital stocks must be understood as socio-natural activities over interlocking and over-lapping spaces and sub-space of dualities and polarities. The basic foundations are established by two inter-connected spaces. They is the space of actual-potential dualities and polarities for direct natural activities transforming the variety actuals to variety potentials and variety potentials to variety actuals. There is also the space of imagination-reality dualities and polarities for indirect social activities transforming variety reality to variety potential and variety imagination to variety reality. In these situations, we have the space of imagination-reality dualities and polarities as a sub-space of the space of actual potential dualities. The analytical structures are such that the potential problem–solution duality $(\mathbf{D}_{\mathfrak{U}})$ resides in the potential solution pole $(\mathbf{P}_{\mathfrak{U}_{\mathfrak{S}}})$ and the actual problem–solution duality $(\mathbf{D}_{\mathfrak{A}})$ resides in the actual problem pole $(\mathbf{P}_{\mathfrak{A}_{\mathfrak{P}}})$ for natural activities. Corresponding to these spaces are the sub-spaces for neuro-decision-choice activities. Corresponding to the potential problem–solution duality is imagination problem–solution duality $(\mathbf{D}_{\mathfrak{G}}) \subset (\mathbf{D}_{\mathfrak{U}})$ which resides in the imagination solution pole $(\mathbf{P}_{\mathfrak{G}_{\mathfrak{S}}}) \subset (\mathbf{P}_{\mathfrak{U}_{\mathfrak{P}}})$ and the real problem–solution duality $(\mathbf{D}_{\mathfrak{H}}) \subset (\mathbf{D}_{\mathfrak{A}})$ which resides in the real problem pole $(\mathbf{P}_{\mathfrak{H}_{\mathfrak{P}}}) \subset (\mathbf{P}_{\mathfrak{A}_{\mathfrak{P}}})$. The dualistic-polar conditions are such that $(\mathbf{D}_{\mathfrak{U}} \cup \mathbf{D}_{\mathfrak{A}}) = (\mathbf{D}_{\mathfrak{U}} \otimes \mathbf{D}_{\mathfrak{A}})$. $(\mathbf{P}_{\mathfrak{A}_{\mathfrak{P}}} \cup \mathbf{P}_{\mathfrak{U}_{\mathfrak{S}}}) = (\mathbf{P}_{\mathfrak{A}_{\mathfrak{P}}} \otimes \mathbf{P}_{\mathfrak{U}_{\mathfrak{S}}})$ with fuzzy-stochastic conditionality $(\mathbf{D}_{\mathfrak{U}} \cap \mathbf{D}_{\mathfrak{A}}) \neq \emptyset$ and $(\mathbf{P}_{\mathfrak{A}_{\mathfrak{P}}} \cap \mathbf{P}_{\mathfrak{U}_{\mathfrak{S}}}) \neq \emptyset$. Similarly, $(\mathbf{D}_{\mathfrak{H}} \cup \mathbf{D}_{\mathfrak{G}}) = (\mathbf{D}_{\mathfrak{H}} \otimes \mathbf{D}_{\mathfrak{G}})$ and $(\mathbf{P}_{\mathfrak{G}_{\mathfrak{S}}} \cup \mathbf{P}_{\mathfrak{H}_{\mathfrak{S}}}) = (\mathbf{P}_{\mathfrak{G}_{\mathfrak{S}}} \otimes \mathbf{P}_{\mathfrak{H}_{\mathfrak{S}}})$ with fuzzy-stochastic conditionality $(\mathbf{D}_{\mathfrak{H}} \cap \mathbf{D}_{\mathfrak{G}}) \neq \emptyset$ and $(\mathbf{P}_{\mathfrak{G}_{\mathfrak{S}}} \cap \mathbf{P}_{\mathfrak{H}_{\mathfrak{S}}}) \neq \emptyset$ where the dualistic-polar conditions and the fuzzy-stochastic conditionality ensure relational continuity and unity with imaginations connected to potentials and realities connected to actuals in knowing and information-knowledge processes. These structures allow us to examine the roles that creativity, curiosity, dreams, perseverance and many such cognitive actions play in knowing and information-knowledge development.

7.4.1 The Categorial Conversion Process and the Problem–Solution Dynamics

The theory that is being discussed in this monograph involves the totality of the human experience in all areas of socio-natural interactive systems of different environmental structures where life finds meaning in the space of problem–solution dualities involving all areas of human existence, and life also finds its essence in the way decision-choice actions relate to the elements in the space of the problem–solution dualities in the variety destruction-replacement processes in the system of socio-natural environmental structures and a system of variety destruction-replacement processes. The variety destruction-replacement processes are the foundation of the cognitive actions under the conditions of preferences and the principle of non-satiation within the space of necessity-freedom dualities all of which find meaning in the space of knowing and information-knowledge dynamics. To understand the dynamics of the general problem–solution process, knowing as an intellectual investment process and information-knowledge dynamics as the intellectual stock-flow process, consider Fig. 7.8, which presents the interactive relation of a variety problem (actual) and a number of variety solutions (potentials) where $(\mathbb{V}_{\mathfrak{A}_{\mathfrak{P}}})$ is the actual

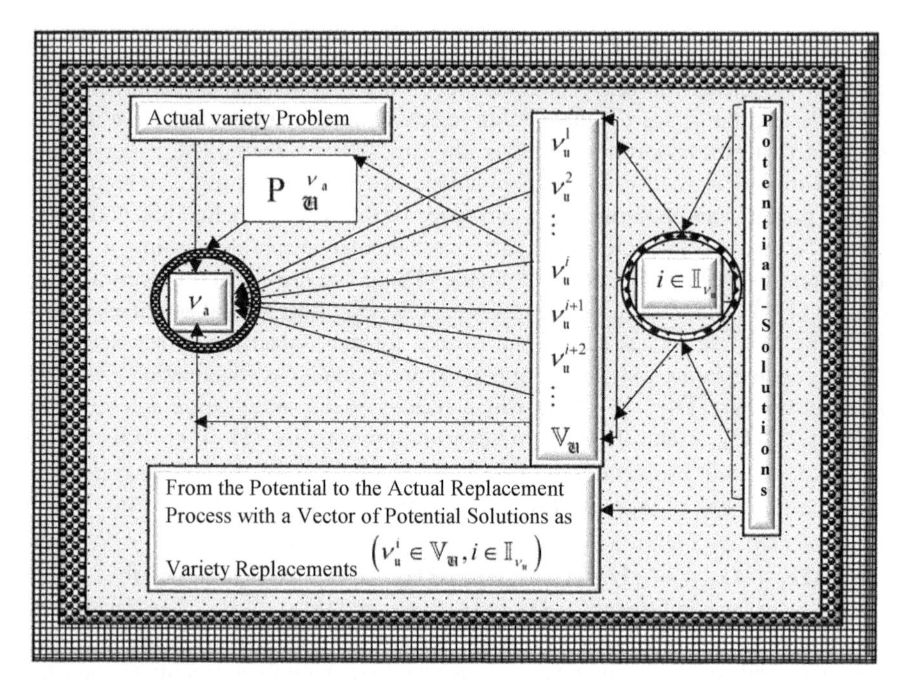

Fig. 7.8 A destruction-replacement geometry from the potential problem–solution polarities to the actual problem–solution polarities to the space of imagination-reality dualities, where the actual-potential conditions must be related to imagination-reality condition in the space of identification-transformation polarities. In the actual pole resides the actual duality and in the potential pole resides the potential duality and similarly, in the reality pole resides the reality duality and in the imagination pole reside the imagination duality

variety problem and corresponding to it is a set of potential variety solutions. Let there be four variety solutions of the form $\left\{\mathbb{V}^1_{\mathfrak{U}\mathfrak{S}}, \mathbb{V}^2_{\mathfrak{U}\mathfrak{S}}, \mathbb{V}^3_{\mathfrak{U}\mathfrak{S}}, \mathbb{V}^4_{\mathfrak{U}\mathfrak{S}}\right\}$.

The concern here is not about the techniques and methods of finding the variety solution to the variety problem in the problem–solution replacement process but about the conceptual framework of the problem–solution games. The techniques and methods will belong to the game theory and the theory of optimization. There are four actual-potential problem–solution polarities in relation to the actual variety problem of the form $\left\{\left(\mathbb{V}_{\mathfrak{A}\mathfrak{P}}, \mathbb{V}^1_{\mathfrak{U}\mathfrak{S}}\right), \left(\mathbb{V}_{\mathfrak{A}\mathfrak{P}}, \mathbb{V}^2_{\mathfrak{U}\mathfrak{S}}\right), \left(\mathbb{V}_{\mathfrak{A}\mathfrak{P}} \mathbb{V}^3_{\mathfrak{U}\mathfrak{S}}\right), \left(\mathbb{V}_{\mathfrak{A}\mathfrak{P}} \mathbb{V}^4_{\mathfrak{U}\mathfrak{S}}\right), i = 1, 2, 3, 4\right\}$, where there are four potential replacements from which an *optimal variety solution* is to be abstracted. The identities of these polarities are established by the potential cost–benefit dualities relative to the cost–benefit duality of the actual variety problem. Let us keep in mind that each of these polarities has a residing cost–benefit duality that may be represented in the fuzzy-stochastic information space with fuzzy-stochastic hybrid characteristic function and presented as a fuzzy decision problem with fuzzy optimization tools. The four actual-potential problem–solution polarities in relation to the actual variety problem may also be expressed in imagination-reality problem–solution polarities in relation to variety problem in the space of reality of the form $\left\{\left(\mathbb{V}_{\mathfrak{H}_{\mathfrak{P}}}, \mathbb{V}^{1}_{\mathfrak{G}_{\mathfrak{S}}}\right), \left(\mathbb{V}_{\mathfrak{H}_{\mathfrak{P}}}, \mathbb{V}^{2}_{\mathfrak{G}_{\mathfrak{S}}}\right), \left(\mathbb{V}_{\mathfrak{H}_{\mathfrak{P}}} \mathbb{V}^{3}_{\mathfrak{G}_{\mathfrak{S}}}\right), \left(\mathbb{V}_{\mathfrak{H}_{\mathfrak{P}}} \mathbb{V}^{4}_{\mathfrak{G}_{\mathfrak{S}}}\right), i = 1, 2, 3, 4\right\}$, where there are four variety imagination replacements from which an *optimal variety solution* is to be abstracted. The identities of these polarities are established by the potential cost–benefit dualities relative to the cost–benefit duality of the actual variety problem.

The relational structure of the actual-potential problem–solution polarity, as geometrically illustrated, showed in Fig. 7.1, is alternatively and explicitly amplified and illustrated with a cognitive geometry in Fig. 7.2 with an illustrative foundation of Fig. 7.1. The possible actual-potential problem–solution polarities are represented algebraically as:

$$\mathbf{P}^{\nu_{\mathfrak{a}}}_{\mathfrak{U}} = \left(\mathbf{P}_{\mathfrak{A}_{\mathfrak{P}}}, \mathbf{P}_{\mathfrak{U}_{\mathfrak{S}}}\right) = \left\{\nu_{\mathfrak{S}} = \left(\nu_{\mathfrak{a}}, \nu^{i}_{\mathfrak{u}}\right), |\nu_{\mathfrak{a}} \in \mathbb{V}_{\mathfrak{A}}, \nu^{i}_{\mathfrak{u}} \in \mathbb{V}_{\mathfrak{U}}, i \in \mathbb{I}_{\nu_{\mathfrak{u}}}\right\} \quad (7.3.1.1)$$

The corresponding imagination-reality problem–solution polarities as derivative from the possible actual-potential polarities are represented algebraically as:

$$\mathbf{P}^{\nu_{\mathfrak{h}}}_{\mathfrak{G}} = \left(\mathbf{P}_{\mathfrak{H}_{\mathfrak{P}}}, \mathbf{P}_{\mathfrak{G}_{\mathfrak{S}}}\right) = \left\{\nu_{\mathfrak{S}} = \left(\nu_{\mathfrak{h}}, \nu^{i}_{\mathfrak{g}}\right), |\nu_{\mathfrak{h}} \in \mathbb{V}_{\mathfrak{H}}, \nu^{i}_{\mathfrak{g}} \in \mathbb{V}_{\mathfrak{G}}, i \in \mathbb{I}_{\nu_{\mathfrak{g}}}\right\} \quad (7.3.1.2)$$

In the general decision-choice analytics, the dualistic-polar dynamics is such that every polarity is composed of a problem pole, defined by a problem duality representing the variety problem and a solution pole defined by a solution duality representing a variety solution where the problem and solution dualities have residing cost–benefit dualities as performance indexes. The variety problem is always one and known while the variety solution is unknown and many, from which a best selection is to be made to replace the variety problem. The decision-choice analytics are in relational connectivity to problem–solution analytics which are connected to cost–benefit analytics that are also connected to information-knowledge analytics and the dynamics of the time trinity of the past-present-future structure of the sankofa-anoma tradition that presents us with three dualities of past-present duality, present-future duality and past-future duality in relational continuum and unity in knowing and information-knowledge production. The number of problem-soluti on polarities for any given actual problem variety is the same as $\left(\#\mathbb{I}_{\nu_{\mathfrak{u}}}\right)$ which specifies the number of elements in the admissible set of decision-choice actions over the space of possibility-probability dualities and its derived space of necessity-freedom dualities.

7.4.2 Necessity-Freedom Analytics and the Categorial Conversion Process

The cognitive actions over the space of problem–solution polarities require the development of categorial converters that convert variety problems into variety solutions and variety solutions into variety problems over the primary-derived dualities under parent-solution dynamics where the rise of variety problems are usually unintentional and hold conditions of necessity while the variety solutions are intentional and hold conditions of freedom. The structure of any categorial convertor whether a solution convertor or a problem convertor depends on the characteristic disposition of the problem–solution duality while the corresponding categorial moment depends on the social Philosophical Consciencism that controls the socio-cultural dimensions of *curiosity*, freedom, variety preferences, information-knowledge search and the degrees of individual social participation, collective decision-choice activities over the action space, social constraints on curiosity and freedom of knowing, and the efficiency of the social variety categorial conversions.

Any categorial problem–solution convertor is an actual-potential variety transformation function that relates a variety solution to a variety problem. Over the epistemological space, the convertor activities are cognitive in structure which creates variety solutions in replacement of variety problems as well as conditions for the rise of new problems from solutions in the *transformation-substitution dynamics* where there are always transformation games between the existing variety as a problem–solution duality and several potential problem–solution varieties as contestants to replace the existing variety which is considered as a problem. Any variety problem creates an opportunity for a new variety as a solution while any variety solution creates an opportunity for the rise of a new problem in space of problem–solution dualities.

General analytics in terms of progress, qualitative and quantitative motions, synergy, complexity and transformations suggest that the solutions to variety identification problems initialize cognitive searches for the information-knowledge solution conditions of variety transformation problems. The transformation-substitution dynamics are also the problem–solution dynamics and ignorance-knowledge dynamics over the space of problem–solution dualities, as well as the space of cost–benefit dynamics in the social decision-choice space relative to the necessity-freedom dynamics over the space of possibility-probability dualities contained in the space of imagination-reality dualities. These convertor activities are not limited to one area of knowing. The convertor activities of the replacement process of variety solutions to variety problems and variety problems to variety solutions are the unifying force of all areas of knowing irrespective of the methods and judgmental acquaintance activities to establish knowledge by acquaintance and knowledge by description where information-knowledge conditions exist in sectorial relational continua and unity as input–output in social disequilibrium dynamics over the space of production-consumption dualities under the conditions of the necessity-freedom principles of transformation.

The necessity-freedom principle must be viewed in terms of universal conditions over the space of neuro-decision-choice activities over the space of imagination-reality dualities, also defines the categorial-conversion-philosophical consciencism principle where the sub-structure of the categorial conversion establishes necessary conditions for variety transformation while the sub-structure of the philosophical consciencism establishes the freedom and sufficient conditions for variety transformations in the sense that the philosophical consciencism provide the decision-choice power for neuro-decision action.

Different areas of knowing involve different set of problem–solution dualities as defined by their corresponding variety characteristic dispositions, and hence will require different sets of *problem–solution convertors* with different sets of categorial conversion moments and different social efficiencies due to the nature of the variety characteristic-signal dispositions that present different sets of problem–solution dualities. It is the same as saying that different decision-choice actions, transformation technologies and mathematical representations will be required for decision-choice activities in different areas of knowing since their different characteristic dispositions will influence the choice of the categorial convertor. All the actual and potential areas of knowing are united by the *principle of problem–solution dualities* with relational continua and unity. The knowledge by acquaintance or the experiential information will always be affected by the nature of the variety signal disposition that corresponds to the variety characteristic dispositions as well as the instrument of acquaintance or observation in the space of source–destination dualities, where the knowledge by acquaintance and knowledge by description are abstraction from the system of signal disposition under a paradigm of thought. The structure of the dynamics of actual-potential polarities for the problem-solving processes as variety destruction-replacement processes is geometrically given in Fig. 7.8, where the actual variety problem is destroyed leaving behind the information knowledge conditions of its birth and existence and replaced with new actual variety solution from the space of imagination as a sub-space of the potential space with new information-knowledge conditions as additions to information-knowledge stocks.

Some analytical observations may be made at this point. The internal duality, composed of negative and positive characteristic sub-sets as the variety duals, defines the identity of any variety and hence every variety presents itself as either a problem or a solution through the corresponding cost–benefit structures relative to variety characteristic-dispositions. The contents as a variety problem or a variety solution are established by the relative characteristic dispositions while the variety problem or the variety solution is known through the corresponding signal dispositions by neuro-decision-choice actions through cognitive abstractions from the space of imaginations. The designation of a variety as either a problem or a solution is established by cognitive agents by transforming the received problem–solution signal dispositions into cost–benefit dispositions in terms of the cost–benefit relation relative to a goal-objective element. The external duality is a dualistic-polar relationship between the actual (reality) variety problem and a potential (imagination) variety solution to form a problem–solution polarity $\left(\mathbf{P}_{\mathfrak{A}}^{\nu_{\mathfrak{a}}}\right)$ or $\left(\mathbf{P}_{\mathfrak{G}}^{\nu_{\mathfrak{h}}}\right)$. The cognitive activities over the space of problem–solution dualities with dualistic-polar conditions unite all areas of knowing including categories of science and non-science into one information-knowledge system in relational intersectionality and unionization of diversities.

From the definitions of the concepts of problem–solution polarity and duality, we may now state the principles of the unity of knowing and science. There are two inter-supportive principles for the unity of knowing and the unity of science. There is the *primary principle* and the *derived principle* based on the *general principle of opposites* where the universe is seen as an integrated matter-energy-information which contains varieties and categorial varieties on the basis of which all languages are formed and developed as the size of the information-knowledge accumulation expands with increasing cognition in the space of acquaintances and improvements in the paradigms of thought in the space descriptions. These varieties and categorial varieties are in diversities, relational interdependence and unity that constitute the universal system of ontological existence under the principle of opposites for which the epistemological existence is a simple derivative by the collective decision-choice actions.

Epilogue

The Epistemic Directions of Future Research on Problem–Solution Dualities and Polarities

Abstract The epilogue concludes the monograph with views and reflections for further research and development in the understanding of the interactions of the spaces of neuro-decision-choice actions, problem–solution dualities, knowing and information-knowledge stock-flow disequilibrium dynamics relative to matter-energy stock-flow equilibrium dynamics under the principles of diversity, unity and opposites with relational continua. The directions of the role that institutions and organizations play in knowing and the knowledge processes are discussed in relation to information-knowledge processes as a systems of input–output processes and construction-destruction processes under a system of neuro-decision-choice actions with *cognitive capacity limitations* composed of *information-knowledge capacity limitations* and *paradigmatic capacity limitations*, where goals, objectives and visions are defined and expressed in the space of imagination-reality dualities as well as the manner in which the space of imagination-reality dualities is contained in the space of actual-potential dualities for the understanding of the general human behavior and cognitive actions over the space of production-consumption dualities, where preferences are governed by the principle of non-satiation over the space of varieties relative to the space of imagination-reality dualities. The epilogue is concluded with discussions on the directions for the examination of imagination, reality, belief, hope, goals, objectives, visions, curiosity and the relevant methodology in knowing and information-knowledge stock-flow dynamics through the space of problem–solution dualities under the principle of philosophical consciencism with the discussions on categorial inter-relationalities for the development of future research universities.

K. K. Dompere, *The Theory of Problem-Solution Dualities and Polarities*, Studies in Systems, Decision and Control 405,
https://doi.org/10.1007/978-3-030-90279-7

Concluding Reflections

The monograph is concluded with a general framework for utilizing the theory of problem–solution dualities and polarities in understanding the stock-flow conditions of information-knowledge disequilibrium dynamics, principles of diversity and the unity of knowing and science from the logic of economic theory of production. The conditions of the principles of diversity and unity of knowing point to the directions for the establishment of two sets of information-knowledge systems for the fundamental and applied knowing and knowledge, and corresponding to them are the two sets of institutions for the development and maintenance of the social organization for neuro-decision-choice activities over the space of production-consumption dualities. The two sets of institutions are the set of institutions for decision-choice actions on the space of fundamental problem–solution dualities to produce the fundamental information-knowledge system, and a set of institutions for neuro-decision-choice actions on the space of applied problem–solution dualities to produce the applied information-knowledge system. This approach of viewing the social organism in terms of information-knowledge systems allows one to see the dualistic-polar conditions of institutional diversity and unity, where the institutional diversities are directed to deal with categorial problem–solution dualities in fundamental and applied neuro-decision-choice actions, and where the unity is directed to pull together categorial diversities for decision-choice actions on the space of categorial unity problem–solution dualities into instruments for managing the controls and the directions of variety transformations of the social organism.

The unity, further, allows us to investigate and understand the elemental neuro-decision-choice behaviors of individual-community dualities (anoma-kokone-kone problem) over the space of problem–solution dualities, where the individual is simultaneously a real cost and a real benefit to the community and the community is simultaneously a real cost and a real benefit to the individual to maintain the principle of relational cost–benefit duality in zones of conflicts over the space of freedom in individual-community dualities. In the respect of human general production activities in all sectors of the social set-up, there are individual skills and abilities and collective skills and abilities in neuro-decision-choice actions over the space of the problem–solution dualities with differential outcomes and results. The unity finds expressions in organization and sub-organizations which are used to exploit the categorial diversities in the general neuro-decision-choice actions over the space of problem–solution dualities. The diversities find expressions in the distribution of the socio-natural characteristic dispositions over the space of varieties. Unity in the intellectual space is seen as relaxing the individual cognitive capacity limitations and expanding the contributory zone of individual actions while expanding the collective intelligence in knowing relative to social rationality over the space of problem–solution dualities. At the level of the space of problem–solution dualities, the concept of diversity finds expression and meaning in *categorial specializations* while the concept of unity finds expression and meaning in progressively inter-categorial supports, dependencies and inter-changes to bring about types of categorial specializations into the relational give-and-take sharing modes to maintain the conditions of internal and external mutual existence and destructions in the space of variety construction-destruction dualities, where over the epistemological space the elements in the destruction-construction dualities obey the social law of cost–benefit dynamics in continuity and at the ontological space the elements in the space of destruction-construction dualities obey the laws of continuity, ecological stability and creative dynamics.

In all of these processes in the infinite universe, given the matter-energy stock-flow equilibrium as establishing the permanency of the universe, and information-knowledge stock-flow disequilibrium as established by the continual variety transformations within the matter-energy statics and information-knowledge dynamics, nothing is created without something being destroyed, nothing is destroyed without something being created, there is no real benefit without real cost, there is no knowledge without ignorance, there is no change without conflict, there is no conflict without tension, there is no tension without force, there is no force without energy and there is no energy without matter in the space of static-dynamic dualities defined over the space of quality-quantity dualities. The permanency of the universe finds definition and expressions within the matter-energy existence which provides input–output varieties for its self-maintenance, while change in the universal existence finds definition and expressions with information-knowledge existence which provides input–output characteristic dispositions for its continual disequilibrium maintenance over the spaces of varieties and the information-knowledge stock-flows, while a change in ignorance finds definition and expressions over the space of expanding varieties in the space of destruction-construction dualities over the space of dividedness-consolidation dualities and polarities.

A Note on Institutions and the Organization of Knowing

The production of information-knowledge intellectual stocks supported by knowing as intellectual investment must be organized where the production effectiveness will depend on institutions and continual institutional creations to service the multiplicity of complex neuro-decision-choice actions. The institutions through which individual-community neuro-decision-choice actions are organized for fundamental problem–solution dualities to produce the fundamental information-knowledge system are housed in the educational sector which derives its power of creation from the governmental power system operating with rules and regulations as socially established in the legal structure and maintained by a system of enforcement processes by the political processes. Similarly, the institutions through which individual-community neuro-decision-choice actions are organized for applied problem–solution dualities to produce the applied information-knowledge system are housed in the business sector which also derives its social power of creation from the governmental power system operating with rules and regulations as socially established in the legal structure and maintained by a system of enforcement processes by the same political process.

The character of any organization depends on the creative interactions of leadership and institutions through which decision-choice actions are implemented while the strength of social institutions depends on the leadership, its personality and neuro-decision-choice actions over the space of thinking and reasoning as applied over the space of problem–solution dualities. The outcomes of the decision-choice actions and the rise of new problem–solution dualities depend on information-knowledge inputs, the logic of thought, the collective personality of the leadership and the administration staff, and their integrity with individual and collective actions.

The social unity of the progress of knowing and information-knowledge dynamics in the space of stability requires a continual organization and reorganization casting institutions within the space of construction-destruction dualities of cognitive diversities in the social space of input–output resource distribution and to develop organizational laws and rules of fairness relative to the elements in the space of categories of problem–solution dualities and the management of intra-relationships and inter-relationships of contents and social varieties. The organizational laws and rules of fairness, equity and equality are related to the conditions over the space of real cost–benefit dualities. There are three important mega-sectorial organizations contained in the basic institutional structures of economics, politics and law. These mega-organizations are the *educational organization*, the *business organization* and the *governmental organization* which are all involved in knowing and information-knowledge production in support of the management and control of the social set-up.

The educational organization produces fundamental knowledge broadly defined as inputs. The business organization produces applied knowledge by using the fundamental knowledge as input to generate outputs. The governmental organization produces rules and laws to coordinate and manage the organs with inputs from the fundamental and applied knowledge systems as well as provides help to the organs whenever there are instabilities to maintain the information-knowledge stock-flow disequilibrium processes relative to production-consumption disequilibrium processes which are also relative to input–output disequilibrium processes. Each mega-sector is directed to deal with a family of categories of problem–solution dualities which constrains the individual and collective neuro-decision-choice actions over the space of individual-collective dualities in relation to the general space of problem–solution dualities and information-knowledge processes, where the information-knowledge contents are additions to the general information-knowledge contents of the philosophical consciencism which harbors systemic good and evil in a dualistic-polar structure.

It is through the ruling philosophical consciencism at any generation timepoint that one can find explanations to social decay and social progress over the space of imagination-reality dualities, where cognitive agents operate where the space of imagination-reality dualities is a subspace of actual-potential dualities where nature operates. It is through the ruling philosophical consciencism that one can find the relevance and non-relevance of critical social theories that have given rise to some theories such as the critical race theory, the critical feminist theory as the historic understandings from the telescopic past to the telescopic present to shape the telescopic future over the space of epistemic feedback processes in relation to social systems as self-exiting and self-correcting living entities induced by neuro-decision-choice systems operating under cognitive capacity limitations. The critical theory is embedded in the Sankofa tradition that connects telescopic past to the telescopic present and provides information-knowledge inputs to construct the telescopic future. The feedback and self-correction process require that the information knowledge inputs must be correct as to the best of cognitive capacity.

The system of variety transformations is such the every variety in it is simultaneously an output from a primary input as well as an input into a process to generate an output. In other words, the system of varieties is infinitely closed under transformation and identification. Every social progress depends on the ability to produce and use information-knowledge inputs, preference ordering of the leadership, its personality, decision-choice actions, and the people's collective support under systems organizations with its institutional configuration for problem–solution dualities where variety solutions are set against variety problems in the space of chain problem–solution interconnectedness where problem–solution mimics input–output dynamics. Under the organizational considerations, the epilogue is used to discuss the frameworks for which the social setup may be organized to provide the best institutional arrangements to serve the information-knowledge needs for individual and social neuro-decision-choice actions over the space of problem–solution dualities to bring about optimal destruction-replacement processes over the intergenerational enveloping paths for executing the infinite production-consumption and input–output processes over the fundamental and applied information-knowledge systems as a system of a family of variety production-consumption dynamics.

Production-Consumption and Input–Output Processes of an Information-Knowledge System

The governmental neuro-decision-choice actions on the space of politico-legal problem–solution dualities will affect both the efficiencies of all activities in the education and business sectors and their interdependencies towards the path of stock-flow disequilibrium dynamics of the general information-knowledge system through their effects on the boundaries of neuro-decision-choice behavior in the space of individual-collective dualities and variety transformations defined by changes in the characteristic-signal dispositions. Both the fundamental and applied information-knowledge systems are socio-economic productions, the results of which are generated by individual and collective neuro-decision-choice actions working in diversity and unity over the economic structure to become socio-economic varieties of elements in the space of socio-economic problem–solution dualities guided by cost–benefit processes.

The educational production of socio-economic varieties involves the creation of the fundamental information-knowledge systems as outputs which become inputs into the production of applied information-knowledge systems. The business production of socioeconomic varieties involves the creation of applied information-knowledge systems as outputs which then become inputs into the production of the fundamental information-knowledge systems and variety production-consumption systems in the space change of real-nominal dualities. The information-knowledge outputs become inputs into the governmental decision-choice processing system, the outcomes of which become inputs into educational, business and governmental activities in the action space of variety identification-transformation dualities. The efficiencies of the neuro-decision-choice actions in the educational production and the business production depend on the general creative understanding and interdependent use of the education sector, business sector and judicious creation of rules, laws, regulations and policies of the decision-choice activities from the governmental sector to control and manage the social organism with elements from the intellectual capital accumulation given the population structure since these rules, laws, regulations and policies are constraints on the action space of thoughts and practices given the space of all preferences. It is here, in the space of production-consumption dualities of individual and social existence, that governmental governance in the framework of a social organism may work to either facilitate or retard social progress over the space of input–output dualities relative to the decision-choice actions over the space of problem–solution dualities and polarities in reference to research, teaching, learning, knowing and information-knowledge disequilibrium processes as they affect the behavior of economic, political and legal structures in human pursuit of happiness locked in the space of necessity-freedom dualities generated by the space of possibility-probability dualities..

The critical understanding of knowing and stock-flow dynamics of information-knowledge production is also a critical understanding of **the theory** and applied economic processes in the space of input–output dynamics with information-knowledge stock-flow disequilibrium dynamics under the spaces of variety problem–solution dualities and destruction-construction dualities relative to the space of paradigms of thought that harbors created logics of thought in developing thinking algorithms which become guides to reasoning over the space of variety problem–solution dualities, setting the variety solutions against the variety problems relative to the individual and/or collective visions with goal-objective elements in the space of goals and objectives under social and inter-generational preference orderings that are defined over the space of cost–benefit dualities. All neuro-decision-choice actions are variety productions through transformations over the space of quality-quantity dualities where things give way as costs for the rise of other things as benefits. The construction-destruction dynamics are such that variety costs may become variety benefits as time moves on and information-knowledge expands with variety destructions and constructions. It is also here that thinking, reasoning and research must examine the relationships between info-statics and info-dynamics on one hand with thermostatics and thermodynamics on the other hand.

The space of variety problem–solution dualities may be related to our understanding of medical sciences and the development of a theory of medical decisions based on the principles of opposites, fuzzy information-knowledge structure and the fuzzy paradigm of thought. Medical decisions are about neuro-decision-choice actions on the sub-space of variety problem–solution dualities that relates to conditions of human health. The basic structure of the theory relates to the idea that the human body is conceived as a macro-duality which is composed of two systems of micro-dualities, where each micro-duality in each of the system is under continual process of destruction-construction duality of deterioration-rejuvenation dynamics. Furthermore, the micro-dualities may also be grouped under a sub-system of key parts, and a sub-system of supporting parts. Each micro-duality is composed of a negative micro-dual with a corresponding characteristic disposition and a positive micro-dual with a corresponding characteristic disposition creating internal micro-conflicts in destruction-construction processes in give-and-take sharing modes for mutual existence and non-existence. The internal negative micro-dual is in a destructive and deterioration mode of the part and the internal positive dual is in a construction and rejuvenation mode of the same part.

The collection of all the negative micro-duals constitutes the negative macro-dual, while the collection of all the positive micro-duals constitutes the positive macro-dual in the space of life-death dualities. The positive macro dual and the negative macro dual constitute the duality of human individuality with deferential distributions of strengths, endowments, energies and immune systems of the macro-dualities to establish a unique set of characteristics in response to the common socio-natural environment. The concepts of negative and positive used to qualify dual of duality must be distinguished from an applicational test of infections where negative and positive are indications of the status of infectious entry into the body. The understanding of the difference of the use of these concepts is important fuzzy decision-choice analytics. This is part of the theory of recycling where old forms give way to new forms and new forms give way to old forms in the space of production-consumption dualities in variety disequilibrium dynamics.

This unique set of characteristics is shared by different people in a group which allows the population to be portioned into categories with similar genetic characteristics (coding) in their responses to the common socio-natural environment. The elements in the genetic categories have similar fuzzy socio-natural responses to similar illnesses, deceases, viruses and other identifiable ailments. Each genetic category has differential responses to medical treatments as well as pharmaceutical treatments in that one size fits all medical practices must be abolished under information-knowledge rationality. It is this possible genetic partition of the population that suggests the path for the development of a *theory of targeted medical practices* and neuro-decision-choice actions over the space of medical problem–solution dualities. The input–output processes of the neuro-decision-choice actions are generated by fuzzy-stochastic information-knowledge systems on a system of dualities with relational continuum and unity under give-and-take sharing processes constrained by cost–benefit conditions over the space of relativity with the use of the fuzzy-paradigm of thought to abstract cost–benefit balances for variety destruction-replacement actions over the space of problem–solution dualities.

The development of a theory of *targeted* medical neuro-decision-choice actions over the space of medical problem–solution dualities with the fuzzy-stochastic information and the use of fuzzy paradigm may have some difficulties, but it is hoped to revolutionize medical and pharmaceutical practices over the space of real cost–benefit dualities where side effects are real costs and health improvements are benefits. In the targeted medical neuro-decision-choice actions, the population is classified in accord with their genetic and other characteristic dispositions. I hope this theoretical framework will be carried on in another monograph that may be devoted to a fuzzy theory of medical decisions. We must understand the concepts of ontological past, present and future information processes as inputs into neuro-decision-choice action. For example, let us examine the history of natural events of pandemics, where the historical accuracy may be useful in dealing with future pandemics by increasing the benefits of neuro-decision-choice actions, while historical inaccuracies and the development of conspiracy theories may increase the costs of neuro-decision-choice actions in dealing with future pandemics.

A Prelude to the Theory of Neuro-Decision-Choice Systems

The general explanatory and prescriptive processes of all human actions must be developed as the theory of neuro-decision-choice systems which will explain, predict and prescribe human actions over the space of problem–solution dualities where the problem–solution dynamics are directed towards achieving greater net benefit. This will affect national and international knowing and information-knowledge systems broadly defined to include theory and applications. It is from this general position that we should understand diversities and the unity of human decision-choice activities of our social universe and the problem–solution dynamics under the principle of opposites for continual dualistic-polar transformations of war-peace dualities, production mimicries of natural dualities, love-hate dualities, life-death dualities and many other dualistic-polar structures, where human neuro-decision-choice actions are constrained by conditions of cognitive capacity limitations, limitativeness and limitationality in relation to information and the paradigm of thought, and where social tensions are generated by distributional conflicts of cost–benefit conditions of outcomes, where matter and energy may be interchangeably cost and benefit in the space of transformations and in the space of knowing and information-capital accumulation under principles of categorial conversion and philosophical consciencism.

Definition IVA: Neuro-Decision-Choice and Neuro-Decision-Choice Space Let the general *neuro-decision-choice space* be represented by ($\mathfrak{D}$) with a generic element ($\mathfrak{d} \in \mathfrak{D}$), the general problem space be ($\mathfrak{P}$) with a generic element ($\mathfrak{p} \in \mathfrak{P}$), a general solution space ($\mathfrak{S}$) with a generic element ($\mathfrak{s} \in \mathfrak{S}$) and a neurological space ($\mathfrak{N}$) with a generic element ($\mathfrak{n} \in \mathfrak{N}$). Then the general *neuro-decision-choice space* is defined as the union of the general problem space, the general solution space and the general neurological space such that $\{(\mathfrak{D} = \mathfrak{P} \cup \mathfrak{S} \cup \mathfrak{N})\}$, and if $(\mathfrak{p} \in \mathfrak{P}|\mathfrak{N})$, then the solution belongs to the intersection of the problem space and the solution space in the form $\{(\mathfrak{s} \in \mathfrak{P} \cap \mathfrak{S})|\mathfrak{n} \in \mathfrak{N}\}$. Similarly, if $(\mathfrak{s} \in \mathfrak{S})$ then the problem belongs to the intersection of the problem space and the solution space such that $\{(\mathfrak{p} \in \mathfrak{P} \cap \mathfrak{S})|\mathfrak{n} \in \mathfrak{N}\}$. In this respect, given the space of real cost–benefit relativity conditions ($\mathfrak{r} \in \mathfrak{R}$) the *neuro-decision-choice space* is analytically $\mathfrak{D} = \{\mathfrak{d}(\mathfrak{p}, \mathfrak{s}, \mathfrak{n}|\mathfrak{r})|\mathfrak{p} \in \mathfrak{P}, \mathfrak{s} \in \mathfrak{S}, \mathfrak{n} \in \mathfrak{N}, \mathfrak{r} \in \mathfrak{R}\}$ and the space of problem–solution dualities ($\mathcal{R}$) is analytically the union of the *problem space* ($\mathfrak{P}$) and the *solution space* ($\mathfrak{S}$), which is ($\mathcal{R} = (\mathfrak{P} \cup \mathfrak{S}) \subseteq \mathfrak{C}$) with a fuzzy conditionality of the form ($\mathfrak{P} \cap \mathfrak{S} \neq \emptyset$) which is a non-zero intersectionality and ($\mathfrak{R}$) is the *relative cost–benefit space.*

Note IVA: Definition of Neuro-Decision-Choice Action A decision-choice activity is a mental action over the problem–solution space to reconcile the conflicts over the real cost and benefit information structures as revealed by the inherent negative–positive characteristic dispositions in an alternative actual and potential varieties to solve the ranking and selection problems in the cost–benefit and criterion spaces to bring about transformations and new information outputs. It is also a cognitive action in the problem–solution space which encompasses all aspects of human existence over the actual and potential spaces in the ontological space through the epistemological space. The mental actions are derived through a complex neurological system from the neurological space that allows comparative analytics of qualitative and quantitative characteristics for assessments and ranking of varieties. The collection of all the neuro-decision-choice actions on negative–positive dualities from the space of cost–benefit dualities constitutes the general neuro-decision-choice space which is a union of the *general problem space, the general solution space* and the *neurological space* given the *criterion* and *vision spaces*. The problems, solutions, preferences, and decision-choice activities are all defined in the space of varieties where each variety is simultaneously a cost and a benefit, and where the relative cost–benefit information becomes an information input in the space of variety assessments and ranking under an individual, collective and social preference schemes with decision-choice activities under deferential cognitive capacity limitations.

The neuro-decision-choice space is not only a collection of decision-choice actions but a collection of action processes that define the path of social histories under individual and collective cognitive capacity limitations. The questions that arise are simply about what cognitive capacity limitations are, and what are their manifestations, and how can they be reduced or eliminated to increase the efficiencies of decision-choice outcomes relative to either the goal-objective elements or a vision where the establishment of a vision and goal-objective elements belong to the space of problem–solution dualities. The cognitive capacity limitations are seen in terms of information-knowledge deficiencies and/or paradigm deficiencies in thinking and reasoning over the space of problem–solution dualities with a set of goal-objective elements toward a vision. The information-knowledge deficiency is an input limitation, while paradigmatic deficiency is a processing limitation creating thinking and reasoning difficulties. The directional dispositions to these questions will be provided through definitions which will extend the vocabulary of the neuro-decision-choice language for exposition and communication.

Definition IVB: Cognitive Capacity Limitation A neuro-decision-choice system, ($\mathfrak{D}$), is said to be under cognitive capacity limitations if the expansion of the supporting information-knowledge structure in the *information field* is necessary and sufficient to obtain the desired goal-objective element and changes of the paradigm of thought are necessary and sufficient to arrive at the zone of correctness and absoluteness of the claims without epistemic fuzzy-stochastic conditionality where the information field is defined to be consistent with characteristic-signal disposition.

Definition IVC: Information-Knowledge Capacity Limitations A neuro-decision-choice system, ($\mathfrak{D}$), under cognitive capacity limitations is said to be information-knowledge limitational (limitative) if the expansion of the supporting information structure from the information field is necessary (sufficient) to obtain the desired goal-objective element toward a vision through correct paradigmatic application.

Definition IVD. Paradigmatic Capacity Limitations A neuro-decision-choice system, ($\mathfrak{D}$), under cognitive capacity limitations is said to be *paradigmatic limitational (limitative)* if changes in the paradigm of thought for information-knowledge processing in decision-choice actions are necessary (sufficient) to arrive at the zone of correctness and absoluteness of the claims where the paradigm of thought is equivalent to methodology containing techniques and methods.

Note IVB: On Definitions of Cognitive Capacity Limitations There are two ways of looking at the concept of cognitive limitations on the space of decision-choice actions. The two ways relate to the necessary and sufficient conditions of outcomes, where necessity relates to initial conditions and anticipations over the possibility space and sufficiency relates to freedom and expectations over the probability space while necessity-freedom dualities relate to possibility-probability dualities in decision-choice actions with outcomes constrained by fuzzy-stochastic conditionality in the space of relativity. The fuzzy-stochastic conditionality indicates the applicable and acceptable zone of claims with error conditionality established by fuzzy-stochastic conditionality with defined conditions of epistemic transversality conditions. The limitationality indicates requirements of necessity, and the limitativeness indicates requirements for sufficiency in relation to inputs and processing to achieve outcomes toward a vision in the neuro-decision-choice system. In some situations, the neuro-decision-choice system may be information-knowledge complete but paradigm deficient. It may also be paradigmatically complete but information-knowledge deficient. It may even be both information-knowledge and paradigm deficient. These conditions must be considered in the development of the theory of neuro-decision-choice systems as general theory of human actions with subjective judgements over the space of problem–solution dualities towards the understanding of the knowing process and information-knowledge accumulation in the general stock-flow disequilibrium dynamics and matter-energy equilibrium states over the space of ontological-epistemological dualities. The knowing process and information-capital accumulation must be placed in the general space of economic production-consumption dualities and then the epistemic need to study the socio-natural variety dynamics, categorial dynamics and the theory of dividedness in relation to the dynamics of intra-generational and inter-generation preferences must be made clear.

The theory of human actions is the theory of neuro-decision-choice actions over the space of problem–solution dualities relative to the space of variety cost–benefit dualities, where every variety resides in the space of negative–positive dualities which is embedded in the space of cost–benefit dualities. The space of cost–benefit dualities is the space of incentives for neuro-decision-choice actions over the space of identification-transformation dualities. All human actions over the space of static-dynamic dualities are seen in the two states of motions as changes in relations or stay in relations, and changes in properties or stay in properties induced by decision-choice activities under some preference ordering and real cost–benefit (pain-pleasure) conditions in duality and polarity. Both variety relations and properties within the space of the variety static-dynamic dualities characterize all decision-choice activities of variety destruction-replacement processes. Variety identification is a change in relations between ignorance and knowing that presents neuro-decision-choice actions in the space of variety identification problem–solution dualities, while variety transformation is a change in properties between ignorance and knowing that present decision-choice actions in the space of variety transformation problem–solution dualities.

All the knowing processes as intellectual investment flows and all the information-knowledge accumulations as intellectual capital stocks are about conditions of variety identities and variety transformations. In other words, the theories of knowing and information-knowledge accumulation are about *what there was, what there is* (past-present dualities) and what would be (present-future dualities) of information-knowledge stocks The space of variety identification-transformation structures of problem–solution dualities is central in knowing and information-knowledge accumulation under the neuro-decision-choice activities, where outcomes establish the enveloping of *past social history, current social history and future social history* in all areas of human actions where history is seen within the space of success-failure dualities in relation to the elements in the goal-objective space. The variety real cost–benefit dispositions provide conditions for variety ranking while the neurological structures relate the socio-natural preferences to variety real cost–benefit dispositions relative to the elements in the space of goals and objectives on the path to a vision on the space of absoluteness under the principle of non-satiation. The variety real cost–benefit dispositions incentivize the variety neuro-decision-choice actions in the space of relativity of dualistic-polar conditions over the epistemological space.

In decision-choice processes, the negative dual and the positive dual reside in the space cost–benefit dualities relative to the elements in the space of goals and objectives which provide dualistic-polar conditions to establish individual and social visions. The conditions of the real cost–benefit duality point to the idea that a new reflection must be brought regarding the way we conceptualize problems and solutions as well as decision-choice actions, the theory of knowing and the theory of knowledge. The concept of the real cost–benefit duality must be broadly interpreted where for example, failure-success duality is seen in the same light as pain-pleasure duality or gain–loss duality and many others. This approach allows us to view human decision-choice actions in terms of problem–solution interplays where a problem is a net cost and a solution is a net benefit. The development of the neuro-decision-choice theory must, therefore, see all human decision-choice actions in terms of variety destruction-replacement processes under economic conditions of production-consumption and reproduction of varieties from the variety space. This is also applicable to knowing, science, non-science, our understanding of knowledge and the theory of cognition in the areas of theory and application in the space of production-consumption dualities.

At the level of neuro-decision-choice, variety knowing is an intellectual investment process and information-knowledge accumulation is an intellectual capital accumulation process which constitutes accumulated knowledge about variety identities [671], and the accumulated technological transformation processes [670, 672] in unity and diversity over the space of socio-economic activities. Every neuro-decision-choice activity is an action on variety problem–solution duality. Every problem–solution variety is either an identification problem or a transformation problem or both over the space of static-dynamic dualities. The identification problem is a property-identification problem or relation-identification problem or both. The transformation problem is a change in the property problem or a change in the relation problem or both. These changes involve real cost–benefit comparative actions over the space of actual-potential dualities where the actual variety problem–solution duality is compared to potential problem–solution dualities in terms of relative cost–benefit standing. This relational structure gives some ideas for the development of the theory of neuro-decision-choice systems that encompasses all areas of human decision-choice actions and actions over the space of variety problem–solution dualities under cost–benefit conditions as incentive over the space of neuro-decision-choice dynamics.

We have provided definitions for some essential concepts used in the monograph wherever it is thought that there is some deficiency in clarity of their uses in technical communications. However, over the space of understanding of decision-choice actions and problem–solution dualities, there is some confusion among the concepts of goals, objectives, mission and vision. The concepts of goals, objectives and goal-objective space have, therefore, been defined. What are the concepts of mission and vision and what are the similarities and differences and how do they relate to the concepts of goals and objectives and knowing in the conceptualized system? It is also maintained that every definition must obey the principle of identity over the space of characteristics relative to the real-nominal existence.

Definition IVE: Vision Vision is a desirably potentially organic variety as currently conceived to be an ultimate variety relative to the current state, to which there is a sequence of action drives to overcome a series of problem–solution dualities in its way by conducting a sequence of missions to conquer the goals and objectives that will lead to the vision as conceived. It is a conceptualization of variety in the potential space to be actualized in the space of actuals through a series of decision-choice actions over the space of the problem–solution dualities relative to the space of imagination-reality dualities passing through the space of possibility-probability dualities. The collection of all goal-objective enablers and the visions is the *visionary structure.*

Definition IVF: The Spaces of the Actual and the Potential The space of actuals (potentials) is a collection of varieties where the actual (potential) characteristic dispositions dominate the potential (actual)characteristic dispositions in dualistic-polar structures, where every actual variety is destructible and turned into a potential variety and every potential variety is constructible and turned into an actual variety through transformation processes such that each variety is seen as residing in the space of actual-potential dualities.

Definition IVG: The Spaces of the Imagination and the Reality The space of reality (imagination) is a collection of varieties where the reality (imagination) characteristic dispositions dominate the imagination (reality) characteristic dispositions in dualistic-polar structures, where every reality variety is destructible and turned into an imagination variety and every imagination variety is constructible and turned into a reality variety through a system of transformation processes such that each variety is seen as residing in the space of imagination-reality dualities.

Note IVC: On Definitions IVE, IVF & IVG It is useful to keep in mind the definitional structures involving the similarities and differences of real and nominal varieties in the dualistic-polar space in terms of domination of characteristic dispositions as seen in the space of relativity and not in the space of absoluteness. To explain the epistemic relations among vision, goal, objective, mission, problem–solution duality, the brain and neuro-decision-choice actions, it will be useful to visit the work on the theory of the knowledge square that shows the complex relational structure from potential to the possible, from the possible to the probable, from the probable to actual and from the actual to the potential [R17.]. Here the ontological space is taken as representing the primary identity, while the epistemological space represents the derived identity of knowing and information-knowledge production and social variety creativity. In fact, at the level of cognitive actions in all areas of existance, it is proper to see the epistemic journeys of cognitive agents within the epistemological space as journeys between the space of imaginations and the space of realities. The cognitive journeys are from the space of imaginations contained in the space of the potentials to the space of possibility, from the space of possibility to the space of probability and from the space of probability to the space of realities containes in the space of the actuals. The cognitive journeys between the space of imaginations and the space of realities are also journeys between the space of potentials and the space actuals in the sense that the space of imagination-reality dualities is a sub-space of the space of actual-potential dualities, where the space of imagination-reality dualities contains sub-space of spaces of possibility, probability, necessity and freedom which relate to cognitive capacity limitations in knowing and the development of information-knowledge stocks.

At the level of socio-natural existence, varieties resides in the space of actual-potential dualities with dualistic-polar characteristic dispositions for natural transformation processes over the space of the actual-potential dualities where an actual variety is transformed by a destruction process and hence enters into the potential space and potential variety changes characteristic disposition by a construction process into the actual and hence enters into the space of the actuals. The destruction of the variety actual or variety reality is a real opportunity cost to receive the real benefit of the variety potential or variety imagination to be actualized or realized.

The natural transformation processes take place over the space of actual-potential dualities moving directly from actual to potential or from potential to actual through one dualistic-polar space the elements of which are cognitively independent of the existence of other varieties in the space of actual-potential dualities. and hence the processes are independent of neuro-decision-choice actions of cognitive agents. The reason for this is that all-natural transformations and processes take place with perfect information inputs without uncertainties and risks since they are produced by nature and used by the same nature in the action processes. There are no cognitive capacity limitations in the sense that nature produces its own information, where the information is also the knowledge input into the natural decision-choice actions with no imperfections to produce other information-knowledge types as further input into the natural transformation processes. At the level of natural processes, information is knowledge and knowledge is information with no uncertainties and no risk of the corresponding natural outcomes. There is no need for information-knowledge quality control for acceptance by nature as the identity relative to social. In general, uncertainties arise where and when there are disparities between information and knowledge and where there are an observational deficiencies over the space of acquaintance to create defective experiential information which may be amplified by paradigmatic conditions as one moves to the space of knowledge by descriptions [670–672]. The concepts of imagination-reality duality, possibility-impossibility duality, possibility-probability duality, necessity-freedom duality and transversality condition do not arise for nature operating over the ontological space.

The question of whether nature has a vision with goals and objectives is not our concern here. At the level of neuro-decision-choice actions, the visionary structure is a family of inter-supportive variety characteristic dispositions established as a hierarchy of goals and objectives over the space of goals and objectives and then mapped onto the space of problem–solution dualities for sequential mission actions. The conditions of the principle of non-satiation point to a series of visionary structures without an end. For the individual, the sequence of the visionary structures ends with the finality of life while the sequence of generational visionary structures is endless with forward–backward telescopic processes with a continual creation of information-knowledge structures. In the process of knowing, it must be observed that methods and techniques are simply to help reconcile the variety problem–solution dualities in the visionary process based on real cost–benefit conditions over the actual-potential transformation process.

The information-knowledge structures from science, non-science or any category of knowing are merely inputs into neuro-decision-choice actions in moving from one step to another, where decision-choice actions are impossible without information-knowledge input and where the outputs of decision-choice actions are information-knowledge items. In this respect, the information-knowledge processes are decision-choice activities over the space of production-consumption dualities with or without markets. The whole process is that the space of real cost–benefit dualities constitutes an incentive space with or without markets for its understanding and analytical utilization in line with dualistic-polar conditions. The cost–benefit characteristics may be reduce/d to monetary units by appropriate price constructions as have been done in [219].

At the level of social knowing, the individual, collective and social neuro-decision-choice actions over the spaces of acquaintance and description create imperfect information-knowledge structures with fuzziness and volume incompleteness due to cognitive capacity limitations with the interpretations of qualitative and quantitative signal dispositions leading to various degrees of uncertainties, where outcomes pass through the complex space of doubt-surety dualities. The neuro-decision-choice processes over the space of actual-potential dualities are always indirect. The existence of cognitive capacity limitations introduces imperfect information structures as inputs into the variety knowing processes, general information-knowledge accumulation processes and all neuro-decision-choice actions at both the levels of individual, collective and social activities. The individual, collective and social neuro-decision-choice processes involving all endeavors must connect the actual and potential varieties and the connections are done through the space of possibility-probability dualities.

At the point of knowing and social creations, the neuro-decision-choice actions proceed from the potential space to the possibility space, and from the possibility space to the probability space and from the probability space to the space of actual varieties and from the space of actuals to the space of potential. The transformation from the space of actual to potential involves a process of destruction. The transformation from the potential through the possible to the probable to the actual involves construction by variety identification to establish necessity and transformation to establish freedom. The neuro-decision-choice processes go through interconnected spaces of dualities of the spaces of potential-possibility dualities, the possibility-probability dualities, the probability-actual dualities and the actual-potential dualities in dealing with knowing, identification and transformation under dualistic-polar conditions of varieties.

The space of visions and the space of goals and objectives are improper subspaces of the potential space where vision is related to an aspiration containing goal-objective varieties as problem–solution dualities that must be overcome with a series of missions if the vision is to be accomplished. A vision is an imagined organic variety to be actualized from the potential space. Any vision is a mental image of a desirable variety that may or may not be reached by cognitive actions and journey to it. When a vision is set, it remains as a mental image that is taken from the potential space to the possibility space by establishing conditions of its actualization. These conditions are cognitively abstracted as characteristic dispositions from the potential space as the required problem–solution dualities that must be carefully overcome by the neuro-decision-choice actions. The required problem–solution dualities are the goals and objectives with their characteristic dispositions also from the potential space standing as impediments that must be overcome on the way to reach the vision. The goals and objectives relative to the selected vision with information and paradigmatic constraints are moved unto the probability space for mission action, while the vision remains in the possibility space with possibilistic and probabilistic uncertainties. The goals and objectives become probabilistic varieties under mission actions with both possibilistic and probabilistic uncertainties or fuzzy-stochastic uncertainties.

Action analytics are required over the space of possibility-probability dualities to establish variety visions and the supporting goals and objectives with a well-defined characteristic disposition. At the possibility space, a vision is a *necessity* defined by a possibilistic disposition with possibilistic uncertainties where the necessity is related to anticipation. The goals, objectives nd mission are established by *freedom* constrained by probabilistic uncertainties in the actualization processes given the possibilistic necessity and anticipation. Let us keep in mind that there is an intimate relationship between the space of possibility-probability dualities and those of necessity-freedom and anticipation-expectation dualities in all neuro-decision-choice processes. Possibility, necessity and anticipation initialize the neuro-decision-choice process while probability, freedom and expectation carry on the actionable missions to the results of transformations.

The essential features of the theory of neuro-decision-choice system are attempts to present a general theory of decision-choice system that is applicable to all areas of human endeavors and provide channels to internalize the users into the space of static-dynamic processes as well as encompass sub-theories involving the conditions of quality-quantity dualities and balances of the input–output structures, the corresponding cost–benefit dispositions and socio-economic transformations where variations in areas of problem–solution structures are due to information-knowledge differences and specificities. Let us keep in mind that every human activity involves decision and choice where this decision-choice action is expressed over an element in the space of problem–solution dualities setting the solution against the problem that leads to either a change in relation or a change in property or both. In this respect, the theory of knowing and the theory of information-knowledge accumulation may be seen in terms of relational connections among the elements in the potential space, the possibility space, the probability space and the space of the actuals under individual, collective and social preferences. The connectivity links are shown in Fig. A1.

The theory of neuro-decision-choice systems as may be conceived is to provide an explanatory and prescriptive links from the space of the potentials to the possibility space through the space of imaginations. The explanatory-prescriptive links

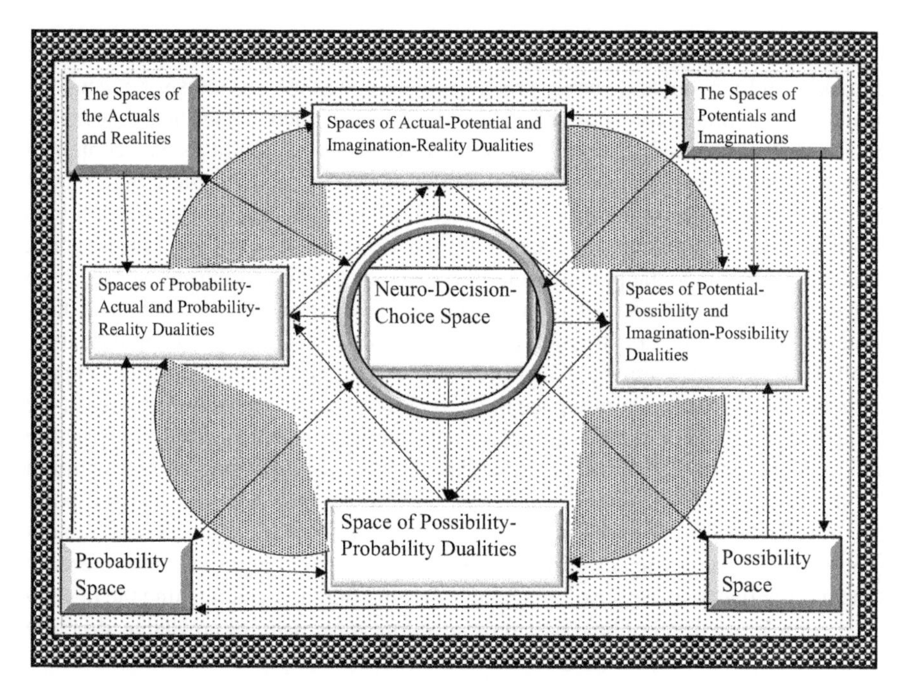

Fig. A1 A cognitive geometry of dualities for neuro-decision-choice actions in the space of imagination-reality dualities as a sub-space of actual-potential dualities

proceed from the space of potential-imagination dualities to the space of actual-reality dualities then from the space of imagination-reality dualities to the space of possibility-probability dualities, with cognitive movements from the possibility space to the probability space, from the space of possibility-probability dualities to the space of probability-actual dualities, from the probability space to the space of reality contained in the space of actuals. From the space of probability-reality dualities the explanatory-prescriptive links proceed to the space of actual-potential dualities, from the space of actuals to the space of potentials and from the space of the probability-reality dualities to the space of actual-potential dualities in the information-knowledge-square process providing an indirect journey by cognitive agents from the space of the potentials to the space of the actuals. This indirect pathways from the potential to the actual and from the actual to the potential is due to information-knowledge disparities and cognitive capacity limitations generating input–output uncertainty and benefit-risk conditions in the space of imagination-reality dualities. At the level of natural operations, however, the natural transformations proceed from the space of the potentials directly to the space of the actuals and from the space of the actuals directly to the space of the potentials under information-knowledge equality with continual accumulations of information-knowledge structures in disequilibrium dynamics relative to the matter-energy disequilibrium dynamics.

At level of society, the social transformations are carried by neuro-decision-choice dynamics in knowing and variety creations, where the direct processes between the actual and potential are not possible. The indirect way is such that the neuro-decision-choice system proceeds from the potential space to the possibility space contained in the space of imaginations to establish the possibility set and the conditions of necessity, then to the probability space to establish the probability set and conditions of the degrees of freedom and then to the space of realities contained in the space of actual to actualize the potential into the space of realities as the sub-space of the space of actuals. The cognitive journey in the knowing and intellectual stock-flow system is through the construction processes, while the intellectual capital quality-control process is through the reduction processes over the space of constructionism-reductionism dualities. At the space of the realities as a sub-space of the actuals, the undesirable actualized realities are selected to enter the potential space by destruction and replaced by an actualized potential based on the *asantrofi-anoma* rationality in the space of cost–benefit dualities under the sankofa-anoma telescopic dualistic-polar conditions of information-knowledge system. This is the destruction-replacement processes of varieties under cost–benefit dualistic-polar conditions and preference schemes relative to the space of imagination-reality dualities establishing the interrelated complexity of telescopic past-present-future information-knowledge structure of input–output elements into the neuro-decision-choice space.

Imagination, Reality, Belief and the Space of Imagination-Reality Dualities

Over the space of possibility-probability dualities arise several conceptual and epistemic difficulties of concepts regarding their statue in the space of knowing and information-knowledge production as intellectual investment-capital accumulation processes. These concepts include imagination, belief, vision, dream, hope, will and reality and their relationships to the knowing process and information-knowledge production. The most disturbing of the use of these concepts in thinking and reasoning is the use of the concept of belief to define knowledge as "justified true belief" (JTB) without any connection to information which is not defined and sometimes taken to be the same thing as knowledge as well as taken to be the same as either fact and evidence in the confused interdependencies violating laws in the theory definitions and variety identity over the space of real-nominal dualities. Again, what is reality and how does reality differ from the actual in the space of knowing in the understanding of *what there was, what there is* and *what there would be*? How do these interchangeable concepts meet the conditions of identity as well as help in communication in the space of source–destination dualities? The direction suggested here is through the principle of characteristic disposition as the content of information and the principle of signal disposition as the message of information in communication and the processes of knowing. In traditional epistemology, the concept of "justified true belief" seems to have no analytical meaning. What are the intrinsic meanings of the sub-concepts of belief, true and justified and how are they related in the spaces of quality-quantity dualities and imagination-reality dualities? These concepts are defined in the space of vagueness with shades of meaning that belong to the fuzzy set characterizations with fuzzy conditionality abstracted from the space of relativity and not in the space of absoluteness for the use of the classical paradigm of thought. The concept of "justified true belief" belongs to a phantom space if it is not related to some form of information as seen in the space of matter-energy dualistic-polar structures. As it stands, the *justified true belief* is disconnected from information and the principle of variety identity that makes it possible to construct a dictionary of terms or glossary of terms. It may be noted that the variety *true* resides in the space of reality. The variety *belief* resides in the space of imagination while *justified* resides in the space of characteristic disposition to allow the conceptual connectivity to make sense of (JTB) in the spaces of production-consumption and input–output processes. Any belief is justified as long as it is related subjectivity of emotion such as love, fear, hate, death and others of such subjective fillings.

The resolution of the phantom nature of the classical definition of knowledge is resolved by relating the concept of knowledge to the concept of information defined as characteristic-signal disposition. In this respect, justification will be related to comparative characteristic dispositions as defining either a nominal or real variety. The implication of the characteristic deposition relates to the theory of definitions where distinctions are made between nominal definition and real definition where the principle of definitions is defined over the space of relativity under fuzziness that allows subjective interpretation in propositions. In fact, ideas, goals, objectives, and visions, must be examined within the space of imaginations contained in the space of potential-possible dualities and show how they relate to information, facts, evidence, evidential things and knowledge which must be examined in the space of reality which is contained in the space actual-reality dualities. Part of the complex conceptual relations is discussed in this monograph requiring more discussions. All conflicts arising from disagreements are simply disagreements over variety characteristic dispositions which are amplified by subjective phenomena in the space of quality-quantity dualities.

Methodology, Information and Problem–Solution Dualities

The set of interactive conditions of problem–solution dualities, information and methodology has been discussed in the monograph. Additionally, there are important points that must be kept in mind. The variety problem–solution dualities are made available indirectly to cognitive agents becourse of their interpretations of signal dispositions that reveal their characteristic dispositions at the level of acquaintance. The variety characteristic-signal dispositions play important roles at the epistemological space where there are complex interactions between possibility and possibility, necessity and freedom. First, the set of the variety characteristic dispositions forms the experiential information, also called knowledge by acquaintance, as the first cognitive step of knowing. Then, the next step is to relate the experiential information as inputs for cognitive processing. The experiential information may have different subjective interpretations leading to different qualitative forms of inputs into a paradigm of thought to abstract knowledge, also called the knowledge by description. The need to have consistency and collective agreement leads to the development of a language and a paradigm of thought with rules for collective use to claim agreeable conclusions. The nature of the paradigm depends on the conditions of vagueness and completeness of experiential information inputs as part of the knowing process. The selected paradigm of thought will control thinking and reasoning with all information structures over the space of the problem–solution dualities. The destruction-construction processes with variety problem–solution replacement dynamics will also be affected in the paradigmatic framework.

The paradigm of thought defines the framework of all decision-choice actions, nature of information representation, operations on qualitative and quantitative structures in static and dynamic domains, the nature of thinking and processing of qualitative and quantitative characteristics, mathematics, logic, and boundaries of pure and applied mathematics, where the applied mathematics must be interpreted in a broad general way in the representations of algorithms of thinking in support of reasoning. The computability-uncomputability duality of mathematical representations of problem–solution duality will depend on Brouwer-[875]vagueness conditions and Gödel completeness conditions of the paradigm of thought, and whether one is operating in the space of relativity or the space of absoluteness. The paradigm of thought that one uses as the guide to thinking is important in all epistemic designs over the space of the problem–solution dualities, variety solutions and the subsequent variety problems that the solution may generate. The nature of the paradigm of thought may constrain the progress of knowing and scientific discoveries.

One important element in the principle of opposites in relation to knowing is that paradigms of thought must be able to establish dynamic connectivity in the parent–offspring process of problem–solution dualities, where relationality is an essential element in changes in relations and properties on the basis of internal or external forces, and that our qualitative and quantitative mathematics must reveal these relations if contradictions, paradoxes and ill-posed problems are to be avoided. Let us keep in mind that the central element in all human actions is the mind that is transformed by the bioneural system to deal with a multiplicity of different challenges in socio-natural environments which are presented as problem–solution dualities that constrain needs, wants, desires, and progresses in all attempts to balance duals of dualities and poles of polarities in all socio-natural relations under a variety of preference relations induced by neurological systems of individuality and collectivity in the conflict zone of preferences. It is the same mind that develops information structures as inputs into decision-choice actions, the results of which become outputs and are then transformed into inputs for further decision-choice actions over the space of problem–solution dualities in the infinite processes of human existence. The space of the neuro-decision-choice actions, its interactions with the space of problem–solution dualities and the space of source–destination dualities in a system of variety dynamics over the space of identification-transformation dualities constitute the information-knowledge economy which is then translated into the variety economy.

Nothing is known in the universal system except in the space of variety acquaintance and with variety signal dispositions that are sent by variety characteristic dispositions linking the ontological space to the epistemological space for neuro-decision-choice recording and interpretation for variety knowing as an intellectual investment under flow conditions and an intellectual accumulation under stock conditions. The space of acquaintances may be multiplicity with or without awareness where perception of our acquaintances may not reveal all the variety characteristic dispositions due to our cognitive limitations that generate vagueness and volume limitation to create qualitative and quantitative uncertainties over the space of possibility-probability dualities, where the theory of knowledge may be seen as a theory of an intellectual capital accumulation under stock conditions over the space of doubt-surety dualities for which the conditions of entropy may help to resolve the dualistic-polar conflicts with neuro-decision-choice actions in the space of doubt-surety dualities of varieties where variety sureties are set against variety doubts or vice versa with an entropic conditionality.

In viewing the neuro-decision-choice processes over the socio-natural varieties, several questions arise as to (1) what is information? (2) What is knowledge and its relationship to information? (3) What is data and its relationship to information and knowledge? (4) What is fact and its relationships to data, information and knowledge? (5) what is an evidence and how is it related to information, knowledge, data, fact and varieties? (6) What is the relational connectivity of these concepts to each other, to thinking, reasoning and neuro-decision-choice input–output processes over the spaces of variety problem–solution dualities and variety production-consumption dualities? (7) What are the differences between information-knowledge production and information-knowledge transmission? These questions are challenges to traditional theories of knowledge where the theories must relate to all areas of knowing and the traditional theory of information that concentrate on forms of storage, datamatics, communication and applications. The most important thing in conceptualizing the system of knowing, knowledge and the unity of science is the question regarding how knowing and knowledge are related to each other and the decision-choice actions over the space of problem–solution dualities as a collection of categorial problem–solution dualities. The discussions in this epilogue are the points of departure and entry points for future research on knowing and information-knowledge production over the space of fundamental-applied dualities.

References

Category Theory in Mathematics, Logic and Sciences

1. Awodey, S.: Structure in mathematics and logic: a categorical perspective. Philos. Math. **3**, 209–237 (1996)
2. Bell, J.L.: Category theory and the foundations of mathematics. British J. Sci. **32**, 349–358 (1981)
3. Bell, J.L.: Categories, toposes and sets. Synthese **51**, 337–393 (1982)
4. Black, M.: The Nature of Mathematics. Adams and Co., Totowa, N.J., Littlefield (1965)
5. Blass, A.: The interaction between category and set theory. Math. Appl. Category Theor. **30**, 5–29 (1984)
6. Brown, B., Woods, J. (eds.): Logical Consequence; Rival Approaches and New Studies in exact Philosophy: Logic, Mathematics and Science, vol. II. Oxford, Hermes (2000)
7. Domany, J.L., et al.: Models of Neural Networks III. Springer, New York (1996)
8. Feferman, S.: Categorical foundations and foundations of category theory. In: Butts, R. (ed.) Logic, Foundations of Mathematics and Computability, pp. 149–169. Reidel, Boston, Mass. (1977)
9. Glimcher, P.W.: Decisions, Uncertainty, and the Brain: The Science of Neoroeconomics. MIT Press, Cambridge, Mass (2004)
10. Gray, J.W. (ed.) Mathematical Applications of Category Theory (American Mathematical Society Meeting 89th Denver Colo. 1983), Providence, R.I., American Mathematical Society (1984)
11. Johansson, I., Investigations, O.: An Inquiry into the Categories of Nature, Man, and Society. Routledge, New York (1989)
12. Kamps, K.H., Pumplun, D., Tholen, W. (eds.): Category Theory: Proceedings of the International Conference, Gummersbach, July 6–10, Springer, New York (1982)
13. Landry, E.: Category theory: the language of mathematics. Philos. Sci. **66**(Supplement), S14–S27
14. Landry, E., Marquis, J.P.: Categories in context: historical, foundational and philosophical. Philios. Math. **13**, 1–43 (2005)
15. Marquis, J.-P.: Three kinds of universals in mathematics. In: Brown, B., Woods, J. (eds.) Logical Consequence; Rival Approaches and New Studies in Exact Philosophy: Logic, Mathematics and Science, vol. II. Oxford, Hermes, pp. 191–212 (2000)
16. McLarty, C.: Category theory in real time. Philos. Math. **2**, 36–44 (1994)

K. K. Dompere, *The Theory of Problem-Solution Dualities and Polarities*, Studies in Systems, Decision and Control 405,
https://doi.org/10.1007/978-3-030-90279-7

17. McLarty, C.: Learning from questions on categorical foundations. Philos. Math. **13**, 44–60 (2005)
18. Ross, D.: Economic Theory and Cognitive Science; Microexplanation. MIT Press, Cambridge, Mass. (2005)
19. Rodabaugh, S., et al. (eds.): Application of Category Theory to Fuzzy Subsets. Kluwer, Boston, Mass. (1992)
20. Sieradski, A.J.: An Introduction to Topology and Homotopy. PWS-KENT Pub. Boston (1992)
21. Taylor, J.G.: Mathematical Approaches to Neural Networks. North-Holland, New York (1993)

Concepts of Information, Fuzzy Probability, Fuzzy Random Variable and Random Fuzzy Variable

22. Van Benthem, J., et al. (eds.): The Age of Alternative Logics: Assessing Philosophy of Logic and Mathematics Today. Springer, New York (2006)
23. Bandemer, H.: From fuzzy data to functional relations. Math. Model. **6**, 419–426 (1987)
24. Bandemer, H., et al.: Fuzzy Data Analysis. Mass, Kluwer, Boston (1992)
25. Kruse, R., et al.: Statistics with Vague Data. D. Reidel Pub. Co., Dordrecht (1987)
26. El Rayes, A.B., et al.: Generalized possibility measures. Inf. Sci. **79**, 201–222 (1994)
27. Dumitrescu, D.: Entropy of a fuzzy process. Fuzzy Sets Syst. **55**(2), 169–177 (1993)
28. Delgado, M., et al.: On the concept of possibility-probability consistency. Fuzzy Sets Syst. **21**(3), 311–318 (1987)
29. Devi, B.B., et al.: Estimation of fuzzy memberships from histograms. Inf. Sci. **35**(1), 43–59 (1985)
30. Dubois, D., et al.: Fuzzy sets, probability and measurement. Euro. J. Oper. Res. **40**(2), 135–154 (1989)
31. Fruhwirth-Schnatter, S.: On statistical inference for fuzzy data with applications to descriptive statistics. Fuzzy Sets Syst. **50**(2), 143–165 (1992)
32. Gaines, B.R.: Fuzzy and probability uncertainty logics. Inf. Control **38**(2), 154–169 (1978)
33. Geer, J.F., et al.: Discord in possibility theory. Int. J. General Syst. **19**, 119–132 (1991)
34. Geer, J.F., et al.: A Mathematical analysis of information-processing transformation between probabilistic and possibilistic formulation of uncertainty. Int. J. General Syst. **20**(2), 14–176 (1992)
35. Goodman, I.R., et al.: Uncertainty Models for Knowledge Based Systems. North-Holland, New York (1985)
36. Grabish, M., et al.: Fundamentals of Uncertainty Calculi with Application to Fuzzy Systems. Kluwer, Boston, Mass (1994)
37. Guan, J.W., et al.: Evidence Theory and Its Applications, vol. 1. North-Holland, New York (1991)
38. Guan, J.W., et al.: Evidence Theory and Its Applications, vol. 2. North-Holland, New York (1992)
39. Hisdal, E.: Are grades of membership probabilities? Fuzzy Sets Syst. **25**(3), 349–356 (1988)
40. Ulrich, H.:A mathematical theory of uncertainty. In: Yager, R.R. (ed.) Fuzzy Set and Possibility Theory: Recent Developments, pp. 344–355. Pergamon, New York (1982)
41. Kacprzyk, J., Fedrizzi, M. (eds.): Combining Fuzzy Imprecision with Probabilistic Uncertainty in Decision Making. Plenum Press, New York (1992)
42. Kacprzyk, J., et al.: Combining Fuzzy Imprecision with Probabil**istic** Uncertainty in Decision Making. Springer-Verlag, New York (1988)
43. Klir, G.J.: Where do we stand on measures of uncertainty, ambignity, fuzziness and the like? Fuzzy Sets Syst. **24**(2), 141–160 (1987)
44. Klir, G.J., et al.: Fuzzy Sets, Uncertainty and Information. Prentice Hll, Englewood Cliff (1988)

45. Klir, G.J., et al.: Probability-possibility transformations: a comparison. Int. J. General Syst. **21**(3), 291–310 (1992)
46. Kosko, B.: Fuzziness vs probability. Int. J. General Syst. **17**(1–3), 211–240 (1990)
47. Manton, K.G., et al.: Statistical Applications Using Fuzzy Sets. Wiley, New York (1994)
48. Meier, W., et. al.: Fuzzy data analysis: methods and industrial applications. Fuzzy Sets Syst. **61**(1), 19–28 (1994)
49. Nakamura, A., et al.: A logic for fuzzy data analysis. Fuzzy Sets Syst. **39**(2), 127–132 (1991)
50. Negoita, C.V., et al.: Simulation, Knowledge-Based Computing and Fuzzy Statistics. Van Nostrand Reinhold, New York (1987)
51. Nguyen, H.T.: Random sets and belief functions. J. Math. Anal. Appl. **65**(3), 531–542 (1978)
52. Prade, H., et al.: Representation and combination of uncertainty with belief functions and possibility measures. Comput. Intell. **4**, 244–264 (1988)
53. Puri, M.L., et al.: Fuzzy random variables. J. Math. Anal. Appl. **114**(2), 409–422 (1986)
54. Rao, N.B., Rashed, A.: Some comments on fuzzy random variables. Fuzzy Sets Syst. **6**(3), 285–292 (1981)
55. Sakawa, M., et al.: Multiobjective fuzzy linear regression analysis for fuzzy input-output data. Fuzzy Sets Syst. **47**(2), 173–182 (1992)
56. Schneider, M., et al.: Properties of the fuzzy expected values and the fuzzy expected interval. Fuzzy Sets Syst. **26**(3), 373–385 (1988)
57. Stein, N.E., Talaki, K.: Convex fuzzy random variables. Fuzzy Sets Syst. **6**(3), 271–284 (1981)
58. Sudkamp, T.: On probability-possibility transformations. Fuzzy Sets Syst. **51**(1), 73–82 (1992)
59. Walley, P.: Statistical Reasoning with Imprecise Probabilities. Chapman and Hall, London (1991)
60. Wang, G.Y., et al.: The theory of fuzzy stochastic processes. Fuzzy Sets Syst. **51**(2), 161–178 (1992)
61. Zadeh, L.A.: Probability measure of fuzzy event. J. Math. Anal. Appl. **23**, 421–427 (1968)

Exact Science, Inexact Sciences and Information

62. Achinstein, P.: The problem of theoretical terms. In: Brody, B.A. (ed.) Reading in the Philosophy of Science. Prentice Hall, Englewood Cliffs, NJ (1970)
63. Amo Afer, A.G.: The Absence of Sensation and the Faculty of Sense in the Human Mind and Their Presence in our Organic and Living Body, Dissertation and Other essays 1727–1749, Halle Wittenberg, Jena, Martin Luther University Translation (1968)
64. Beeson, M.J.: Foundations of Constructive Mathematics. Springer, Berlin (1985)
65. Benacerraf, P.: God, the devil and Gödel. Monist **51**, 9–32 (1967)
66. Benecerraf, P., Putnam, H. (eds.): Philosophy of Mathematics: Selected Readings. Cambridge University Press, Cambridge (1983)
67. Black, M.: The Nature of Mathematics. Adams and Co., Totowa, Littlefield (1965)
68. Blanche, R.: Contemporary Science and Rationalism. Oliver and Boyd, Edinburgh (1968)
69. Blanshard, B.: The Nature of Thought. Unwin, London Allen (1939)
70. Blauberg, I.V., Sadovsky, V.N., Yudin, E.G.: Systems Theory: Philosophical and Methodological Problems. Progress Publishers, Moscow (1977)
71. Braithwaite, R.B.: Scientific Explanation. Cambridge University Press, Cambridge (1955)
72. Brody, B.A. (ed.): Reading in the Philosophy of Science. Prentice Hall, Englewood Cliffs, N.J. (1970)
73. Brody, B.A.: Confirmation and explanation. In: Brody, B.A. (ed.) Reading in the Philosophy of Science, pp. 410–426. Prentice-Hall, Englewood Cliffs, N.J. (1970)
74. Brouwer, L.E.J.: Intuitionism and formalism. Bull. Am. Math. Soc. **20**, 81–96 (1913); Also in Benecerraf, P., Putnam, H. (eds.) Philosophy of Mathematics: Selected Readings, pp. 77–89. Cambridge University Press, Cambridge (1983)

75. Brouwer, L.E.J.: Consciousness, philosophy, and mathematics. In: Benecerraf, P., Putnam, H. (eds.) Philosophy of Mathematics: Selected Readings, pp. 90–96. Cambridge University Press, Cambridge (1983)
76. Brouwer, L.E.J.: Collected Works, Vol. 1: Philosophy and Foundations of Mathematics. In: Heyting, A. (ed.). Elsevier, New York (1975)
77. Campbell, N.R.: What is Science? Dover, New York (1952)
78. Carnap, R.: Foundations of logic and mathematics. In: International Encyclopedia of Unified Science, pp. 143–211. Chicago, Univ. of Chicago (1939)
79. Carnap, R.: On inductive logic. Philosophy of Science **12**, 72–97 (1945)
80. Carnap, R.: The methodological character of theoretical concepts. In: Feigl, H., Scriven, M. (eds.) Minnesota Studies in the Philosophy of Science, vol. I, pp. 38–76 (1956)
81. Charles, D., Lennon, K. (eds.): Reduction, Explanation, and Realism. Oxford University Press, Oxford (1992)
82. Cohen, Robert S. and Marx W. Wartofsky (eds.), Methodological and Historical Essays in the Natural and Social Sciences. D. Reidel Publishing Co., Dordrecht (1974)
83. van Dalen, D. (ed.): Brouwer's Cambridge Lectures on Intuitionism. Cambridge University Press, Cambridge (1981)
84. Davidson, D.: Truth and Meaning: Inquiries into Truth and Interpretation. Oxford University Press, Oxford (1984)
85. Davis, M.: Computability and Unsolvability. McGraw-Hill, New York (1958)
86. Denonn, L.E. (ed.): The Wit and Wisdom of Bertrand Russell. The Beacon Press, Boston, MA. (1951)
87. Dompere, K.K.: Fuzziness and Foundations of Exact and Inexact Sciences. Springer, New York (2013)
88. Dummett, M.: The philosophical basis of intuitionistic logic. In: Benecerraf, P., Putnam, H. (eds.) Philosophy of Mathematics: Selected Readings, pp. 97–129. Cambridge University Press, Cambridge (1983)
89. Feigl, H., Scriven, M. (eds.): Minnesota Studies in the Philosophy of Science, vol. I (1956)
90. Garfinkel, A.: Forms of Explanation: Structures of Inquiry in Social Science. Yale University Press, New Haven, Conn. (1981)
91. George, F.H.: Philosophical Foundations of Cybernetics. Great Britain, Tunbridge Well (1979)
92. Gillam, B.: Geometrical illusions. Sci. Am. 102–111 (1980)
93. Gödel, K.: What is cantor's continuum problem? In: Benecerraf, P., Putnam, H. (eds.) Philosophy of Mathematics: Selected Readings, pp. 470–486. Cambridge University Press, Cambridge (1983)
94. Gorsky, D.R.: Definition. Progress Publishers, Moscow (1974)
95. Gray, W., Rizzo, N.D. (eds.): Unity Through Diversity. Gordon and Breach, New York (1973)
96. Hart, W.D. (ed.): The Philosophy of Mathematics. Oxford University Press, Oxford (1996)
97. Hartkamper, A., Schmidt, H.: Structure and Approximation in Physical Theories. Plenum Press, New York (1981)
98. Hausman, D.M.: The Exact and Separate Science of Economics. Cambridge University Press, Cambridge (1992)
99. Helmer, O., Rescher, N.: On the Epistemology of the Inexact Sciences, P-1513. Rand Corporation, Santa Monica, CA (1958)
100. Hempel, C.G.: Studies in the logic of confirmation. Mind **54**(Part I), 1–26 (1945)
101. Hempel, C.G.: The theoretician's dilemma. In: Feigl, H., Scriven, M. (eds.) Minnesota Studies in the Philosophy of Science, vol. II, pp. 37–98 (1958)
102. Hempel, C.G., Oppenheim, P.: Studies in the logic of explanation. In: Philosophy of Science, vol. 15, pp. 135–175 (1948). [Also in Brody, B.A. (ed.) Reading in the Philosophy of Science, pp. 8–27. Prentice-Hall, Englewood Cliffs, NJ (1970)
103. Heyting, A.: Intuitionism: An Introduction. North-Holland, Amsterdam (1971)
104. Hintikka, J. (ed.): The Philosophy of Mathematics. Oxford University Press, London (1969)
105. Hockney, D., et al. (eds.): Contemporary Research in Philosophical Logic and Linguistic Semantics. Reidel Pub., Co., Dordrecht-Holland (1975)

106. Hoyninggen-Huene, P., Wuketits, F.M. (eds.): Reductionism and Systems Theory in the Life Science: Some Problems and Perspectives. Kluwer Academic Pub. Dordrencht (1989)
107. Ilyenkov, E.V.: Dialectical Logic: Essays on Its History and Theory. Progress Publishers, Moscow (1977)
108. Kedrov, B.M.: Toward the methodological analysis of scientific discovery. Sov. Stud. Philos. **11962**, 45–65
109. Kemeny, J.G., Oppenheim, P.: On reduction. In: Brody, B.A. (ed.) Reading in the Philosophy of Science, pp. 307–318. Prentice-Hall, Englewood Cliffs, NJ. (1970)
110. Klappholz, K.: Value judgments of economics. British J. Philos. **15**, 97–114 (1964)
111. Kleene, S.C.: On the interpretation of intuitionistic number theory. J. Symb. Log. **10**, 109–124 (1945)
112. Kmita, J.: The methodology of science as a theoretical discipline. In: Soviet Studies in Philosophy, pp. 38–49. Spring (1974)
113. Krupp, S.R. (ed.): The Structure of Economic Science. Prentice-Hall, Englewood Cliff, N. J. (1966)
114. Kuhn, T.: The Structure of Scientific Revolution. University of Chicago Press, Chicago (1970)
115. Kuhn, T.: The function of dogma in scientific research. In: Brody, B.A. (ed.) Reading in the Philosophy of Science, pp. 356–374. Prentice-Hall, Englewood Cliffs, NJ. (1970)
116. Kuhn, T.: The Essential Tension: Selected Studies in Scientific Tradition and Change. University of Chicago Press, Chicago (1979)
117. Lakatos, I. (ed.): The Problem of Inductive Logic. North Holland, Amsterdam (1968)
118. Lakatos, I.: Proofs and Refutations: The Logic of Mathematical Discovery. Cambridge University Press, Cambridge (1976)
119. Lakatos, I.: Mathematics, Science and Epistemology: Philosophical Papers. In: Worrall, J., Currievol, G. (eds.), vol. 2. Cambridge University Press, Cambridge (1978)
120. Lakatos, I.: The Methodology of Scientific Research Programmes, vol. 1. Cambridge University Press, New York (1978)
121. Lakatos, I., Musgrave, A. (eds.): Criticism and the Growth of Knowledge, pp. 153–164. Cambridge University Press, New York, Holland (1979)
122. Lawson, T.: Economics and Reality. Routledge, New York (1977)
123. Lenzen, V.: Procedures of empirical science. In: Neurath, O., et al. (eds.) International Encyclopedia of Unified Science, vol. 1–10, pp. 280–338. University of Chicago Press, Chicago (1955)
124. Levi, I.: Must the scientist make value judgments? In: Brody, B.A. (ed.) Reading in the Philosophy of Science, pp. 559–570. Prentice-Hall, NJ., Englewood Cliffs (1970)
125. Tse-tung, M.: On Practice and Contradiction, in Selected Works of Mao Tse-tung, Piking, 1937. Also, London, Revolutions (2008)
126. Lewis, D.: Convention: A Philosophical Study. Harvard University Press, Cambridge, Mass. (1969)
127. Mayer, T.: Truth Versus Precision in Economics. Edward Elgar, London (1993)
128. Menger, C.: Investigations into the Method of the Social Sciences with Special Reference to Economics. New York University Press, New York (1985)
129. Mirowski, P. (ed.): The Reconstruction of Economic Theory. Mass. Kluwer Nijhoff, Boston (1986)
130. Mueller, I.: Philosophy of Mathematics and Deductive Structure in Euclid's Elements. MIT Press, Cambridge, Mass. (1981)
131. Nagel, E.: Review: Karl Niebyl, modern mathematics and some problems of quantity, quality, and motion in economic analysis. J. Symbol. Logic, 74 (1940)
132. Nagel, E., et al. (eds.): Logic, Methodology, and the Philosophy of Science. Stanford University Press, Stanford (1962)
133. Narens, L.: A theory of belief for scientific refutations. Synthese, **145**, 397–423 (2005)
134. Narskii, I.S.: On the problem of contradiction in dialectical logic. Sov. Stud. Philos. **vi**(4), 3 10 (1965)

135. Neurath, O., et al. (eds.): International Encyclopedia of Unified Science, vol. 1–10. University of Chicago Press, Chicago (1955)
136. Neurath, O.: Unified science as encyclopedic. In: Neurath, O., et al. (eds.) International Encyclopedia of Unified Science, vol. 1–10, pp. 1–27. University of Chicago Press, Chicago (1955)
137. Planck, M.: Scientific Autobiography and Other Papers, Westport, Conn. Greenwood (1971)
138. Planck, M.: The meaning and limits of exact science. In: Planck, M. (ed.) Scientific Autobiography and Other Papers, pp. 80–120. Conn. Greenwood, Westport (1971)
139. Polanyi, M.: Genius in science. In: Cohen, R.S., Wartofsky, M.W. (eds.) Methodological and Historical Essays in the Natural and Social Sciences, pp. 57–71. D. Reidel Publishing Co., Dordrecht (1974)
140. Popper, K.: The Nature of Scientific Discovery. Harper and Row, New York (1968)
141. Putnam, H.: Models and reality. In: Benecerraf, P., Putnam, H. (eds.) Philosophy of Mathematics: Selected Readings, pp. 421–444. Cambridge University Press, Cambridge (1983)
142. Reise, S.: The Universe of Meaning. The Philosophical Library, New York (1953)
143. Robinson, R.: Definition. Clarendon Press, Oxford (1950)
144. Rudner, R.: The scientist qua scientist makes value judgments. Philos. Sci. **20**, 1–6 (1953)
145. Russell, B.: Our Knowledge of the External World. Norton, New York (1929)
146. Russell, B.: Human Knowledge, Its Scope and Limits. Allen and Unwin, London (1948)
147. Russell, B.: Logic and Knowledge: Essays 1901–1950. Capricorn Books, New York (1971)
148. Russell, B.: An Inquiry into Meaning and Truth. Norton, New York (1940)
149. Russell, B.: Introduction to Mathematical Philosophy. George Allen and Unwin, London (1919)
150. Russell, B.: The Problems of Philosophy. Oxford University Press, Oxford (1978)
151. Rutkevih, M.N.: Evolution, progress, and the law of dialectic. Sov. Stud. Philos. **IV**(3), 34–43 (1965)
152. Ruzavin, G.I.: On the problem of the interrelations of modern formal logic and mathematical logic. Sov. Stud. Philos. **3**(1), 34–44 (1964)
153. Scriven, M.: Explanations, predictions, and laws. In: Brody, B.A. (ed.) Reading in the Philosophy of Science, pp. 88–104. Prentice-Hall, Englewood Cliffs, N.J. (1970)
154. Sellars, W.: The language of theories. In: Brody, B.A. (ed.) Reading in the Philosophy of Science, pp. 343–353. Prentice-Hall, NJ. (1970)
155. Sterman, J.: The growth of knowledge: testing a theory of scientific revolutions with a formal model. Technol. Forecast. Soc. Chang. **28**, 93–122 (1995)
156. Tsereteli, S.B. On the concept of dialectical logic. Sov. Stud. Philos. **V**(2), 15–21 (1966)
157. Tullock, G.: The Organization of Inquiry. Indiana, Liberty Fund Inc., Indianapolis (1966)
158. Van Fraassen, B.: Introduction to Philosophy of Space and Time. Random House, New York (1970)
159. Veldman, W.: A survey of intuitionistic descriptive set theory. In: Petkov, P.P. (ed.) Mathematical Logic: Proceedings of the Heyting Conference, pp. 155–174. Plenum Press, New York (1990)
160. Vetrov, A.A.: Mathematical logic and modern formal logic. Sov. Stud. Philos. **3**(1), 24–33 (1964)
161. von Mises, L.: Epistemological Problems in Economics. New York University Press, New York (1981)
162. Wang, H.: Reflections on Kurt Gödel. MIT Press, Cambridge, Mass. (1987)
163. Watkins, J.W.N.: The paradoxes of confirmation. In: Brody, B.A. (ed.) Reading in the Philosophy of Science, pp. 433–438. Prentice-Hall, Englewood Cliffs, NJ. (1970)
164. Whitehead, A.N.: Process and Reality. The Free Press, New York (1978)
165. Wittgenstein, L., Logico-philosophicus, T.: Atlantic Highlands. The Humanities Press Inc., N.J. (1974)
166. Woodger, J.H.: The Axiomatic Method in Biology. Cambridge University Press, Cambridge (1937)
167. Zeman, J.: Information, knowledge and time. In: Kubát, L., Zeman, J. (eds.) Entropy and Information in the Physical Sciences, pp. 245–260. Amsterdam, Elsevier (1975)

Fuzzy Logic, Information and Knowledge-Production

168. Baldwin, J.F.: A new approach to approximate reasoning using a fuzzy logic. Fuzzy Sets Syst. **2**(4), 309–325 (1979)
169. Baldwin, J.F.: Fuzzy logic and fuzzy reasoning. Intern. J. Man-Mach. Stud. **11**, 465–480 (1979)
170. Baldwin, J.F.: Fuzzy logic and its application to fuzzy reasoning. In: Gupta, M.M., et al. (eds.) Advances in Fuzzy Set Theory and Applications, pp. 96–115. North-Holland, New York (1979)
171. Baldwin, J.F., et al.: Fuzzy relational inference language. Fuzzy Sets Syst. **14**(2), 155–174 (1984)
172. Baldsin, J., Pilsworth, B.W.: Axiomatic approach to implication for approximate reasoning with fuzzy logic. Fuzzy Sets Syst. **3**(2), 193–219 (1980)
173. Baldwin, J.F., et al.: The resolution of two paradoxes by approximate reasoning using a fuzzy logic. Synthese **44**, 397–420 (1980)
174. Dompere, K.K.: Fuzzy Rationality: Methodological Critique and Unity of Classical, Bounded and Other Rationalities, (Studies in Fuzziness and Soft Computing, vol. 235). Springer, New York (2009)
175. Dompere Kofi, K.: Epistemic Foundations of Fuzziness, (Studies in Fuzziness and Soft Computing, vol. 236). Springer, New York (2009)
176. Dompere Kofi, K.: Fuzziness and Approximate Reasoning: Epistemics on Uncertainty, Expectation and Risk in Rational Behavior, (Studies in Fuzziness and Soft Computing, vol. 237). Springer, New York (2009)
177. Dompere, K.K.: The Theory of the Knowledge Square: The Fuzzy Rational Foundations of Knowledge-Production Systems. Springer, New York (2013)
178. Dompere, K.K.: Cost-benefit analysis, benefit accounting and fuzzy decisions: part i, theory. Fuzzy Sets Syst. **92**, 275–287 (1997)
179. Dompere, K.K.: The theory of social cost and costing for cost-benefit analysis in a fuzzy decision space. Fuzzy Sets Syst. **76**, 1–24 (1995)
180. Dompere, K.K.: Fuzzy Rational Foundations of Exact and Inexact Sciences. Springer, New York (2013)
181. Gaines, B.R.: Foundations of fuzzy reasoning. Inter. J. Man-Mach. Stud. **8**, 623–668 (1976)
182. Gaines, B.R.: Foundations of fuzzy reasoning. In: Gupta, M.M., et al. (eds.), Fuzzy Information and Decision Processes, pp. 19–75. North-Holland, New York (1982)
183. Gaines, B.R.: Precise past, fuzzy future. Int. J. Man-Mach. Stud. **19**(1), 117–134 (1983)
184. Giles, R.: Lukasiewics logic and fuzzy set theory. Intern. J. Man-Mach. Stud. **8**, 313–327 (1976)
185. Giles, R.: Formal system for fuzzy reasoning. Fuzzy Sets Syst. **2**(3), 233–257 (1979)
186. Ginsberg, M.L. (ed.): Readings in Non-monotonic Reason. Los Altos, Ca., Morgan Kaufman (1987)
187. Goguen, J.A.: The logic of inexact concepts. Synthese **19**, 325–373 (1969)
188. Gottinger, H.W.: Towards a fuzzy reasoning in the behavioral science. Cursos Congr. Univ. Santiago de Compostela **16**(2), 113–135 (1973)
189. Gupta, M.M., et al. (eds.): Approximate Reasoning In Decision Analysis. North Holland, New York (1982)
190. Höhle, U., Klement, E.P.: Non-Clasical Logics and their Applications to Fuzzy Subsets: A Handbook of the Mathematical Foundations of Fuzzy Set Theory. Kluwer, Boston, Mass. (1995)
191. Kaipov, V.K., et al.: Classification in fuzzy environments. In: Gupta, M.M., et al. (eds.) Advances in Fuzzy Set Theory and Applications, pp. 119–124. North-Holland, New York (1979)
192. Kaufman, A.: Progress in modeling of human reasoning of fuzzy logic. In: Gupta, M.M., et al. (eds.) Fuzzy Information and Decision Process, pp. 11–17. North-Holland, New York (1982)

193. Lakoff, G.: Hedges: a study in meaning criteria and the logic of fuzzy concepts. J. Philos. Logic **2**, 458–508 (1973)
194. Lee, R.C.T.: Fuzzy logic and the resolution principle. J. Assoc. Comput. Mach. **19**, 109–119 (1972)
195. LeFaivre, R.A.: The representation of fuzzy knowledge. J. Cybern. **4**, 57–66 (1974)
196. Negoita, C.V.: Representation theorems for fuzzy concepts. Kybernetes **4**, 169–174 (1975)
197. Nowakowska, M.: Methodological problems of measurements of fuzzy concepts in social sciences. Behav. Sci. **22**(2), 107–115 (1977)
198. Skala, H.J.: On many-valued logics, fuzzy sets, fuzzy logics and their applications. Fuzzy Sets Syst. **1**(2), 129–149 (1978)
199. Van Fraassen, B.C.: Comments: Lakoff's fuzzy propositional logic. In: Hockney, D. et al., (eds.) Contemporary Research in Philosophical Logic and Linguistic Semantics, pp. 273–277. Holland, Reild (1975)
200. Yager, R.R., et al. (eds.): An Introduction to Fuzzy Logic Applications in Intelligent Systems. Kluwer, Boston, Mass. (1992)
201. Zadeh, L.A.: Quantitative fuzzy semantics. Inform. Sci. **3**, 159–176 (1971)
202. Zadeh, L.A.: A fuzzy set interpretation of linguistic hedges. J. Cybern. **2**, 4–34 (1972)
203. Zadeh, L.A.: The concept of a linguistic variable and its application to approximate reasoning. In: Fu, K.S., et al. (eds.) Learning Systems and Intelligent Robots, pp. 1–10. Plenum Press, New York (1974)
204. Zadeh, L.A., et al. (eds.): Fuzzy Sets and Their Applications to Cognitive and Decision Processes. Academic Press, New York (1974)
205. Zadeh, L.A.: The birth and evolution of fuzzy logic. Int. J. General Syst. **17**(2–3), 95–105 (1990)

Fuzzy Mathematics and Paradigm of Approximate Reasoning Under Conditions of Inexactness and Vagueness

206. Bellman, R.E.: Mathematics and human sciences. In: Wilkinson, J., et al. (eds.) The Dynamic Programming of Human Systems, pp. 11–18. MSS Information Corp, New York (1973)
207. Bellman, R.E, Glertz, M.: On the analytic formalism of the theory of fuzzy sets. Inf. Sci. **5**, 149–156 (1973)
208. Butnariu, D.: Fixed points for fuzzy mapping. Fuzzy Sets Syst. **7**(2), 191–207 (1982)
209. Butnariu, D.: Decompositions and range for additive fuzzy measures. Fuzzy Sets Syst. **10**(2), 135–155 (1983)
210. Chang, C.L.: Fuzzy topological spaces. J. Math. Anal. Appl. **24**, 182–190 (1968)
211. Chang, S.S.L.: Fuzzy mathematics, man and his environment. In: IEEE Transactions on Systems, Man and Cybernetics, SMC-2, pp. 92–93 (1972)
212. Chang, S.S.: Fixed point theorems for fuzzy mappings. Fuzzy Sets Syst. **17**, 181–187 (1985)
213. Chapin, E.W.: An axiomatization of the set theory of Zadeh. Notices, Am. Math. Soc. 687-02-4, 754 (1971)
214. Chaudhury, A.K., Das, P.: Some results on fuzzy topology on fuzzy sets. Fuzzy Sets Syst. **56**, 331–336 (1993)
215. Chitra, H., Subrahmanyam, P.V.: Fuzzy sets and fixed points. J. Math. Anal. Appl. **124**, 584–590 (1987)
216. Czogala, J., et al.: Fuzzy relation equations on a finite set. Fuzzy Sets Syst. **7**(1), 89–101 (1982)
217. DiNola, A., et al. (eds.): The Mathematics of Fuzzy Systems. Verlag TUV Rheinland, Koln (1986)
218. Dompere, K.K.: Cost-Benefit Analysis and the Theory of Fuzzy Decisions: Identification and Measurement Theory (Series: Studies in Fuzziness and Soft Computing, vol. 158). Springer, Berlin, Heidelberg (2004)

219. Dompere, Kofi K., Cost-Benefit Analysis and the Theory of Fuzzy Decisions: Fuzzy Value Theory (Series: Studies in Fuzziness and Soft Computing, vol. 160). Springer, Berlin, Heidelberg (2004)
220. Dubois, D., Prade, H.: Fuzzy Sets and Systems. Academic Press, New York (1980)
221. Dubois.: Fuzzy real algebra: some results. Fuzzy Sets Syst. **2**(4), 327–348 (1979)
222. Dubois, D., Prade, H.: Possibility Theory: An Approach to Computerized Processing of Uncertainty. Plenum Press, New York (1988)
223. Dubois, D., Prade, H.: Gradual inference rules in approximate reasoning. Inf. Sci. **61**(1–2), 103–122 (1992)
224. Dubois, D., Prade, H.: On the combination of evidence in various mathematical frameworks. In: Flamm, J., Luisi, T. (eds.) Reliability Data Collection and Analysis, pp. 213–241. Kluwer, Boston (1992)
225. Dubois, D., Prade, H.: Fuzzy sets and probability: misunderstanding, bridges and gaps. In: Proceedings of Second IEEE International Conference on Fuzzy Systems, pp. 1059–1068, San Francisco, (1993)
226. Dubois, D., Prade, H.: A survey of belief revision and updating rules in various uncertainty models. Int. J. Intell. Syst. **9**(1), 61–100 (1994)
227. Filev, D.P., et al.: A generalized defuzzification method via bag distributions. Int. J. Intell. Syst. **6**(7), 687–697 (1991)
228. Goetschel, R. Jr., et al.: Topological properties of fuzzy number. Fuzzy Sets Syst. **10**(1), 87–99 (1983)
229. Goodman, I.R.: Fuzzy sets as equivalence classes of random sets. In: Yager, R.R. (ed.) Fuzzy Set and Possibility Theory: Recent Development, pp. 327–343. Pergamon Press, New York (1992)
230. Gupta, M.M., et al. (eds.): Fuzzy Automata and Decision Processes. North-Holland, New York (1977)
231. Gupta, M.M., Sanchez, E. (eds.): Fuzzy Information and Decision Processes. North-Holland, New York (1982)
232. Higashi, M., Klir, G.J.: On measure of fuzziness and fuzzy complements. Int. J. General Syst. **8**(3), 169–180 (1982)
233. Higashi, M., Klir, G.J.: Measures of uncertainty and information based on possibility distributions. Int. J. General Syst. **9**(1), 43–58 (1983)
234. Higashi, M., Klir, G.J.: On the notion of distance representing information closeness: possibility and probability distributions. Int. J. General Syst. **9**(2), 103–115 (1983)
235. Higashi, M., Klir, G.J.: Resolution of finite fuzzy relation equations. Fuzzy Sets Syst. **13**(1), 65–82 (1984)
236. Higashi, M., Klir, G.J.: Identification of fuzzy relation systems. IEEE Trans. Syst. Man Cybern. **14**(2), 349–355 (1984)
237. Jin-wen, Z.: A unified treatment of fuzzy set theory and Boolean valued set theory: fuzzy set structures and normal fuzzy set structures. J. Math. Anal. Appl. **76**(1), 197–301 (1980)
238. Kandel, A., Byatt, W.J.: Fuzzy processes. Fuzzy Sets Syst. **4**(2), 117–152 (1980)
239. Kaufmann, A., Gupta, M.M.: Introduction to Fuzzy Arithmetic: Theory and Applications. Van Nostrand Rheinhold, New York (1991)
240. Kaufmann, A.: Introduction to the Theory of Fuzzy Subsets, vol. 1. Academic Press, New York (1975)
241. Klement, E.P., Schwyhla, W.: Correspondence between fuzzy measures and classical measures. Fuzzy Sets Syst. **7**(1), 57–70 (1982)
242. Klir, G., Yuan, B.: Fuzzy Sets and Fuzzy Logic. Prentice Hall, Upper Saddle River, NJ (1995)
243. Kruse, R., et al.: Foundations of Fuzzy Systems. John Wiley and Sons, New York (1994)
244. Lasker, G.E. (ed.): Applied Systems and Cybernetics, vol. VI: Fuzzy Sets and Systems. Pergamon Press, New York (1981)
245. Lientz, B.P.: On time dependent fuzzy sets. Inf. Sci. **4**, 367–376 (1972)
246. Lowen, R.: Fuzzy uniform spaces. J. Math. Anal. Appl. **82**(2), 367–376 (1981)
247. Michalek, J.: Fuzzy topologies. Kybernetika **11**, 345–354 (1975)

248. Negoita, C.V., et al.: Applications of Fuzzy Sets to Systems Analysis. Wiley and Sons, New York (1975)
249. Negoita, C.V.: Representation theorems for fuzzy concepts. Kybernetes **4**, 169–174 (1975)
250. Negoita, C.V., et al.: On the state equation of fuzzy systems. Kybernetes **4**, 231–214 (1975)
251. Netto, A.B.: Fuzzy classes. Notices Am. Math. Soc. **68T-H28**, 945 (1968)
252. Pedrycz, W.: Fuzzy relational equations with generalized connectives and their applications. Fuzzy Sets Syst. **10**(2), 185–201 (1983)
253. Raha, S., et al.: Analogy between approximate reasoning and the method of interpolation. Fuzzy Sets Syst. **51**(3), 259–266 (1992)
254. Ralescu, D.: Toward a general theory of fuzzy variables. J. Math. Anal. Appl. **86**(1), 176–193 (1982)
255. Rodabaugh, S.E.: Fuzzy arithmetic and fuzzy topology. In: Lasker, G.E. (ed.) Applied Systems and Cybernetics, vol. VI: Fuzzy Sets and Systems, pp. 2803–2807. Pergamon Press, New York (1981)
256. Rosenfeld, A.: Fuzzy groups. J. Math. Anal. Appl. **35**, 512–517 (1971)
257. Ruspini, E.H.: Recent developments in mathematical classification using fuzzy sets. In: Lasker, G.E. (ed.) Applied Systems and Cybernetics, vol. VI: Fuzzy Sets and Systems, pp. 2785–2790. Pergamon Press, New York (1981)
258. Santos, E.S.: Fuzzy algorithms. Inform. Control **17**, 326–339 (1970)
259. Stein, N.E., Talaki, K.: Convex fuzzy random variables. Fuzzy sets and Syst. **6**(3), 271–284 (1981)
260. Triantaphyllon, E., et al.: The problem of determining membership values in fuzzy sets in real world situations. In: Brown, D.E., et al. (eds.) Operations Research and Artificial Intelligence: The Integration of Problem-solving Strategies, pp. 197–214. Kluwer, Boston, Mass (1990)
261. Tsichritzis, D.: Participation measures. J. Math. Anal. Appl. **36**, 60–72 (1971)
262. Turksens, I.B.: Four methods of approximate reasoning with interval-valued fuzzy sets. Int. J. Approx. Reason. **3**(2), 121–142 (1989)
263. Turksen, I.B.: Measurement of membership functions and their acquisition. Fuzzy Sets Syst. **40**(1), 5–38 (1991)
264. Wang, P.P. (ed.): Advances in Fuzzy Sets, Possibility Theory, and Applications. Plenum Press, New York (1983)
265. Wang, Z., Klir, G.: Fuzzy Measure Theory. Plenum Press, New York (1992)
266. Wang, P.Z., et al. (eds.): Between Mind and Computer: Fuzzy Science and Engineering. World Scientific Press, Singapore (1993)
267. Wang, S.: Generating fuzzy membership functions: a monotonic neural network model. Fuzzy Sets Syst. **61**(1), 71–82 (1994)
268. Wong, C.K.: Fuzzy points and local properties of fuzzy topology. J. Math. Anal. Appl. **46**, 19874, 316–328
269. Wong, C.K.: Categories of fuzzy sets and fuzzy topological spaces. J. Math. Anal. Appl. **53**, 704–714 (1976)
270. Yager, R.R., et al. (eds.): Fuzzy Sets, Neural Networks, and Soft Computing. Nostrand Reinhold, New York (1994)
271. Zadeh, L.A.: A computational theory of decompositions. Int. J. Intell. Syst. **2**(1), 39–63 (1987)
272. Zimmerman, H.J.: Fuzzy Set Theory and Its Applications. Mass, Kluwer, Boston (1985)

Fuzzy Optimization, Information, Decision-Choice Theory and The Science of Knowing

273. Bose, R.K., Sahani, D.: Fuzzy mappings and fixed point theorems. Fuzzy Sets Syst. **21**, 53–58 (1987)
274. Butnariu, D.: Fixed points for fuzzy mappings. Fuzzy Sets Syst. **7**, 191–207 (1982)

275. Dompere, K.K.: Fuzziness, rationality, optimality and equilibrium in decision and economic theories. In: Lodwick, W.A., Kacprzyk, J. (eds.) Fuzzy Optimization: Recent Advances and Applications (Series: Studies in Fuzziness and Soft Computing, vol. 254) Springer, Berlin, Heidelberg (2010)
276. Eaves, B.C.: Computing Kakutani fixed points. J. Appl. Math. **21**, 236–244 (1971)
277. Heilpern, S.: Fuzzy mappings and fixed point theorem. J. Math. Anal. Appl. **83**, 566–569 (1981)
278. Kacprzyk, J., et al. (eds.): Optimization Models Using Fuzzy Sets and Possibility Theory. D. Reidel, Boston, Mass. (1987)
279. Kaleva, O.: A note on fixed points for fuzzy mappings. Fuzzy Sets Syst. **15**, 99–100 (1985)
280. Lodwick, W.A., Kacprzyk, J. (eds.): Fuzzy Optimization: Recent Advances and Applications, (Studies in Fuzziness and Soft Computing, Vol. 254). Springer, Berlin, Heidelberg (2010)
281. Negoita, C.V.: The current interest in fuzzy optimization. Fuzzy Sets Syst. **6**(3), 261–270 (1981)
282. Negoita, C.V., et al.: On fuzzy environment in optimization problems. In: Rose, J., et al. (eds.) Modern Trends in Cybernetics and Systems, pp. 13–24. Springer, Berlin (1977)
283. Zimmerman, H.-J.: Description and optimization of fuzzy systems. Int. J. Gen. Syst. **2**(4), 209–215 (1975)

Ideology, Disinformation, Misinformation and Propaganda in Intra- and Inter-Epistemological Masseging Systems

284. Abercrombie, N., et al.: The Dominant Ideology Thesis, London, Allen and Unwin (1980)
285. Abercrombie, N.: Class, Structure, and Knowledge: Problems in the Sociology of Knowledge. New York University Press, New York (1980)
286. Aron, R.: The Opium of the Intellectuals. MD, University Press of America, Lanham (1985)
287. Aronowitz, S.: Science as Power: Discourse and Ideology in Modern Society. University of Minnesota Press, Minneapolis (1988)
288. Barinaga, M., Marshall, E.: Confusion on the cutting edge. Science **257**, 616–625 (1992)
289. Barnett, R.: Beyond All Reason: Living with Ideology in the University. Society for Research into Higher Education and Open University Press, Philadelphia, PA. (2003)
290. Barth, H.: Truth and Ideology. University of California Press, Berkeley (1976)
291. Basin, A., Verdie, T.: The economics of cultural transmission and the dynamics of preferences. J. Econ. Theor. **97**, 298–319 (2001)
292. Bikhchandani, S., et al.: A theory of fads, fashion, custom, and cultural change. J. Polit. Econ. **100**, 992–1026 (1992)
293. Robert, B., Richerson, P.J.: Culture and Evolutionary Process. University of Chicago Press, Chicago (1985)
294. Buczkowski, P., Klawiter, A.: Theories of Ideology and Ideology of Theories. Rodopi, Amsterdam (1986)
295. Chomsky, N.: Manufacturing Consent. Pantheo Pess, New York (1988)
296. Chomsky, N.: Problem of Knowledge and Freedom. Collins, Glasgow (1972)
297. Cole, J.R.: Patterns of intellectual influence in scientific research. Soc. Educ **43**, 377–403 (1968)
298. Cole, J.R., Cole, S.: Social Stratification in Science. University of Chicago Press, Chicago (1973)
299. Debackere, K., Rappa, M.A.: Institutioal variations in problem choice and persistence among scientists in an emerging fields. Res. Policy **23**, 425–441 (1994)
300. Dompere, K.K.: Ideology and decision-choice rationalities. In: Dompere, K.K. (ed.) Fuzzy Rationality: Methodological Critique and Unity of Classical, Bounded and Other Rationalities, (Studies in Fuzziness and Soft Computing, vol. 235), pp. 143–165. Springer, New York (2009)

301. Dompere, K.K.: Culture, ideology and categorial conversion of social polarities in systems dynamics. In: Dompere, K.K. (ed.) The Theory of Philosophical Consciencism, pp. 124–158. Adonis & Abbey Pub, Practice Foundations of Nkrumaism in Social Systemicity, London (2017)
302. Dompere, K.K.: Abstract ideas and practice of ideas in social settings extensions and reflections on Nkrumah and Africa under systems thinking. In: Dompere, K.K. (ed.) The Theory of Categorial Conversion: Rational Foundations of Nkrumaism in Socio-Natural Systemicity and Complexity, pp. 61–85. Adonis & Abbey Pub, London (2017)
303. Dompere, K.K.: Zones of Thought: Reflections on the theories of Thought. In: Dompere, K.K. (ed.) Fuzziness and Foundations of Exact and Inexact Sciences, (Studies in Fuzziness and Soft Computing, vol. 290), pp. 103–116. Springer, New York (2013)
304. Fraser, C., Gaskell, G. (eds.): The Social Psychological Study of Widespread Beliefs. Clarendon Press, Oxford (1990)
305. Gieryn, T.F.: Problem retention and problem change in science. Sociol. Inquiry **48**, 96–115 (1978)
306. Harrington, J.E.J.: The rigidity of social systems. J. Polit. Econ. **107**, 40–64
307. Hinich, M., Munger, M.: Ideology and the Theory of Political Choice, Ann Arbor University of Michigan Press (1994)
308. Hull, D.L.: Science as a Process: An Evolutionary Account of the Social and Conceptual Development of Science. University of Chicago Press, Chicago (1988)
309. Marx, K., Engels, F.: The German Ideology. International Pub, New York (1970)
310. Mészáros, I.: Philosophy, Ideology and Social Science: Essay in Negation and Affirmation, Brighton, Sussex, Wheatsheaf (1986)
311. Mészáros, I.: The Power of Ideology. New York University Press, New York (1989)
312. Newcomb, T.M., et al.: Persistence and Change. Wiley, New York (1967)
313. Pickering, A.: Science as Practice and Culture. University of Chicago Press, Chicago (1992)
314. Therborn, G.: The Ideology of Power and the Power of Ideology. NLB Publications, London (1980)
315. Thompson, K.: Beliefs and Ideology. Tavistock Publication, New York (1986)
316. Ziman, J.: The problem of 'problem choice.' Minerva **25**, 92–105 (1987)
317. Ziman, J., Knowledge, P.: An Essay Concerning the Social Dimension of Science. Cambridge University Press, Cambridge (1968)
318. Zuckerman, H.: Theory choice and problem choice in science. Sociol. Inq. **48**, 65–95 (1978)

Information, Thought, Knowing and Knowledge

319. Aczel, J., Daroczy, Z.: On Measures of Information and their Characterizations. Academic Press, New York (1975)
320. Afanasyev, Social Information and Regulation of Social Development, Moscow, Progress (1978)
321. Anderson, J.R.: The Architecture of Cognition. Harvard University Press, Cambridge, Mass. (1983)
322. Angelov, S., Georgiev, D.: The problem of human being in contemporary scientific knowledge. Sov. Stud. Philos. Summer, 49–66 (1974)
323. Ash, R.: Information Theory. Wiley, New York (1965)
324. Bergin, J.: Common knowledge with monotone statistics. Econometrica **69**, 1315–1332 (2001)
325. Bestougeff, H., Ligozat, G.: Logical Tools for Temporal Knowledge Representation. Ellis Horwood, New York (1992)
326. Brillouin, L.: Science and information Theory. Academic Press, New York (1962)
327. Bruner, J.S., et al.: A Study of Thinking. Wiley, New York (1956)

328. Brunner, K., Meltzer, A.H. (eds.): Three Aspects of Policy and Policy Making: Knowledge, Data and Institutions, Carnegie-Rochester Conference Series, vol. 10. North-Holland, Amsterdam (1979)
329. Burks, A.W.: Chance, Cause, Reason: An Inquiry into the Nature of Scientific Evidence. University of Chicago Press, Chicago (1977)
330. Calvert, R.: Models of Imperfect Information in Politics. Hardwood Academic Publishers, New York (1986)
331. Cornforth, M.: The Theory of Knowledge. International Pub, New York (1972)
332. Cornforth, M.: The Open Philosophy and the Open Society. International Pub, New York (1970)
333. Coombs, C.H.: A Theory of Data. Wiley, New York (1964)
334. Dretske, F.I.: Knowledge and the Flow of Information. MIT Press, Cambridge, Mass., (1981)
335. Dreyfus, H.L.: A framework for misrepresenting knowledge. In: Ringle, M. (ed.) Philosophical Perspectives in Artificial Intelligence. Humanities press, Atlantic Highlands, N.J. (1979)
336. Fagin, R., et al.: Reasoning About Knowledge. MIT Press, Cambridge, Mass (1995)
337. Geanakoplos, J.: Common knowledge. J. Econ.Perspect. **6**, 53–82 (1992)
338. George, F.H.: Models of Thinking. Allen and Unwin, London (1970)
339. George, F.H.: Epistemology and the problem of perception. Mind **66**, 491–506 (1957)
340. Harwood, E.C.: Reconstruction of Economics. American Institute for Economic Research, Great Barrington, Mass (1955)
341. Hintikka, J.: Knowledge and Belief. Cornell University Press, Ithaca, N.Y. (1962)
342. Hirshleifer, J.: The private and social value of information and reward to inventive activity. Am. Econ. Rev. **61**, 561–574 (1971)
343. Kapitsa, P.L.: The influence of scientific ideas on society. Sov. Stud. Philos. Fall 52–71 (1979)
344. Kedrov, B.M.: The road to truth. Sov. Stud. Philos. **4**, 3–53 (1965)
345. Klatzky, R.L.: Human Memory: Structure and Processes. W. H. Freeman Pub, San Francisco, Ca. (1975)
346. Kreps, D., Wilson, R.: Reputation and imperfect information. J. Econ. Theor. **27**, 253–279 (1982)
347. Kubát, L., Zeman, J. (eds.): Entropy and Information. Elsevier, Amsterdam (1975)
348. Kurcz, G., Shugar, W., et al. (eds.): Knowledge and Language. North-Holland, Amsterdom (1986)
349. Lakemeyer, G., Nobel, B. (eds.): Foundations of Knowledge Representation and Reasoning. Springer, Berlin (1994)
350. Lektorskii, V.A.: Principles involved in the reproduction of objective in knowledge. Sov. Stud. Philos. **4**(4), 11–21 (1967)
351. Levi, I.: The Enterprise of Knowledge. MIT Press, Cambridge, Mass. (1980)
352. Levi, I.: Ignorance, probability and rational choice. Synthese **53**, 387–417 (1982)
353. Levi, I.: Four types of ignorance. Soc. Sci. **44**, 745–756
354. Marschak, J.: Economic Information, Decision and Prediction: Selected Essays, vol. II, Part II, Dordrecnt-Holland, Boston, Mass. (1974)
355. Menges, G. (ed.): Information, Inference and Decision. D. Reidel Pub., Dordrecht, Holland (1974)
356. Masuch, M., Pólos, L. (eds.): Knowledge Representation and Reasoning Under Uncertainty. Springer, New York (1994)
357. Moses, Y. (ed.): Proceedings of the Fourth Conference of Theoretical Aspects of Reasoning about Knowledge. Morgan Kaufmann, San Mateo (1992)
358. Nielsen, L.T., et al.: Common knowledge of aggregation expectations. Econometrica **58**, 1235–1239 (1990)
359. Newell, A.: Unified Theories of Cognition. Harvard University Press, Cambridge, Mass (1990)
360. Newell, A., Simon, H.A.: Human Problem Solving. Prentice-Hall, Englewood Cliff, N.J. (1972)
361. Ogden, G.K., I.A.: The Meaning of Meaning. Harcourt-Brace Jovanovich, New York (1923)
362. Planck, M.: Scientific Autobiography and Other Papers, Westport, Conn., Greenwood (1968)

363. Pollock, J.: Knowledge and Justification. Princeton, Princeton University Press (1974)
364. Polanyi, M.: Personal Knowledge. Routledge and Kegan Paul, London (1958)
365. Popper, K.R.: Objective Knowledge. Macmillan, London (1949)
366. Popper, K.R.: Open Society and it Enemies, vols. 1 and 2. Princeton University Press, Princeton (2013)
367. Popper, K.R.: The Poverty of Historicism. Taylor and Francis, New York (2002)
368. Price, H.H.: Thinking and Experience. Hutchinson, London (1953)
369. Putman, H.: Reason, Truth and History. Cambridge University Press, Cambridge (1981)
370. Putman, H.: Realism and Reason. Cambridge University Press, Cambridge (1983)
371. Putman, H.: The Many Faces of Realism. Open Court Publishing Co., La Salle (1987)
372. Russell, B.: Human Knowledge, its Scope and Limits. Allen and Unwin, London (1948)
373. Russell, B.: Our Knowledge of the External World. Norton, New York (1929)
374. Samet, D.: Ignoring ignorance and agreeing to disagree. J. Econ. Theor. **52**, 190–207 (1990)
375. Schroder, H.M., Suedfeld, P. (eds.): Personality Theory and Information Processing. Ronald Pub. New York (1971)
376. Searle, J.: Minds, Brains and Science. Harvard University Press, Cambridge, Mass. (1985)
377. Shin, H.: Logical structure of common knowledge. J. Econ. Theor. **60**, 1–13 (1993)
378. Simon, H.A.: Models of Thought. Yale University Press, New Haven, Conn. (1979)
379. Smithson, M.: Ignorance and Uncertainty, Emerging Paradigms. Springer, New York (1989)
380. Sowa, J.F.: Knowledge Representation: Logical, Philosophical, and Computational Foundations. Pacific Grove, Brooks Pub. (2000)
381. Stigler, G.J.: The economics of information. J. Polit. Econ. **69**, 213–225 (1961)
382. Tiukhtin, V.S.: How reality can be reflected in cognition: reflection as a property of all matter. Sov. Stud. Philos. **3**(1), 3–12 (1964)
383. Tsypkin, Y.Z.: Foundations of the Theory of Learning Systems. Academic Press, New York (1973)
384. Ursul, A.D.: The problem of the objectivity of information. In: Kubát, L., Zeman, J. (eds.) Entropy and Information, Amsterdam, pp. 187–230. Elsevier (1975)
385. Vardi, M. (ed.): Proceedings of Second Conference on Theoretical Aspects of Reasoning about Knowledge. Asiloman, Ca., Los Altos, Ca, Morgan Kaufman (1988)
386. Vazquez, M., et al.: Knowledge and reality: some conceptual issues in system dynamics modelling. Syst. Dyn. Rev. **12**, 21–37 (1996)
387. Zadeh, L.A.: A theory of commonsense knowledge. In: Skala, H.J., et al. (eds.) Aspects of Vagueness, pp. 257–295. D. Reidel Co., Dordrecht (1984)
388. Zadeh, L.A.: The concept of linguistic variable and its application to approximate reasoning. Inf. Sci. **8**, 199–249 (1975). (Also in vol. 9, pp. 40–80)

Languages, Information and Representation

389. Agha, A.: Language and Social Relations. Cambridge University Press, Cambridge (2006)
390. Aitchison, J. (ed.): Language Change: Progress or Decay? Cambridge University Press, Cambridge, New York, Melbourne (2001)
391. Anderson, S.: Languages: A Very Short Introduction. Oxford University Press, Oxford (2012)
392. Aronoff, M., Fudeman, K.: What is Morphology. Wiley, New York (2011)
393. Bauer, L. (ed.): Introducing Linguistic Morphology. Georgetown University Press, Washington, D.C. (2003)
394. Barber, A., Stainton, R.J. (eds.): Concise Encyclopedia of Philosophy of Language and Linguistics. Elsevier, New York (2010)
395. Brown, K., Ogilvie, S. (eds.): Concise Encyclopedia of Languages of the World. Elsevier Science, New York (2008)

396. Campbell, L. (ed.): Historical Linguistics: an Introduction Cambridge. MIT Press, MASS (2004)
397. Chomsky, N.: Syntactic Structures. The Hague, Mouton (1957)
398. Chomsky, N.: The Architecture of Language. Oxford University Press, Oxford (2000)
399. Clarke, D.S.: Sources of Semiotic: Readings with Commentary from Antiquity to the Present. Southern Illinois University Press, Carbondale (1990)
400. Collinge, N.E. (ed.): An Encyclopedia of Language. Routledge, London, New York (1989)
401. Comrie, B. (ed.): Language Universals and Linguistic Typology: Syntax and Morphology. Oxford, Blackwell (1989)
402. Comrie, B. (ed.): The World's Major Languages. Routledge, New York (2009)
403. Coulmas, F., Systems, W.: An Introduction to Their Linguistic Analysis. Cambridge University Press, Cambridge (2002)
404. Croft, W., Cruse, D.A.: Cognitive Linguistics. Cambridge University Press, Cambridge (2004)
405. Croft, W.: Typology. In: Aronoff, M., Rees-Miller, J. (eds.) The Handbook of Linguistics, pp. 81–105. Oxford, Blackwell (2001)
406. Crystal, D. (ed.): The Cambridge Encyclopedia of Language. Cambridge University Press, Cambridge (1997)
407. Deacon, T.: The Symbolic Species: The Co-evolution of Language and the Brain. Norton & Company, New York: W.W. (1997)
408. Deming, W.E.: On errors in surveys. Am. Soc. Rev. **IX**, 359–369 (1944)
409. Devitt, M., Sterelny, K.: Language and Reality: An Introduction to the Philosophy of Language. MIT Press, Boston (1999)
410. Duranti, A.: Language as culture in U.S. anthropology: three paradigms. Curr. Anthropol. **44**(3), 323–348 (2003)
411. Evans, N., Levinson, S.C.: The myth of language universals: language diversity and its importance for cognitive science. Behav. Brain Sci. **32**(5), 429–492 (2009)
412. Fitch, W.T.: The Evolution of Language. Cambridge University Press, Cambridge (2010)
413. Foley, W.A., Linguistics, A.: An Introduction. Blackwell, Oxford (1997)
414. Ginsburg, S.: Algebraic and Automata-Theoretic Properties of Formal Languages. North-Holland, New York (1973)
415. Goldsmith, J.A.: The Handbook of Phonological Theory: Blackwell Handbooks in Linguistics. Blackwell Publishers, Oxford (1995)
416. Greenberg, J.: Language Universals: With Special Reference to Feature Hierarchies. The Hague, Mouton & Co. (1966)
417. Hauser, M.D., Chomsky, N., Fitch, W.T.: The faculty of language: what is it, who has it, and how did it evolve? Science, 22 **298**(5598), 1569–1579 (2002)
418. Harz, H.: Information, sign, image. In: Kubát, L., Zeman, J. (eds.) Entropy and Information in Science and Philosophy. Elsevier, New York (1975)
419. International Phonetic Association.: Handbook of the International Phonetic Association: A guide to the use of the International Phonetic Alphabet. Cambridge University Press, Cambridge (1999)
420. Katzner, K.: The Languages of the World. Routledge, New York (1999)
421. Labov, W.: Principles of Linguistic Change vol. I Internal Factors. Blackwell, Oxford (1994)
422. Labov, W.: Principles of Linguistic Change vol. II Social Factors. Blackwell, Oxford (2001)
423. Levinson, S.C.: Pragmatics. Cambridge University Press, Cambridge (1983)
424. Lewis, M.P. (ed.) Ethnologue: Languages of the World. SIL International, Dallas, Tex. (2009)
425. Lyons, J.: Language and Linguistics. Cambridge University Press, Cambridge (1981)
426. MacMahon.: April M.S., Understanding Language Change. Cambridge University Press, Cambridge (1994)
427. Matras, Y., Bakker, P. (eds.) The Mixed Language Debate: Theoretical and Empirical Advances. Walter de Gruyter, Berlin (2003)
428. Moseley, C. (ed.): Atlas of the World's Languages in Danger, Paris. UNESCO Publishing (2010)

429. Nerlich, B.: History of pragmatics. In: Cummings, L. (ed.) The Pragmatics Encyclopedia, pp. 192–93. Routledge, New York (2010)
430. Newmeyer, F.J.: The History of Linguistics. Linguistic Society of America (2005)
431. Newmeyer, F.J.: Language Form and Language Function (PDF). MIT Press, Cambridge, MA (1998)
432. Nichols, J.: Linguistic Diversity in Space and Time. University of Chicago Press, Chicago (1992)
433. Nichols, J.: Functional theories of grammar. Annu. Rev. Anthropol. **13**, 7–117 (1949)
434. Sandler, W., Lillo-Martin, D.: Natural sign languages. In: Aronoff, M., Rees-Miller, J. (eds.) The Handbook of Linguistics, pp. 533–563. Blackwell, Oxford (2001)
435. Swadesh, M.: The phonemic principle. Language **10**(2), 117–129 (1934)
436. Tomasello, M.: The cultural roots of language. In: Velichkovsky, B., Rumbaugh, D. (eds.) Communicating Meaning: The Evolution and Development of Language, pp. 275–308. Psychology Press, New York (1996)
437. Tomasello, M.: Origin of Human Communication. MIT Press, Cambridge Mass. (2008)
438. Thomason, S.G.: Language Contact—An Introduction. Edinburgh University Press, Edinburgh (2001)
439. Ulbaek, I.B.: The origin of language and cognition. In: Hurford, J.R., Knight, C. (eds.) Approaches to the Evolution of Language, pp. 30–43. Cambridge University Press, Cambridge (1998)
440. Van Valin, jr, R.D.: Functional linguistics. In: Aronoff, M., Rees-Miller, J. (eds.) The Handbook of Linguistics, pp. 319–337. Blackwell, Oxford (2001)

Language, Knowledge-Production Process and Epistemics

441. Aho, A.V.: Indexed grammar—an extension of context-free grammars. J. Assoc. Comput. Mach. **15**, 647–671 (1968)
442. Black, M. (ed.): The Importance of Language. Prentice-Hall, Englewood Cliffs, N.J (1962)
443. Carnap, R.: Meaning and Necessity: A Study in Semantics and Modal Logic. University of Chicago Press, Chicago (1956)
444. Chomsky, N.: Linguistics and philosophy. In: Hook, S. (ed.) Language and Philosophy, pp. 51–94. New York University Press, New York (1968)
445. Chomsky, N.: Language and Mind. Harcourt Brace Jovanovich, New York (1972)
446. Cooper, W.S.: Foundations of Logico-Linguistics: A Unified Theory of Information, Language and Logic. D. Reidel, Dordrecht (1978)
447. Cresswell, M.J.: Logics and Languages. Methuen Pub, London (1973)
448. Dilman, I.: Studies in Language and Reason. Barnes and Nobles, Books, Totowa, N.J. (1981)
449. Fodor, J.A.: The Language and Thought. New York, Thom as Y. Crowell Co (1975)
450. Givon, T.: On Understanding Grammar. Academic Press, New York (1979)
451. Gorsky, D.R.: Definition. Progress Publishers, Moscow (1974)
452. Hintikka, J.: The Game of Language. D. Reidel Pub, Dordrecht (1983)
453. Johnson-Lair, P.N., Models, M.: Toward Cognitive Science of Language, Inference and Consciousness. Harvard University Press, Cambridge, Mass (1983)
454. Kandel, A.: Codes over languages. IEEE Trans. Syst. Man Cybern. **4**, 135–138 (1975)
455. Keenan, E.L., Faltz, L.M.: Boolean Semantics for Natural Languages. D. Reidel Pub, Dordrecht (1985)
456. Lakoff, G.: Linguistics and natural logic. Synthese **22**, 151–271 (1970)
457. Lee, E.T., et al.: Notes on fuzzy languages. Inf. Sci. **1**, 421–434 (1969)
458. Mackey, A., Merrill, D. (eds.): Issues in the Philosophy of Language. Yale University Press, New Haven, CT (1976)

459. Nagel, T.: Linguistics and epistemology. In: Hook, S. (ed.) Language and Philosophy, pp. 180–184. New York University Press, New York (1969)
460. Pike, K.: Language in Relation to a Unified Theory of Structure of Human Behavior. Mouton Pub, The Hague (1969)
461. Quine, W.V.O.: Word and Object. MIT Press, Cambridge, Mass (1960)
462. Russell, B.: An Inquiry into Meaning and Truth. Penguin Books (1970)
463. Tarski, A.: Logic, Semantics and Metamathematics. Clarendon Press, Oxford (1956)
464. Whorf, B.L. (ed.): Language, Thought and Reality. Humanities Press, New York (1956)

Language, Knowledge-Production Process, Knowing, Ontic, Imagination, Reality and Epistemics

465. Atmanspacher, H., Kronz, F.: Many realisms. Acta Polytechnica Scandinavica, Ma-91, 31–43 (1198)
466. Atmanspacher, H., Wiedenmann, G., Amann, A.: Descartes revisited—the endo/exo-distinction and its relevance for the study of complex systems. Complexity **1**(3), 15–21 (1995)
467. Atmanspacher, D.H., Primas, H.: Epistemic and ontic quantum realities. In: Khrennikov, A. (ed.) Foundations of Probability and Physics, pp. 49–61. American Institute of Physics (2005). Originally published in Time, Quantum and Information, Lutz Castell and Otfried Ischebeck (Eds.), pp. 301–32.Springer, Berlin (2003)
468. Atmanspacher, H.: Determinism is Ontic, Determinability is Epistemic, University of Pittsburg Archives, Pittsburgh, Pa (2001)
469. Berger, P.L., Luckmann, T.: The Social Construction of Reality: A Treatise in the Sociology of Knowledge, p. 1966. Anchor Books, New York (1966)
470. Bohm, D.: Wholeness and the Implicate Order. Routledge and Kegan Paul, London (1980)
471. Boyd, R.: Determinism, laws and predictability in principle. Philosophy of Science **39**, 431–450 (1972)
472. Campbell, J.: Past, Space, and Self. MIT Press, Mass., Cambridge (1994)
473. Carruthers, P.: Phenomenal Consciousness, p. 2000. Cambridge University Press, Mass., Cambridge (2000)
474. Caves, C.M.: Information and entropy. Phys. Rev. A **47**(1993), 4010–4017 (1993)
475. Caves, C.: Information, entropy, and chaos. In: Halliwell, J.J., Pérez-Mercader, J., Zurek, W.H. (eds.) Physical Origins of Time Asymmetry, pp. 47–89. Cambridge University Press, Cambridge, (1994)
476. Chalmers, D.: The Conscious Mind. Oxford University Press, Oxford, UK (1996)
477. Chalmers, D.: Facing up to the problem of consciousness. J. Conscious. Stud. **2**, 200–2019 (1995)
478. Crutchfield, J.P.: Observing complexity and the complexity of observation. In: Atmanspacher, H., Dalenoort, G.J. (eds.) Inside Versus Outside, pp. 235–272. Springer, Berlin (1994)
479. Einstein, A., Podolsky, B., Rosen, N.: Can quantum-mechanical description of physical reality be considered complete? Phys. Rev. **47**, 777–780 (1935)
480. Elskens, Y., Prigogine, I.: From instability to irreversibility. Proc. Natl. Acad. Sci. USA, **83**, 5756–5760.198 (1986)
481. Faisal, A.: Computer Science: visionary of virtual reality. Nature **551**(7680), 298–299 (2017)
482. Farmer, D.: Information dimension and the probabilistic structure of chaos. Z. Naturforsch **37a**, 1304–1325 (1982)
483. Fetzer, J.H., Almeder, R.F.: Glossary of Epistemology/Philosophy of Science, p. 100f. Paragon House, New York (1993)
484. Lycan, W.: Consciousness. MIT Press, Cambridge, MASS (1987)
485. Lycan, W.: Consciousness and Experience. MIT Press, Cambridge, MASS (1996)

486. Mensah, T.O.: Fiber Optics Engineering: Processing and Applications (Aiche Symposium Series), Amer Inst. Of Chemical Engineers, New York, N.Y (1987)
487. Miller, A.: What is the Manifestation argument? In: Pacific Philosophical Quarterly, vol. 83, pp. 352–383 (2002)
488. Miller, A.: The significance of semantic realism. Synthese **136**, 191–217 (2003)
489. Miller, A.: What is the acquisition argument? In: Barber, A. (ed.) Epistemology of Language, pp. 459–495. Oxford University Press, Oxford (2003)
490. Miller, A.: Realism and antirealism. In: Leporc, E., Smith, B. (eds.) A Handbook of Philosophy of Language, pp. 983–1005. Oxford University Press, Oxford (2006)
491. Miller, A.: Rule-following, error theory and eliminativism. Int. J. Philos. Stud. **23**, 323–336 (2015)
492. Putnam, H.: Reason, Truth and History. Cambridge University Press, Cambridge (1981)
493. Putnam, H.: Realism and Reason. Cambridge University Press, Cambridge (1983)
494. Railton, P.: Moral realism. Philos. Rev. **95**, 163–207 (1986)
495. Searle, J.: The Rediscovery of the Mind. MIT Press, Cambridge, MASS. (1992)
496. Seager, W.: Consciousness, information, and panpsychism. J. Conscious. Stud. **2**, 272–288 (1995)
497. Seigel, S.: The Contents of Visual Experience. Oxford University Press, Oxford (2010)
498. Siewert, C.: The Significance of Consciousness. Princeton University Press, Princeton, NJ (1998)
499. Shallice, T.: From Neuropsychology to Mental Structure. Cambridge University Press, Cambridge, MASS (1988)
500. Shear, J.: Explaining Consciousness: The Hard Problem. MIT Press, Cambridge, MASS (1997)
501. Wiebers, D.: Theory of Reality: Evidence for Existence Beyond the Brain and Tools for Your Journey. Amazon, New York (2012)
502. Wright, C.: Frege's Conception of Numbers as Objects. Aberdeen University Press, Aberdeen (1983)
503. Wright, C.: Truth and Objectivity. Harvard University Press, Cambridge, MASS. (1992)
504. Wright, C.: Realism, Meaning, and Truth, 2nd edn. Oxford, Blackwell (1993)
505. Wright, C.: Saving the Differences: Essays on Themes from Truth and Objectivity. Harvard University Press, Cambridge MASS. (2003)
506. Wright, C.: Rule-following without reasons: Wittgenstein's Quietism and the constitutive question. Ratio (New Series) **20**, 481–502 (2007)
507. Wright, C.: The ontic conception of scientific explanation. Stud. History Philos. Sci. **54**(4), 20–30 (2015)

Philosophy of Information and Semantic Information

508. Aisbett, J., Gibbon, G.: A practical measure of the information in a logical theory. J. Exp. Theor. Artif. Intell. **11**(2), 201–218 (1999)
509. Badino, M.: An application of information theory to the problem of the scientific experiment. Synthese **140**, 355–389 (2004)
510. Bar-Hillel, Y. (ed.): Language and Information: Selected Essays on Their Theory and Application. Addison-Wesley, Reading (1964)
511. Bar-Hillel, Y., Carnap, R.: An outline of a theory of semantic information, (1953). In: Bar-Hillel, Y. (ed.) Language and Information: Selected Essays on Their Theory and Application, pp. 221–274. Addison-Wesley, Reading (1964)
512. Barwise, J., Seligman, J.: Information Flow: The Logic of Distributed Systems. University Press, Cambridge, Cambridge (1997)
513. Braman, S.: Defining information. Telecommun. Policy **13**, 233–242 (1989)

514. Bremer, M.E.: Do logical truths carry information? Mind. Mach. **13**(4), 567–575 (2003)
515. Bremer, M., Cohnitz, D.: Information and Information Flow: an Introduction. Ontos Verlag, Frankfurt, Lancaster (2004)
516. Chaitin, G.J.: Algorithmic Information Theory. Cambridge University Press, Cambridge (1987)
517. Chalmers, D.J.: The Conscious Mind: In Search of a Fundamental Theory. Oxford University Press, New York (1996)
518. Cherry, C.: On Human Communication: A Review, a Survey, and a Criticism. MIT Press, Cambridge (1978)
519. Colburn, T.R.: Philosophy and Computer Science. Sharpe, Armonk, M.E. (2000)
520. Cover, T.M., Thomas, J.A.: Elements of Information Theory. Wiley, New York (1991)
521. Dennett, D.C.: Intentional systems. J. Philos. **68**, 87–106 (1971)
522. Deutsch, D., The Fabric of Reality. London Penguin (1997)
523. Devlin, K.J.: Logic and Information. Cambridge University Press, Cambridge (1991)
524. Fetzer, J.H.: Information, misinformation, and disinformation. Mind. Mach. **14**(2), 223–229 (2004)
525. Floridi, L.: Philosophy and Computing: An Introduction. Routledge, London (1999)
526. Floridi, L.: What is the philosophy of information? Metaphilosophy **33**(1–2), 123–145 (2002)
527. Floridi, L.: Two approaches to the philosophy of information. Mind. Mach. **13**(4), 459–469 (2003)
528. Floridi, L.: Open problems in the philosophy of information. Metaphilosophy **35**(4), 554–582 (2004)
529. Floridi, L.: Outline of a theory of strongly semantic information. Mind. Mach. **14**(2), 197–222 (2004)
530. Floridi, L.: Is information meaningful data? Philos. Phenomen. Res. **70**(2), 351–370 (2005)
531. Fox, C.J.: Information and Misinformation: An Investigation of the Notions of Information. Informing, and Misinforming, Westport Greenwood Press, Misinformation (1983)
532. Frieden, B.R.: Science from Fisher Information: A Unification. Cambridge University Press, Cambridge (2004)
533. Golan, A.: Information and entropy econometrics—editor's view. J. Econ. **107**(1–2), 1–15 (2002)
534. Graham, G.: The Internet: A Philosophical Inquiry. Routledge, London (1999)
535. Grice, H.P.: Studies in the Way of Words. Harvard University Press, Cambridge (1989)
536. Hanson, P.P. (ed.): Information, language, and cognition. University of British Columbia Press, Vancouver (1990)
537. Harms, W.F.: The use of information theory in epistemology. Philos. Sci. **65**(3), 472–501 (1998)
538. Heil, J.: Levels of reality. Ratio **16**(3), 205–221 (2003)
539. Hintikka, J., Suppes, P. (eds.): Information and Inference. Reidel, Dordrecht (1970)
540. Kemeny, J.: A logical measure function. J. Symb. Log. **18**, 289–308 (1953)
541. Kolin, K.K.: The nature of information and philosophical foundations of informatics. Open Educ. **2**, 43–51 (2005)
542. Kolin, K.K.: The evolution of informatics. Inf. Technol. **1**, 2–16 (2005)
543. Kolin, K.K.: The formation of informatics as basic science and complex scientific problems. In: Kolin, K. (ed.) Systems and Means of Informatics. Special Issue. Scientific and Methodological Problems of Informatics, pp. 7–57. IPI RAS, Moscow (2006)
544. Kolin, K.K.: Fundamental Studies in informatics: a general analysis, trends and prospects. Sci. Tech. Inf. **1**(7), 5–11 (2007)
545. Kolin, K.K.: Structure of reality and the phenomenon of information. Open Educ. **5**, 56–61 (2008)
546. Losee, R.M.: A discipline independent definition of information. J. Am. Soc. Inf. Sci. **48**(3), 254–269 (1997)
547. Lozinskii, E.: Information and evidence in logic systems. J. Exp. Theor. Artif. Intell. **6**, 163–193 (1994)

548. Machlup, F., Mansfield, U. (eds.): The Study of Information: Interdisciplinary Messages. Wiley, New York (1983)
549. MacKay, D.M.: Information, Mechanism and Meaning. MIT Press, Cambridge (1969)
550. Marr, D.: Vision: A Computational Investigation into the Human Representation and Processing of Visual Information. W.H. Freeman, San Francisco (1982)
551. Mingers, J.: The nature of information and its relationship to meaning. In: Winder, R.L., et al. (eds.) Philosophical Aspects of Information Systems, pp. 73–84. Taylor and Francis, London (1997)
552. Nauta, D.: The Meaning of Information. Mouton, The Hague (1972)
553. Newell, A.: The knowledge level. Artif. Intell. **18**, 87–127 (1982)
554. Newell, A., Simon, H.A.: Computer science as empirical inquiry: symbols and search. Commun. ACM **19**, 113–126 (1976)
555. Pierce, J.R.: An Introduction to Information Theory: Symbols, Signals and Noise. Dover Publications, New York (1980)
556. Poli, R.: The basic problem of the theory of levels of reality. Axiomathes, **12**, 261–283 (2001)
557. Sayre, K.M.: Cybernetics and the Philosophy of Mind. Routledge and Kegan Paul, London (1976)
558. Simon, H.A.: The Sciences of the Artificial. MIT Press, Cambridge (1996)
559. Smokler, H.: Informational content: a problem of definition. J. Philos. **63**(8), 201–211 (1966)
560. Ursul, A.D.: The Nature of the Information. Philosophical Essay. Politizdat, Moscow (1968)
561. Ursul, A.D.: Information. Methodological Aspects. Nauka, Moscow (1971)
562. Ursul, A.D.: Reflection and Information. Nauka, Moscow (1973)
563. Ursul, A.D.: The Problem of Information in Modern Science: Philosophical Essays. Nauka, Moscow (1975)
564. Ursul, A.D.: The problem of the objectivity of information. In: Kubatir, L., Zeman, J. (eds.) Entropy and Information in Science and Philosophy. Elsevier, New York (1975)
565. Weaver, W.: The mathematics of communication. Sci. Am. **181**(1), 11–15 (1949)
566. Winder, R.L., Probert, S.K., Beeson, I.A.: Philosophical Aspects of Information Systems. Taylor & Francis, London (1997)

Planning, Prescriptive Science and Information in Cost-Benefit Analytics

567. Alexander, E.R.: Approaches to Planning. Gordon and Breach, Philadelphia, Pa (1992)
568. Bailey, J.: Social Theory for Planning. Routledge and Kegan Paul, London (1975)
569. Burchell, R.W., Sternlieb, G. (eds.): Planning Theory in the 1980's: A Search for Future Directions. Rutgers University Center for Urban and Policy Research, New Brunswick, N.J (1978)
570. Camhis, M.: Planning Theory and Philosophy. Tavistock Publicationa, London (1979)
571. Chadwick, G.: A Systems View of Planning. Pergamon, Oxford (1971)
572. Cooke, P.: Theories of Planning and Special Development. Hutchinson, London (1983)
573. Dompere, K.K., Lawrence, T.: Planning. In: Hussain, S.B. (ed.) Encyclopedia of Capitalism, vol. II, pp. 649–653. Facts On File, Inc., New York (2004)
574. Dompere, K.K.: Social Goal-Objective Formation, Democracy and National Interest: A Theory of Political Economy under Fuzzy Rationality, (Studies in Systems, Decision and Control, vol. 4). Springer, New York (2014)
575. Dompere, K.K.: Fuzziness, Democracy, Control and Collective Decision-Choice System: A Theory on Political Economy of Rent-Seeking and Profit-Harvesting, (Studies In Systems, Decision and Control, vol. 5). Springer, New York (2014)
576. Dompere, K.K.: The Theory of Aggregate Investment in Closed Economic Systems. Greenwood Press, Westport (1999)

577. Dompere, K.K.: The Theory of Aggregate Investment and Output Dynamics in Open Economic Systems. Greenwood Press, Westport (1999)
578. Faludi, A.: Planning Theory. Pergamon, Oxford (1973)
579. Faludi, A. (ed.): A Reader in Planning Theory. Pergamon, Oxford (1973)
580. Harwood, E.C. (ed.),: Reconstruction of economics, american institute for economic research, Great Barrington, Mass, 1955., Also in John Dewey and Arthur Bently, 'Knowing and the known', Boston, Beacon Press, p. 269 (1949)
581. Kickert, W.J.M.: Organization of Decision-Making A Systems-Theoretic Approach. North-Holland, New York (1980)
582. Knight, F.H.: Risk, Uncertainty and Profit. University of Chicago Press, Chicago (1971)
583. Knight, F.H.: On History and Method of Economics. University of Chicago Press, Chicago (1971)

Possible-Actual Worlds and Information Analytics

584. Adams, R.M.: Theories of actuality. Noûs **8**, 211–231 (1974)
585. Allen, S. (ed.): Possible Worlds in Humanities, Arts and Sciences, Proceedings of Nobel Symposium, vol. 65. Walter de Gruyter Pub. New York (1989)
586. Armstrong, D.M.: A Combinatorial Theory of Possibility. Cambridge University Press (1989)
587. Armstrong, D.M.: A World of States of Affairs. Cambridge University Press, Cambridge (1997)
588. Bell, J.S.: Six possible worlds of quantum mechanics. In: Allen, S. (ed.) Possible Worlds in Humanities, Arts and Sciences, Proceedings of Nobel Symposium, vol. 65, pp. 359–373. Walter de Gruyter Pub., New York (1989)
589. Bigelow, J.: Possible Worlds foundations for probability. J. Philos. Log. **5**, 299–320 (1976)
590. Bradley, R., Swartz, N.: Possible World: An Introduction to Logic and its Philosophy. Bail Blackwell, Oxford (1997)
591. Castañeda, H.-N.: Thinking and the structure of the world. Philosophia **4**, 3–40 (1974)
592. Chihara, C.S.: The Worlds of Possibility: Modal Realism and the Semantics of Modal Logic. Clarendon (1998)
593. Chisholm, R.: Identity through possible worlds: some questions. Noûs 1, 1–8; reprinted in Loux, The Possible and the Actual (1967)
594. Divers, J.: Possible Worlds. Routledge, London (2002)
595. Forrest, P.: Occam's Razor and possible worlds. Monist **65**, 456–464 (1982)
596. Forrest, P., Armstrong, D.M.: An argument against David Lewis' theory of possible worlds. Aust. J. Philos. **62**, 164–168 (1984)
597. Grim, P.: There is no set of all truths. Analysis **46**, 186–191 (1986)
598. Heller, M.: Five Layers of interpretation for possible worlds. Philos. Stud. **90**, 205–214 (1998)
599. Herrick, P.: The Many Worlds of Logic. Oxford University Press, Oxford (1999)
600. Krips, H.: Irreducible probabilities and indeterminism. J. Philos. Log. **18**, 155–172 (1989)
601. Kuhn, T.S.: Possible worlds in history of science. In: Allen, S. (ed.) Possible Worlds in Humanities, Arts and Sciences, Proceedings of Nobel Symposium, vol. 65, pp. 9–41. Walter de Gruyter Pub., New York (1989)
602. Kuratowski, K., Mostowski, A.: Set Theory: With an Introduction to Descriptive Set Theory. North-Holland, New York (1976)
603. Lewis, D.: On the Plurality of Worlds. Basil Blackwell, Oxford (1986)
604. Loux, M.J. (ed.): The Possible and the Actual: Readings in the Metaphysics of Modality. Cornell University Press, Ithaca & London (1979)
605. Parsons, T.: Nonexistent Objects. Yale University Press, New Haven (1980)
606. Perry, J.: From worlds to situations. J. Philos. Log. **15**, 83–107 (1986)

607. Rescher, N., Brandom, R.: The Logic of Inconsistency: A Study in Non-Standard Possible-World Semantics and Ontology. Rowman and Littlefield (1979)
608. Skyrms, B.: Possible worlds, physics and metaphysics. Philos. Stud. **30**, 323–332 (1976)
609. Stalmaker, R.C.: Possible world. Nous **10**, 65–75 (1976)
610. Quine, W.V.O.: Word and Object. Press, M.I.T (1960)
611. Quine, W.V.O.: Ontological relativity. J. Philos. **65**, 185–212 (1968)

Rationality, Information, Games, Conflicts and Exact Reasoning

612. Border, K.: Fixed Point Theorems with Applications to Economics and Game Theory. Cambridge University Press, Cambridge (1985)
613. Brandenburger, A.: Knowledge and equilibrium games. J. Econ. Perspect. **6**, 83–102 (1992)
614. Campbell, R., Sowden, L.: Paradoxes of Rationality and Cooperation: Prisoner's Dilemma and Newcomb's Problem. University of British Columbia Press, Vancouver (1985)
615. Scott, G., Humes, B.: Games, Information, and Politics: Applying Game Theoretic Models to Political Science. University of Michigan Press, Ann Arbor (1996)
616. Gjesdal, F.: Information and incentives: the agency information problem. Rev. Econ. Stud. **49**, 373–390 (1982)
617. Harsanyi, J.: Games with incomplete information played by 'Bayesian' players I: the basic model. Manage. Sci. **14**, 159–182 (1967)
618. Harsanyi, J.: Games with incomplete information played by 'Bayesian' players II: Bayesian equilibrium points. Manage. Sci. **14**, 320–334 (1968)
619. Harsanyi, J.: Games with incomplete information played by 'Bayesian' players III: the basic probability distribution of the game. Manage. Sci. **14**, 486–502 (1968)
620. Harsanyi, J.: Rational Behavior and Bargaining Equilibrium in Games and Social Situations. Cambridge University Press, New York (1977)
621. Krasovskii, N.N., Subbotin, A.I.: Game-Theoretical Control Problems. Springer, New York (1988)
622. Lagunov, V.N.: Introduction to Differential Games and Control Theory. Heldermann Verlag, Berlin (1985)
623. Maynard Smith, J.: Evolution and the Theory of Games. Cambridge University Press, Cambridge (1982)
624. Myerson, R., Theory, G.: Analysis of Conflict. Harvard University Press, Mass (1991)
625. Rapoport, A., Chammah, A.: Prisoner's Dilemma: A Study in Conflict and Cooperation. University of Michigan Press, Ann Arbor (1965)
626. Roth, A.E.: The economist as engineer: game theory, experimentation, and computation as tools for design economics. Econometrica **70**, 1341–1378 (2002)
627. Shubik, M.: Game Theory in the Social Sciences: Concepts and Solutions. MIT Press, Mass. (1982)

Social Sciences, Mathematics and the Problems of Exact and Inexact Information

628. Ackoff, R.L.: Scientific Methods: Optimizing Applied Research Decisions. Wiley, New York (1962)
629. Angyal, A.: The structure of wholes. Philos. Sci. **6**(1), 23–37 (1939)
630. Bahm, A.J.: Organicism: The philosophy of interdependence. Int. Philos. Q. **VII**(2) (1967)

631. Bealer, G.: Quality and Concept. Clarendon Press, Oxford (1982)
632. Black, M., Thinking, C.: Englewood Cliffs. Prentice-Hall, N.J. (1952)
633. Brewer, M.B., Collins, B.E. (eds.): Scientific Inquiry and Social Sciences. Jossey-Bass Pub., San Francisco, Ca (1981)
634. Campbell, D.T.: On the conflicts between biological and social evolution and between psychology and moral tradition. Am. Psychol. **30**, 1103–1126 (1975)
635. Churchman, C.W., Ratoosh, P. (eds.): Measurement: Definitions and Theories. John Wiley, New York (1959)
636. Foley, D.: Problems versus conflicts economic theory and ideology. Am. Econ. Assoc. Papers Proc. **65**, 231–237 (1975)
637. Garfinkel, A.: Forms of Explanation: Structures of Inquiry in Social Science. Yale University Press, Conn. (1981)
638. Georgescu-Roegen, N.: Analytical Economics. Harvard University Press, Cambridge (1967)
639. Gillespie, C.: The Edge of Objectivity. Princeton, Princeton University Press (1960)
640. Hayek, F.A.: The Counter-Revolution of Science. Free Press of Glencoe Inc., New York (1952)
641. Laudan, L.: Progress and Its Problems: Towards a Theory of Scientific Growth. University of California Press, Berkeley (1961)
642. Marx, K.: The Poverty of Philosophy. International Pub, New York (1971)
643. Phillips, D.C.: Holistic Thought in Social Sciences. Stanford University Press, Stanford (1976)
644. Popper, K.: Objective Knowledge. Oxford University Press, Oxford (1972)
645. Rashevsky, N.: Organismic sets: outline of a general theory of biological and social organism. General Syst. **XII**, 21–28 (1967)
646. Roberts, B., Holdren, B.: Theory of Social Process. Iowa University Press, Ames (1972)
647. Rudner, R.S.: Philosophy of Social Sciences. Prentice Hall, Englewood Cliff, N.J. (1966)
648. Simon, H.A.: The structure of ill-structured problems. Artif. Intell. **4**, 181–201 (1973)
649. Toulmin, S.: Foresight and Understanding: An Enquiry into the Aims of Science. Harper and Row, New York (1961)
650. Winch, P.: The Idea of a Social Science. Humanities Press, New York (1958)

Tranformations, Decisions, Polarity, Duality and Conflicts

651. Amen, R.U.N. Metu Neter, vol. 1, Khamit Corp., Bronx, NY (1990)
652. Anovsky, M.E.: Linin and Modern Natural Science. Progress Pub., Moscow (1978)
653. Arrow, K.J.: Limited knowledge and economic analysis. Am. Econ. Rev. **64**, 1–10 (1974)
654. Berkeley, G.: Treatise Concerning the Principles of Human Knowledge, Works. In: Fraser, A. (ed.), vol. I. Oxford University Press, Oxford (1871–1814)
655. Berkeley, G.: Material Things are Experiences of Men or God, in [R1.5], pp. 658–668 (1967)
656. Boulding, K.E.: A New Theory of Societal Evolution. Sage Pub. Beverly Hills Ca (1978)
657. Brody, B.A. (ed.): Readings in the Philosophy of Science. Prentice-Hall Inc., Englewood Cliffs, NJ (1970)
658. Brouwer, L.E.J.: Consciousness, philosophy, and mathematics. In: Benecerraf, P., Putnam, H. (eds.) Philosophy of Mathematics: Selected Readings, pp. 90–96. Cambridge University Press, Cambridge (1983)
659. Brown, B., Woods, J. (eds.): Logical Consequence; Rival Approaches and New Studies in exact Philosophy: Logic, Mathematics and Science, vol. II. Oxford, Hermes (2000)
660. Cornforth, M.: Dialectical Materialism and Science. International Pub, New York (1960)
661. Cornforth, M.: Materialism and Dialectical Method. International Pub, New York (1953)
662. Cornforth, M.: Science and Idealism: An Examination of "Pure Empiricism. International Pub., New York (1947)
663. Cornforth, M.: The Open Philosophy and the Open Society: A Reply to Dr. Karl Popper's Refutations of Marxism. International Pub., New York (1968)

664. Cornforth, M.: The Theory of Knowledge. International Pub, New York (1960)
665. Diop, C.A.: The African Origins of Civilization: Myth or Reality. Lawrence Hill, Brooklyn, New York (1974)
666. Diop, C.A.: Civilization or Barbarism. Lawrence Hill, Brooklyn, New York (1991)
667. Dompere, K.K.: On epistemology and decision-choice rationality. In: Trapple, R. (ed.) Cybernetics and System Research, pp. 219–228. North Holland, New York (1982)
668. Dompere, K.K., Ejaz, M.: Epistemics of Development Economics: Toward a Methodological Critique and Unity. Greenwood Press, Westport, CT (1995)
669. Dompere, K.K.: The Theory of Categorial Conversion: Rational Foundations of Nkrumaism in Socio-Natural Systemicity and Complexity. Adonis-Abbey Pubs., London (2016–2017)
670. Dompere, K.K.: The Theory of Philosophical Consciencism: Practice Foundation of Nkrumaism in Social Systemicity. Adonis-Abbey Pubs., London (2016–2017)
671. Dompere, K.K.: The Theory of Info-Statics: Conceptual Foundations of Information and Knowledge. Springer, New York. (Series: Studies in Systems, Decision and Control, vol. 112) (2017)
672. Dompere, K.K.: The Theory of Info-Dynamics: Rational Foundations of Information-Knowledge Dynamics. Springer, New York. (Series: Studies in Systems, Decision and Control, vol. 114) (2017)
673. Dompere, K.K.: Polyrhythmicity: Foundations of African Philosophy. Adonis and Abbey Pub, London (2006)
674. Engels, F.: Dialectics of Nature. International Pub, New York (1971)
675. Engels, F.: Origin of the Family, Private Property and State. International Pub, New York (1971)
676. Ewing, A.C.: A Reaffirmation of Dualism, in [R1.5], pp. 454–461
677. Fedoseyer, P.N., et al.: Philosophy in USSR: Problems of Dialectical Materialism. Progress Pub, Moscow (1977)
678. Kedrov, B.M.: On the dialectics of scientific discovery. Sov. Stud. Philos. **6**, 16–27 (1967)
679. Lenin, V.I.: Materialism and Empirio-Criticism: Critical Comments on Reactionary Philosophy. International Pub, New York (1970)
680. Lenin, V.I.: Collected Works Vol. 38: Philosophical Notebooks. International Pub., New York (1978)
681. Lenin, V.I.: On the National Liberation Movement. Foreign Language Press, Peking (1960)
682. Hegel, G.: Collected works, Berlin, Duncher und Humblot, 1832–1845 [also Science of Logic, translated by W. H. Johnston and L. G. Struther, London, 1951]
683. Hempel, C.G., Oppenheim, P.: Studies in the logic of explanation. in [R15.5], pp. 8–27
684. Ilyenkov, E.V.: Dialectical Logic: Essays on its History and Theory. Progress Pub, Moscow (1977)
685. Keirstead, B.S.: The conditions of survival. Am. Econ. Rev. **40**(2), 435–445
686. Kühne, K.: Economics and Marxism, vol. I: The Renaissance of the Marxian System. St Martin's Press, New York (1979)
687. Kühne, K.: Economics and Marxism, vol. II: The Dynamics of the Marxian System. St Martin's Press, New York (1979)
688. March, J.C.: Bounded rationality, ambiguity and engineering of choice. Bell J. Econ. **9**(2) (1978)
689. Marx, K.: Contribution to the Critique of Political Economy. Charles H. Kerr and Co., Chicago (1904)
690. Marx, K.: Economic and Philosophic Manuscripts of 1884. Progress Pub, Moscow (1967)
691. Marx, K.: The Poverty of Philosophy. International Publishers, New York (1963)
692. Marx, K.: Economic and Philosophic Manuscripts of 1844. Progress Pub, Moscow, 196
693. Massey, G. A Book of the Beginnings, vols. 1–2. William and Norgate, London (1881)
694. Massey, G.: Pyramid Text, vols. 1–4. Longmans Green, New York (1952)
695. Massey, G.: The Natural Genesis, vols. 1 and 2. Black Classic Press, Baltimore, MD (1998) (First published 1883)
696. Massey, G., Egypt, A.: The Light of the World. MD, Black Classic Press, Baltimore (1992)

697. Niebyl, K.H.: Modern mathematics and some problems of quantity, quality and motion in economic analysis. Philos. Sci. **7**(1), 103–120 (1940)
698. Nkrumah, K.: Consciencism. Heinemann, London (1964)
699. Obenga, T.T.: African Philosophy During the Period of the Pharaohs, 2800–300BC. Per Ankh Pub., Popenguine, Senegal, W.A. (2006)
700. Obenga, T.T.: African Philosophy in World History. Per Ankh Pub., Popenguine, Senegal, W.A. (1998)
701. Price, H.H.: Thinking and Experience. Hutchinson, London (1953)
702. Putman, H.: Reason, Truth and History. Cambridge University Press, Cambridge (1981)
703. Putman, H.: Realism and Reason. Cambridge University Press, Cambridge (1983)
704. Quiggin, J., Economics, Z.: How Dead Ideas Still Walk Among Us. Princeton University Press, Prince ton, NJ (2010)
705. Robinson, J.: Economic Philosophy. Anchor Books, New York (1962)
706. Robinson, J.: Freedom and Necessity: An Introduction to the Study of Society. Vintage Books, New York (1971)
707. Robinson, J., Heresies, E.: Some Old-Fashioned Questions in Economic Theory. Basic Books, New York (1973)
708. Schumpeter, J.A.: The Theory of Economic Development. Harvard University Press, Cambridge, Mass (1934)
709. Schumpeter, J.A.: Capitalism, Socialism and Democracy. Harper & Row, New York (1950)
710. Schumpeter, J.A.: March to socialism. Am. Econ. Rev. **40**, 446–456 (1950)
711. Schumpeter, J.A.: Theoretical problems of economic growth. J. Econ. History **8**(Supplement), 1–9 (1947)
712. Schumpeter, J.A.: The analysis of economic change. Rev. Econ. Stat. **17**, 2–10 (1935)
713. Schwaller de Lubicz, R.A.: The Egyptian Miracle: An Introduction to the Wisdom of the Temple. (Vermont) Inner Traditions International, Rochester (1985)
714. Schwaller de Lubicz, R.A.: A Study of Numbers: A Guide to The Constant Creation of The Universe. (Vermont) Inner Traditions International, Rochester (1986)
715. Schwallier de Lubicz, R.A.: The Temple in Man: The Secrets of Ancient Egypt. Autumn Press, Brookline, (Massachusetts) (1977)
716. Schaller de Lubicz, R.A.: The Temple in Man: Sacred Architecture and the Perfect Man. Inner Traditions Published, Rochester, (Vermont) (1981)
717. Schwaller de Lubicz, R.A.: Symbol and the Symbolic: Egypt, Science, and The Evolution of Consciousness. Autumn Press, Brookline, (Massachusetts) (1978)
718. Schwaller de Lubicz, R.A.: The Temple of Man: Apet of The South at Luxor. Rochester, (Vermont) Inner Traditions 91998)
719. Wright, R.A. (ed.): African Philosophy: An Introduction. University Press of America, New York (1984)
720. Weber, L.E., Duderswtadt, J.J. (eds.): Reinventing the Research University. Economica, London (2004)

Vagueness, Thinking, Approximation and Reasoning in the Information-Knowledge Process

721. Adams, E. w., and H. F. Levine, "On the Uncertainties Transmitted from Premises to Conclusions in deductive Inferences," *Synthese* Vol. 30, 1975, pp. 429 – 460.
722. Arbib, M.A.: The Metaphorical Brain. McGraw-Hill, New York (1971)
723. Bečvář, J.: Notes on Vagueness and Mathematics. In: Skala, H.J., et al., (eds.) Aspects of Vagueness, pp. 1–11. D. Reidel Co., Dordrecht (1984)
724. Black, M.: Vagueness: an exercise in logical analysis. Philos. Sci. **17**, 141–164 (1970)
725. Black, M.: Reasoning with loose concepts. Dialogue **2**, 1–12 (1973)

726. Black, M.: Language and Philosophy. Cornell University Press, Ithaca, N.Y. (1949)
727. Black, M.: The analysis of rules. In: Black, M. [R18.8] Models and Metaphors: Studies in Language and Philosophy, pp. 95–139. Cornell University Press, Ithaca, New York (1962)
728. Black, M.: Models and Metaphors: Studies in Language and Philosophy. Cornell University Press, Ithaca, New York (1962)
729. Black, M.: Margins of Precision. Cornell University Press, Ithaca (1970)
730. Boolos, G.S., Jeffrey, R.C.: Computability and Logic. Combridge University Press, New York (1989)
731. Cohen, P.R.: Heuristic Reasoning about uncertainty: An Artificial Intelligent Approach. Pitman, Boston (1985)
732. Darmstadter, H.: Better theories. Philos. Sci. **42**, 20–27 (1972)
733. Davis, M.: Computability and Unsolvability. McGraw-Hill, New York (1958)
734. Dummett, M.: Wang's Paradox. Synthese **30**, 301–324 (1975)
735. Dummett, M.: Truth and Other Enigmas. Harvard University Press, Cambridge, Mass (1978)
736. Endicott, T.: Vagueness in the Law. Oxford University Press, Oxford (2000)
737. Evans, G.: Can there be vague objects? Analysis **38**, 208 (1978)
738. Fine, K.: Vagueness, truth and logic. Synthese **54**, 235–259 (1975)
739. Gale, S.: Inexactness, fuzzy sets and the foundation of behavioral geography. Geogr. Anal. **4**(4), 337–349 (1972)
740. Ginsberg, M.L. (ed.): Readings in Non-monotonic Reason. Morgan Kaufman, Los Altos, Ca. (1987)
741. Goguen, J.A.: The logic of inexact concepts. Synthese **19**, 325–373 (1968/69)
742. Grafe, W.: Differences in individuation and vagueness. In: Hartkamper, A., Schmidt, H.-J. (eds.) Structure and Approximation in Physical Theories, pp. 113–122. Plenum Press, New York (1981)
743. Goguen, J.A.: The logic of inexact concepts. Synthese **19**, 1968–1969
744. Graff, D., Timothy. (eds.): Vagueness. Ashgate Publishing, Aldershot (2002)
745. Hartka Amper A, Schmidt HJ (eds.) Structure and Approximation in Physical Theories. Plenum Press, New York (1981)
746. Hersh, H.M., et al.: A fuzzy set approach to modifiers and vagueness in natural language. J. Exp. **105**, 254–276 (1976)
747. Hilpinen, R.: Approximate truth and truthlikeness. In: Pprelecki, M., et al. (eds.) Formal Methods in the Methodology of Empirical Sciences, pp. 19–42. Dordrecht and Ossolineum, Wroclaw, Reidel (1976)
748. Hockney, D., et al. (eds.), Contemporary Research in Philosophical Logic and Linguistic Semantics. Reidel Pub. Co., Dordrecht-Holland (1975)
749. Ulrich, H., et al. (eds.) Non-Clasical Logics and their Applications to Fuzzy Subsets: A Handbook of the Mathematical Foundations of Fuzzy Set Theory. Kluwer, Boston, Mass (1995)
750. Katz, M.: Inexact geometry. Notre-Dame J. Formal Logic **21**, 521–535 (1980)
751. Katz, M.: Measures of proximity and dominance. In: Proceedings of the Second World Conference on Mathematics at the Service of Man, pp. 370–377. Universidad Politecnica de Las Palmas (1982)
752. Katz, M.: The logic of approximation in quantum theory. J. Philos. Log. **11**, 215–228 (1982)
753. Keefe, R.: Theories of Vagueness. Cambridge University Press, Cambridge (2000)
754. Keefe, R., Smith, P. (eds.): Vagueness: A Reader. MIT Press, Cambridge (1996)
755. Kling, R.: Fuzzy planner: reasoning with inexact concepts in a procedural problem-solving language. J. Cybern. **3**, 1–16 (1973)
756. Kruse, R.E., et al.: Uncertainty and Vagueness in Knowledge Based Systems: Numerical Methods. Springer-Verlag, New York (1991)
757. Ludwig, G.: Imprecision in physics. In: Hartkamper, A., Schmidt, H.J. (eds.) Structure and Approximation in Physical Theories, pp. 7–19. Plenum Press, New York (1981)
758. Kullback, S., Leibler, R.A.: Information and sufficiency. Annals Math. Stat. **22**, 79–86 (1951)

759. Lakoff, G.: Hedges: a study in meaning criteria and logic of fuzzy concepts. In: Hockney, D., et al. (eds.) Contemporary Research in Philosophical Logic and Linguistic Semantics, pp. 221–271. Reidel Pub. Co., Dordrecht-Holland (1975)
760. Lakoff, G.: Hedges: a study in meaning criteria and the logic of fuzzy concepts. J. Philos. Logic **2**, 458–508 (1973)
761. Levi, I.: The Enterprise of Knowledge. MIT Press, Cambridge, Mass (1980)
762. Lucasiewicz, J.: Selected Works. Studies in the Logical Foundations of Mathematics. North-Holland, Amsterdam (1970)
763. Machina, K.F.: Truth, belief and Vagueness. J. Philos. Logic **5**, 47–77 (1976)
764. Menges, G., et. al.: On the problem of Vagueness in the social sciences. In: Menges, G. (ed.) Information, Inference and Decision, pp. 51–61. D. Reidel Pub., Dordrecht, Holland (1974)
765. Merricks, T.: Varieties of Vagueness. Philos. Phenomenol. Res. **53**, 145–157 (2001)
766. Mycielski, J.: On the axiom of determinateness. Fund. Mathematics **53**, 205–224 (1964)
767. Mycielski, J.: On the axiom of determinateness II. Fund. Math. **59**, 203–212 (1966)
768. Naess, A.: Towards a theory of interpretation and preciseness. In: Linsky, L. (ed.) Semantics and the Philosophy of Language, Urbana, Ill. Univ. of Illinois Press (1951)
769. Narens, L.: The theory of belief. J. Math. Psychol. **49**, 1–31 (2003)
770. Narens, L.: A theory of belief for scientific refutations. Synthese **145**, 397–423 (2005)
771. Netto, A.B.: Fuzzy classes. Notices, Amar, Math. Soc. **68T-H28**, 945 (1968)
772. Neurath, O., et al. (eds.): International Encyclopedia of Unified Science, vol. 1–10. University of Chicago Press, Chicago (1955)
773. Niebyl, K.H.: Modern mathematics and some problems of quantity, quality and motion in economic analysis. Science **7**(1), 103–120 (1940)
774. Orlowska, E.: Representation of vague information. Inf. Syst. **13**(2), 167–174 (1988)
775. Parrat, L.G.: Probability and Experimental Errors in Science. Wiley, New York (1961)
776. Raffman, D.: Vagueness and context-sensitivity. Philos. Stud. **81**, 175–192 (1996)
777. Reiss, S.: The Universe of Meaning. The Philosophical Library, New York (1953)
778. Russell, B.: Vagueness. Aust. J. Philos. **1**, 84–92 (1923)
779. Russell, B.: An Inquiry into Meaning and Truth. Norton, New York (1940)
780. Shapiro, S.: Vagueness in Context. Oxford University Press, Oxford (2006)
781. Skala, H.J.: Modelling Vagueness. In: Gupta, M.M., Sanchez, E. (eds.) Fuzzy Information and Decision Processes, pp. 101–109. Amsterdam North-Holland (1982)
782. Skala, H.J., et al. (eds.) Aspects of Vagueness. D. Reidel Co., Dordrecht (1984)
783. Sorensen, R.: Vagueness and Contradiction. Oxford University Press, Oxford (2001)
784. Tamburrini, G., Termini, S.: Some foundational problems in formalization of Vagueness. In: Gupta, M.M., et al. (eds.) Fuzzy Information and Decision Processes, pp. 161–166. North Holland, Amsterdam (1982)
785. Termini, S.: Aspects of Vagueness and some epistemological problems related to their formalization. In: Skala, H.J., et al. (eds.) Aspects of Vagueness, pp. 205–230. D. Reidel Co, Dordrecht (1984)
786. Tikhonov, A.N., Arsenin, V.Y.: Solutions of Ill-Posed Problems. Wiley, New York (1977)
787. Tversky, A., Kahneman, D.: Judgments under uncertainty: heuristics and biases. Science **185**, 1124–1131 (1974)
788. Ursul, A.D.: The problem of the objectivity of information. In: Kubát, L., Zeman, J. (eds.) Entropy and Information, Amsterdam, pp. 187–230. Elsevier (1975)
789. Vardi, M. (ed.): Proceedings of Second Conference on Theoretical Aspects of Reasoning about Knowledge. Asiloman, Ca, Los Altos, Ca, Morgan Kaufman (1988)
790. Verma, R.R.: Vagueness and the principle of the excluded middle. Mind **79**, 66–77 (1970)
791. Vetrov, A.A.: Mathematical logic and modern formal logic. Sov. Stud. Philos. **3**(1), 24–33 (1964)
792. von Mises, R.: Probability, Statistics and Truth. Dover Pub., New York (1981)
793. Williamson, T.: Vagueness. Routledge, London (1994)
794. Wiredu, J.E.: Truth as a logical constant with an application to the principle of the excluded middle. Philos. Quart. **25**, 305–317 (1975)

795. Wright, C.: On coherence of vague predicates. Synthese **30**, 325–365 (1975)
796. Wright, C.: The epistemic conception of vagueness. South. J. Philos. **33**(Supplement), 133–159 (1995)
797. Zadeh, L.A.: A theory of commonsense knowledge. In: Skala, H.J., et al. (eds.) Aspects of Vagueness, pp. 257–295. Dordrecht, D. Reidel Co. (1984)
798. Zadeh, L.A.: The concept of linguistic variable and its application to approximate reasoning. Inf. Sci. **8**, 199–249 (1975). (Also, in vol. 9, pp. 40–80) (1975)

Vagueness, Disinformation, Misinformatiomn and Fuzzy Game Theory in Socio-natural Transformations

799. Aubin, J.P.: Cooperative fuzzy games. Math. Oper. Res. **6**, 1–13 (1981)
800. Kubat Aubin, J.P.: Mathematical Methods of Game and Economics Theory. North Holland, New York (1979)
801. Butnaria, D.: Fuzzy games: a description pf the concepts. Fuzzy Sets Syst. **1**, 181–192 (1978)
802. Butnaria, D.: Stability and shapely value for a n-persons fuzzy games. Fuzzy Sets Syst. **4**(1), 63–72 (1980)
803. Nurmi, H.: A fuzzy solution to a majority voting game. Fuzzy Sets Syst. **5**, 187–198 (1981)
804. Regade., R.K.: Fuzzy games in the analysis of options. J. Cybern. **6**, 213–221 (1976)
805. Spillman, B., et al.: Coalition analysis with fuzzy sets. Kybernetes **8**, 203–211 (1979)
806. Wernerfelt, B.: Semifuzzy games. Fuzzy Sets Syst. **19**, 21–28 (1986)

Weapon Foundations for Information System

807. Forte, B.: On a system of functional equation in information theory. Aequationes Math. **5**, 202–211 (1970)
808. Gallick, J.: The Information: A History, a Theory, a Flood. Pantheon, New York, NY, (2011)
809. Hopcroft, J.E., Motwani, R., Ljeffrey, D.: Ullman, Introduction to automata Theory, Languages, and Computation. Pearson Education (2000)
810. Howard, N.: Paradoxes of Rationality. MIT Press, Cambridge, Mass. (1972)
811. Ingarden, R.S.: A simplified axiomatic definition of information. Bull. Acad. Polo Sci. Ser, Sci. Math Astronomy Phys. **11**, 209–212 (1963)
812. Lee, P.M.: On the axioms of information theory. Annals Math Stat. **35**, 415–418 (1964)
813. Luce, R.D. (ed.): Development in Mathematical Psychology. Greenwood Press, Westport (1960)
814. Floridi, L.: Is information meaningful data? Philos. Phenomenol. Res. **70**(2), 351–370 (2005)
815. Meyer, L. Meaning in music and information theory. J. Aesthetics Art Criticism **15**, 412–424 (1957)
816. Rich, E.: Automata, Computability, and Complexity: Theory and Applications, Pearson (2008)
817. Shannon, C.E.: The mathematical theory of communication. Bell Syst. Tech. J. **27**(3), 379–423 (1945), **27**(4), 623–666 (1948)
818. Shannon, C.E., Weaver, W.: The Mathematical Theory of Communication. University of Illinois Press (1949)
819. Theil, H.: Statistical Decomposition Analysis. North-Holland, Amsterdam (1974)
820. Vigo, R.: Representational information: a new general notion and measure of information. Inf. Sci. **18**, 4847–4859 (2011)
821. Vigo, R.: Complexity over uncertainty in generalized representational information theory (GRIT): a structure-sensitive general theory of information. Information **4**(1), 1–30 (2013)

822. Vigo, R.: Mathematical Principles of Human Conceptual Behavior: The Structural Nature of Conceptual Representation and Processing. Routledge, New York and London (2014)
823. von Mises, R.: Probability, Statistics and Truth. Dover, New York (1981)
824. Wicker, S.B., Kim, S.: Fundamentals of Codes, Graphs, and Iterative Decoding. Springer, New York (2003)
825. Wiener, N.: Cybernetics. John Wiley and Sons, New York (1948)
826. Wiener, N.: The Human use of Human Beings. Mass, Houghton, Boston (1950)
827. Young, P.: The Nature of Information. Greenwood Publishing Group, Westport, Ct (1987)

Weapon Foundations for Fuzzy Information and Entropy

828. Belis, M., Guiasu, S.: A quantitative–qualitative measure of information in cybernetic systems. IEEE Trans. Inform. Theory **14**, 593–594 (1968)
829. Burillo, P., Bustince, H.: Entropy on intuitionistic fuzzy sets and on interval-valued fuzzy sets. Fuzzy Sets Syst. **78**(3), 305–316 (1996)
830. Ceng, H.D., Chen, Y.H., Sun, Y.: A novel fuzzy entropy approach to image enhancement and thresholding. Signal Process. **75**, 277–301 (1999)
831. De Luca, A.S.T.: A definition of non-probabilistic entropy in setting of fuzzy set theory. Inform. Control. **20**, 301–312 (1972)
832. Dompere, K.K.: A General Theory of Entropy: Rational Foundations of Information-Knowledge Dynamics. Springer, New York. (Series: Studies in Systems, Decision and Control, vol. 114) (2019)
833. Dumitrescu, D.: Fuzzy measures and the entropy of fuzzy partitions. J. Math. Anal. Appl. **176**, 359–373 (1993)
834. Dumitrescu, D.: Entropy of fuzzy process. Fuzzy Sets Syst. **55**, 169–177 (1993)
835. Dumitrescu, D.: Entropy of a fuzzy dynamical system. Fuzzy Sets Syst. **70**, 45–57 (1995)
836. Garbaczewski, P.: Differential entropy and dynamics of uncertainty. J. Stat. Phys. **123**, 315–355 (2006)
837. Hu, Q., Yu, D.: Entropies of fuzzy indiscernibility relation and its operations. Internat. J. Uncertain. Fuzziness Knowledge-Based Systems **12**, 575–589 (2004)
838. Hu, Q., Yu, D., Xie, Z., Liu, J.: Fuzzy probabilistic approximation spaces and their information measures. IEEE Trans. Fuzzy Syst. **14**, 191–201 (2006)
839. Hudetz, T.: Space-time dynamical entropy for quantum systems. Lett. Math. Phys. **16**, 151–161 (1988)
840. Hung, W.L., Yang, M.S.: Fuzzy entropy on intuitionistic fuzzy sets. Int. J. Intell. Syst. **21**(4), 443–451 (2006)
841. Kapur, J.N.: Four families of measures of entropy. Indian J. Pure Appl. Math. **17**, 429–449 (1986)
842. Kapur, J.N.: Measures of Fuzzy Information. Mathematical Sciences Trust Society, New Delhi (1997)
843. Kasko, B.: Fuzzy entropy and conditioning. Inf. Sci. **40**, 165–174 (1986)
844. Klir, G.J.: Generalized information theory: aims, results and open problems. Reliab. Eng. Syst. Safety **85**(1–3), 21–38 (2004)
845. Kolmogorov, A.N.: Foundations of the Theory of Probability. Chelsea Publishing Company, New York, NY (1950)
846. Liu, B., Liu, Y.K.: Expected value of fuzzy variable and fuzzy expected value models. IEEE Trans. Fuzzy Syst. **10**(4), 445–450 (2002)
847. Loo, S.G.: Measures of fuzziness. Cursos Congr. Univ. Santiago de Compostela **20**, 201–210 (1977)
848. Markechová, D.: The entropy of fuzzy dynamical systems and generators. Fuzzy Sets Syst. **48**, 351–363 (1992)

849. Mesiar, R., Rybárik, J.: Entropy of fuzzy partitions: a general model. Fuzzy Sets Syst. **99**, 73–79 (1998)
850. Mesiar, R.: The Bayes principle and the entropy on fuzzy probability spaces. Int. J. Gen. Syst. **20**, 67–72 (1991)
851. Parkash, O.: A new parametric measure of fuzzy entropy. Inform. Process. Manage. Uncertain. **2**, 1732–1737 (1998)
852. Parkash, O., Sharma, P.K.: Measures of fuzzy entropy and their relations. Int. J. Manag. Syst. **20**, 65–72 (2004)
853. Rahimi, M., Riazi, A.: On local entropy of fuzzy partitions. Fuzzy Sets Syst. **234**, 97–108 (2014)
854. Rieˇcan, B.: An entropy construction inspired by fuzzy sets. Soft Comput. **7**, 486–488 (2003)
855. Szmidt, E., Kacprzyk, J.: Entropy for intuitionistic fuzzy sets. Fuzzy Sets Syst. **118**(3), 467–477 (2001)
856. Verma, R., Sharma, B.D.: On generalized exponential fuzzy entropy. Eng. Technol. **5**, 956–959 (2011)
857. Zeng, W., Li, H.: Relationship between similarity measure and entropy of interval valued fuzzy sets. Fuzzy Sets Syst. **157**(11), 1477–1484 (2006)

Weapon Foundations for Knowledge by Acquaintance and Knowledge by Description

858. Balog, K.: Acquaintance and the mind-body problem. In: Gozzano, S., Hill, C. (eds.) New Perspectives on Type Identity: The Mental and the Physical, pp. 16–43. Cambridge University Press, Cambridge (2012)
859. BonJour, L.: Toward a defense of empirical foundationalism. In: DePaul, M.R. (ed.) Resurrecting Old-Fashioned Foundationalism, pp. 21–38. Rowman and Littlefield, Lanham (2001)
860. Brewer, B.: Perception and its Objects. Oxford University Press, Oxford (2011)
861. Chalmers, D.: The Conscious Mind. Oxford University Press, Oxford (1996)
862. Chisolm, R.: Acquaintance and the Mind-Body Problem. Oxford University Press, Oxford (2008)
863. Churchman, C.W., Ratoosh, P.: Measurement. Wiley and Sons, Definitions and Theories (1959)
864. Fales, E.: A Defense of the Given. Rowman and Littlefield, Lanham (1996)
865. Gertler, B.: A defense of the knowledge argument. Philos. Stud. **93**, 317–336 (1999)
866. Gertler, B.: Introspecting phenomenal states. Philos. Phenomenol. Res. **63**, 305–328 (2001)
867. Gertler, B.: Renewed Acquaintance. In: Smithies, D., Stoljar, D. (eds.) Introspection and Consciousness, pp. 93–128. Oxford University Press, Oxford (2012)
868. Harrison, G.W., List, J.A.: Field experiments. J. Econ. Lit. **42** (2004)
869. Hasan, A., Fumerton, R.: Knowledge by acquaintance vs. description. In: Zalta, E.N. (ed.) Stanford Encyclopedia of Philosophy. Spring (2014)
870. Lazerowitz, M.: Knowledge by description. Philos. Rev. **46**(4), 402–415 (1937)
871. Morgenstern, O.: On the Accuracy of Economic Observation. Princeton, Princeton University Press (1973)
872. Parker, D.H.: Knowledge by acquaintance. Philos. Rev. **54**(1), 1–18 (1945)
873. Parker, D.H.: Knowledge by description. Philos. Rev. **54**(5), 458–488 (1945B)
874. Poston, T.: Similarity and acquaintance. Philos. Stud. **147**, 369–378 (2010)
875. Russell, B.: On denoting. Mind **14**, 479–493 (1905)
876. Russell, B.: Knowledge by acquaintance and knowledge by description. Proc. Aristotelian Soc. **11**, 108–128 (1910–1911)

877. Russell, B.: On the nature of acquaintance. In: Marsh, R.C. (ed.) Bertrand Russell: Logic and Knowledge: Essays 1901–1950, pp. 127–174. Capricorn Books, New York (1971)
878. Russell, B.: Logic and Knowledge: Essays 1901–1950. In: Marsh, R.C. (ed.) Capricorn Books, New York (1971)

Written and Audio Languages and Information

879. Agha, A.: Language and Social Relations. Cambridge University Press, Cambridge (2006)
880. Aitchison, J. (ed.): Language Change: Progress or Decay? Cambridge University Press, Cambridge, New York, Melbourne (2001)
881. Allerton, D.J.: Language as form and pattern: grammar and its categories. In: Collinge, N.E. (ed.) An Encyclopedia of Language. Routledge, London, New York (1989)
882. Anderson, S.: Languages: A Very Short Introduction. Oxford University Press, Oxford (2012)
883. Aronoff, M., Fudeman, K.: What is Morphology. John Wiley & Sons, New York (2011)
884. Alex, B., Stainton, R.J. (eds.): Concise Encyclopedia of Philosophy of Language and Linguistics. Elsevier, New York (2010)
885. Bauer, L. (ed.): Introducing Linguistic Morphology. Georgetown University Press, Washington, D.C. (2003)
886. Brown, K., Ogilvie, S. (eds.): Concise Encyclopedia of Languages of the World. New York Elsevier Science (2008)
887. Campbell, L. (ed.): Historical Linguistics: an Introduction. MIT Press, Cambridge, MA (2004)
888. Chao, Y.R.: Language and Symbolic Systems. Cambridge University Press, Cambridge (1968)
889. Chomsky, N.: Syntactic Structures. Mouton, The Hague (1957)
890. Chomsky, N.: The Architecture of Language. Oxford University Press, Oxford (2000)
891. Clarke, D.S.: Sources of Semiotic: Readings with Commentary from Antiquity to the Present. Southern Illinois University Press, Carbondale (1990)
892. Comrie, B. (ed.): Language Universals and Linguistic Typology: Syntax and Morphology. Blackwell, Oxford (1989)
893. Comrie, B. (ed.): The World's Major Languages. Routledge, New York (2009)
894. Coulmas, F. Writing Systems: An Introduction to Their Linguistic Analysis. Cambridge University Press (2002)
895. Croft, W., Cruse, D.A.: Cognitive Linguistics. Cambridge University Press, Cambridge (2004)
896. Crystal, D.: The Cambridge Encyclopedia of Language. Cambridge University Press, Cambridge (1997)
897. Deacon, T.: The Symbolic Species: The Co-evolution of Language and the Brain. W.W. Norton & Company, New York (1997)
898. Devitt, M., Sterelny, K.: Language and Reality: An Introduction to the Philosophy of Language. MIT Press, Boston (1999)
899. Evans, N., Levinson, S.C.: The myth of language universals: Language diversity and its importance for cognitive science. Behav. Brain Sci. **32**(5), 429–492 (2009)
900. Fitch, W.T.: The Evolution of Language. Cambridge University Press, Cambridge (2010)
901. Foley, W.A.: Anthropological Linguistics: An Introduction. Blackwell, Oxford (1997)
902. Greenberg, J., Universals, L.: With Special Reference to Feature Hierarchies. Mouton & Co., The Hague (1966)
903. Hauser, M.D., Chomsky, N., Fitch, W.T.: The Faculty of Language: What Is It, Who Has It, and How Did It Evolve? Science 22, **298**(5598), 1569–1579 (2002)
904. International Phonetic Association: Handbook of the International Phonetic Association: A guide to the use of the International Phonetic Alphabet. Cambridge University Press, Cambridge (1999)
905. Katzner, K.: The Languages of the World. Routledge, New York (1999)
906. Labov, W.: Principles of Linguistic Change vol. I: Internal Factors, Oxford, Blackwell (1994)

907. Labov, W.: Principles of Linguistic Change, vol. II: Social Factors, Oxford, Blackwell (2001)
908. Ladefoged, P., Maddieson, I.: The Sounds of the World's Languages, pp. 329–330. Blackwell, Oxford (1996)
909. Levinson, S.C.: Pragmatics. Cambridge University Press, Cambridge (1983)
910. Lewis, M.P. (ed.): Ethnologue: Languages of the World. SIL International, Dallas, Tex. (2009)
911. Lyons, J.: Language and Linguistics. Cambridge University Press, Cambridge (1981)
912. MacMahon.: April M.S., Understanding Language Change. Cambridge University Press, Cambridge (1994)
913. Matras, Y., Bakker, P. (eds.): The Mixed Language Debate: Theoretical and Empirical Advances. Walter de Gruyter, Berlin (2003)
914. Moseley, C. (ed.): Atlas of the World's Languages in Danger. UNESCO Publishing, Paris (2010)
915. Nerlich, B.: History of pragmatics. In: Cummings, L. (ed.) The Pragmatics Encyclopedia, pp. 192–193. Routledge, New York (2010)
916. Newmeyer, F.J.: The History of Linguistics. Linguistic Society of America (2005)
917. Newmeyer, F.J.: Language Form and Language Function (PDF). MIT Press, Cambridge, MA (1998)
918. Nichols, J.L.: Diversity in Space and Time. University of Chicago Press, Chicago (1992)
919. Nichols, J.: Functional theories of grammar. Ann. Rev. Anthropol. **13**, 97–117 (1984)
920. Senft, G.: Systems of Nominal Classification. Cambridge University Press. (ed.) (2008)
921. Swadesh, M.: The phonemic principle. Language **10**(2), 117–129 (1934)
922. Tomasello, M.: The cultural roots of language. In: Velichkovsky, B., Rumbaugh, D. (eds.) Communicating Meaning: The Evolution and Development of Language, pp. 275–308. Psychology Press, (1996)
923. Tomasello, M.: Origin of Human Communication. MIT Press (2008)
924. Ulbaek, I.: The origin of language and cognition. In: Hurford, J.R., Knight, C. (eds.) Approaches to the evolution of language, pp. 30–43. Cambridge University Press (1998)

Zones of Epistemic Conflicts on Possibility and Probability

925. Alberoni, F.: Contribution to the study of subjective probability. Part I. J. General Psychol. **66**, 241–264 (1962)
926. Berger, J.O.: Statistical Decision Theory and Bayesian Analysis (Springer Series in Statistics). Springer-Verlag, New York (1985)
927. Benacerraf, P.: Mathematical truth. J.Philos. **70**, 661–679 (1973)
928. de Finetti, B.: Theory of Probability. A Critical Introductory Treatment, (translation by A. Machi and AFM Smith of 1970 book) 2 volumes. Wiley, New York (1974–1975)
929. Dompere, K.K.: Savage Axioms, Ellsberg's Paradox and fuzzy optimal decision-choice rationality. Fuzzy Econ. Rev. **17**(1), 15–51
930. DeGroot, M.: Optimal Statistical Decisions. Wiley Classics Library, Wiley, New York (2004)
931. Edwards, W.: Conservatism in human information processing. In: Kleinmuntz, B. (ed.) Formal Representation of Human Judgment, pp. 17–52. Wiley, New York (1968)
932. Eells, E., Sober, E.: Probabilistic causality and the question of transitivity. Philos. Sci. **50**, 35–57 (1983)
933. Feller, W.: An Introduction to Probability Theory and its Applications, vol. 1. Wiley, New York (1968)
934. Fetzer, J.H.: Dispositional probabilities. In: Buck, R.C., Cohen, R.S. (eds.) Boston Studies in the Philosophy of Science VIII, pp. 473–82. Reidel, Dordrecht (1971)
935. Fetzer, J.H.: Statistical probabilities: single case propensities vs. long-run frequencies. In: Leinfellner, W., Kohler, E. (eds.) Developments in the Methodology of Social Science, pp. 387–397. Reidel, Dordrecht (1974)

936. Fetzer, J.H., Nute, D.E.: Syntax, semantics, and ontology: a probabilistic causal calculus. Synthese **40**(1979), 453–495 (1979)
937. Gärdenfors, P., Sahlin, N.-E.: Decision, Probability, and Utility: Selected Readings. Cambridge University Press, Cambridge (1988)
938. Hacking, I.: Slightly more realistic personal probability. Philos. Sci. **34**(4), 311–325 (1967)
939. Hazewinkel, M. (ed.): Bayesian approach to statistical problems. In: Encyclopedia of Mathematics. Springer Science, New York (2001)
940. Fetzer, J.H. (ed.): Probability and Causality. Reidel, Dordrecht (1988)
941. Jeffrey, R.C.: The Logic of Decision. University of Chicago Press, Chicago (1990)
942. Kolmogorov, A.N. Foundations of the Theory of Probability. Chelsea Publishing Company, New York, NY (19500
943. Kyburg, H.E., Jr.: Propensities and probabilities. Br. Jo. Philos. Sci. **25**, 359–375 (1974)
944. Lindley, D.V.: Introduction to Probability and Statistics: From a Bayesian Viewpoint Part I, Probability. Cambridge University Press, Cambridge (1965)
945. Lindley, D.V.: Introduction to Probability and Statistics: From a Bayesian Viewpoint Part 2, Inference. Cambridge University Press, Cambridge (1965)
946. Morgenstern, O.: "Utility". In Andrew Schotter. Selected Economic Writings of Oskar Morgenstern, pp. 65–70. New York University Press, New York (1978)
947. Pfanzagl, J.: Subjective probability derived from the Morgenstern-von Neumann utility theory. In: Shubik, M. (ed.) Essays in Mathematical Economics in Honor of Oskar Morgenstern, pp. 237–251. Princeton University Press, Princeton, N.J. (1967)
948. Popper, K.R.: The propensity interpretation of the calculus of probability, and the quantum theory. In: Korner, S. (ed.) Observation and Interpretation in the Philosophy of Physics, pp. 65–70. Dover Publications, New York (1957)
949. Popper, K.R.: The propensity interpretation of probability. Br. J. Philos. Sci. **10**, 25–42 (1959)
950. Salmon, W.C.: The Foundations of Scientific Inference. University of Pittsburgh Press, Pittsburg (1967)
951. Savage, L.J.: The Foundations of Statistics. Dover Pub, New York (1972)
952. Savage, L.J.: Subjective probability and statistical practice. In: Machol, R.E., Gray, P. (eds.) Recent Developments in Decision and Information Processes. Macmillan, New York (1962)
953. Savage, L.J.: Difficulties in the theory of personal probability. Philos. Sci. **34**, 305–310 (1967)
954. Sklar, L.: Is probability a dispositional property? J. Philos. **67**, 355–366 (1970)
955. Skyrms, B.: Causal Necessity. Yale University Press, New Haven (1980)
956. Stigler, S.M.: The History of Statistics: The Measurement of Uncertainty Before 1900. Belknap Press/Harvard University Press, Cambridge (1990)
957. Suppes, P.: New foundations of objective probability: axioms for propensities. In: Suppe, P., Henkin, L., Joja, A., Moisil G.C. (eds.) Logic, Methodology and Philosophy of Science IV, pp. 515–29, North-Holland, Amsterdam, London (1973)
958. Tversky, A., Kahneman, D.: The belief in the law of small numbers. Psychol. Bull. **76**, 105–110 (1971)
959. Von Mises, R., *Probability, Statistics and Truth,* Macmillan, New York 1957
960. von Neumann, J., Morgenstern, O.: Theory of Games and Economic Behavior. Princeton University Press (1943, 1947)
961. Wagenaar, W.A.: Subjective randomness and the capacity to generate information. In: Sanders, A.F. (ed.) Attention and performance III, Acta Psychologica, vol. 33, pp. 233–242 (1970)

Zones of Epistemic Conflicts in the Unity of Science and Knowing

962. Andler, D.: Unity without myths. In: Symons, J., Torres, J.M., Plomb, O. (eds.) New approaches to the Unity of Science, vol. 1: Otto Neurath and the Unity of Science. Springer, New York (2011)

963. Anton, J. P.: The unity of scientific inquiry and categorial theory in Aristotle. In: Boston Studies in the Philosophy of Science, pp. 121:29–43 (1990)
964. Barrow, J.D.: New Theories of Everything: The Quest for Ultimate Explanation. Oxford University Press, Oxford (2007)
965. Bechtel, W.P., Hamilton, A.: Reduction, integration, and the unity of science: natural, behavioral, and social sciences and the humanities. In: Kuipers, T. (ed.) Philosophy of Science: Focal Issues (Volume 1 of the Handbook of the Philosophy of Science). Elsevier, New York (2007)
966. Boersema, D.: Metaphysics, mind, and the unity of science. Behav. Brain Sci. **27**(5), 627–628 (2004)
967. Briskman, L.: Three views concerning the unity of science. In: Radnitzky, G. (ed.), Centripetal Forces in the Sciences, New York, Paragon House Publishers, 1987. pp. 1--105.details
968. Bunge, M. A. (ed).: The Methodological Unity of Science. Reidel, Boston (1973)
969. Burian, R.M.: Conceptual change, cross-theoretical explanation, and the unity of science. Synthese **32**(1–2), pp. 1–28 (1975)
970. Byrne, L.: An educational application of resources of the unity of science movement. Philos. Sci. **7**(2), 241–262 (1940)
971. Carnap, R.: The Unity of Science. K. Paul, Trench, Trubner & Co., London (1934)
972. Carrier, M.: The unity of science. Int. Stud. Philos. Sci. **4**(1), 17–31 (1990)
973. Causey, R.L.: Unity of science. Philos. Rev. **90**(1), 150–153 (1977)
974. Crave, C.F.R.: Beyond reduction: mechanisms, multifield integration and the unity of neuroscience. Stud. History Philos. Sci. Part C **36**(2), 373–395 (2005)
975. Damasio, A.R. (ed.): Unity of Knowledge: The Convergence of Natural and Human Science. New York Academy of Sciences, New York (2001)
976. Davies, D.: Explanatory disunities and the unity of science. Int. Stud. Philos. Sci. **10**(1), 5–21 (1996)
977. de Santillana, G., Zilsel, E.: Foundations of the unity of science. II 8: the development of rationalism and empiricism. Philos. Rev. **52**(1), 87–87 (1943)
978. Dewey, J.: Foundations of the unity of science, vol. II, No. 4: theory of valuation. Philos. Rev. **50**(4), 443–446. (1941)
979. Donovan, R.: Science without unity. Int. Philos. Q. **30**(1), 122–125 (1990)
980. Dupre, J.: Against scientific imperialism. In: 1994—PSA: Proceedings of the Biennial Meeting of the Philosophy of Science Association, pp. 374–381 (1994)
981. Dupré, J.: The disunity of science. Mind **92**(367), 321–346 (1983)
982. Galison, P., Stump, D.J. (eds.): The Disunity of Science: Boundaries, Contexts, and Power. Stanford University Press, Stanford, Ca. (1996)
983. Neurath, O., et al. (eds.) International Encyclopedia of Unified Science, vol. 1–10. University of Chicago Press, Chicago (1955)
984. Neurath, O.: Unified science as encyclopedic. In: Neurath, O., et al. (eds.) International Encyclopedia of Unified Science, vol. 1–10, pp. 1–27. University of Chicago Press, Chicago (1955)

Zones of Epistemic Conflicts on Decidability, Computability and Predictability in the Space of the Problem-Solution Dualities and Polarities

985. Davis, M., Undecidable, T.: Basic Papers on Undecidable Propositions. Dover, Dover Books, Unsolvable Problems and Computable Functions (2004)
986. Gödel, K.: On Formally Undecidable Propositions of Principia Mathematica and Related Systems. Dover, Dover Pub. (1992)

987. Gödel, K.K.: Gödel, Collected Works, vol. 1-3. Oxford University Press, Oxford (1995)
988. Hartshorne, C.: The Logic of Perfection and Other Essays in Neoclassical Metaphysics. Open Court Press, Lasalle, Illinois (1962)
989. Tarski, A., Mostowski, A., et al.: Undecidable Theories: Studies in Logic and Foundation of Mathematics. Dover Books, Dover (2010)
990. Wang, H.: Reflections on Kurt, Cambridge. MIT Press, Mass (1987)

Index

K. K. Dompere, *The Theory of Problem-Solution Dualities and Polarities*, Studies in Systems, Decision and Control 405,
https://doi.org/10.1007/978-3-030-90279-7

Q

R

S

T

U

V

W

Z

Printed by Books on Demand, Germany